SYNTAX and SEMANTICS

VOLUME 10

A complete list of the contents of previous volumes appears at the end of this book.

SYNTAX and SEMANTICS

VOLUME 10

Selections from the Third Groningen Round Table

Edited by

FRANK HENY

Institute for General Linguistics
Rijksuniversiteit
Groningen, The Netherlands

HELMUT S. SCHNELLE

Ruhruniversität
Bochum, Germany

ACADEMIC PRESS *New York San Francisco London*

A Subsidiary of Harcourt Brace Jovanovich, Publishers

ACADEMIC PRESS, INC.
111 Fifth Avenue, New York, New York 10003

United Kingdom Edition published by
ACADEMIC PRESS, INC. (LONDON) LTD.
24/28 Oval Road, London NW1 7DX

LIBRARY OF CONGRESS CATALOG CARD NUMBER: 72-9423

ISBN 0-12-613510-X

PRINTED IN THE UNITED STATES OF AMERICA

79 80 81 82 9 8 7 6 5 4 3 2 1

CONTENTS

LIST OF CONTRIBUTORS

Numbers in parentheses indicate the pages on which the authors' contributions begin.

JOACHIM BALLWEG (225). *Institut für deutsche Sprache, Universität Manheim, Germany*

RENATE BARTSCH (23), *Instituut voor Filosofie, Centrale Interfaculteit, Universiteit van Amsterdam, Amsterdam, The Netherlands*

ROBIN COOPER (61), *Department of Linguistics, University of Wisconsin—Madison, Madison, Wisconsin 53706*

ROBERT M. HARNISH (315), *Department of Philosophy and Department of Linguistics, University of Arizona, Tucson, Arizona 85721*

ROLAND R. HAUSSER (93). *Institut für deutsche Philologie, Universität München, Germany*

FRANK HENY (1), *Institute for General Linguistics, Rijksuniversiteit Groningen, Groningen, The Netherlands*

H. HIŻ (343), *Department of Linguistics, University of Pennsylvania, Philadelphia, Pennsylvania 19104*

ASA KASHER (257), *Department of Philosophy, Tel-Aviv University, Tel-Aviv, Israel*

HANS-HEINRICH LIEB (353), *Freie Universitat Berlin, Fachbereich Germanistik, Berlin, Germany*

BARBARA H. PARTEE (275), *Departments of Linguistics and Philosophy, University of Massachusetts, Amherst, Massachusetts 01003*

TIMOTHY C. POTTS (141), *Department of Philosophy, The University of Leeds, Leeds, West Yorkshire, England*

DAVID S. SCHWARZ (241), *Departments of Philosophy, University of Connecticut, Storrs, Connecticut 06268*

THOMAS WASOW (199), *Department of Linguistics and Philosophy, Stanford University, Stanford, California 94305*

PREFACE

This volume contains a selection of papers from the Third Groningen Round Table, held in June, 1976. This conference, organized by the Institute for General Linguistics of the Faculty of Letters, Rijksuniversiteit Groningen, Holland, was the third in a series of meetings held there at intervals of roughly two years, each devoted to a distinct topic of interest to linguists. For the Third Round Table, this was the Semantics of Natural Language. The editors of this volume solicited a number of papers for the meeting (Helmut Schnelle having been invited to act as guest president) and a general invitation was then extended to scholars working in the area to submit relevant papers. As far as the present volume goes, the solicited papers were all in principle accepted—though two were not available—and of the twenty or so others, four were selected for inclusion. No attempt has been made to include a transcription of the discussion, though that has clearly influenced the final form of several of the papers. A number of the papers have in fact undergone quite radical revision and appear in every different versions from those read at the conference. For all these reasons, then, this cannot be regarded as the "proceedings" of the Third Round Table, but (as the title implies) is a selection of papers originating in connection with that meeting.

Thanks are due to the Faculty of Letters, Groningen, for bearing the major costs of this meeting and to the university for making facilities available; also to the Ministerie voor Onderwijs en Wetenschappen for a contribution to the costs of the conference, which helped in no small measure to make it the success it was. Without these two sources of financial support this collection would never have materialized. The participants—both those who read papers and those who did not—also deserve thanks; they, too, contributed in many ways to the quality of these papers as well as to the conference itself. Finally, Susan Barnard bore most of the responsibility for the detailed arrangements for the meeting and for the preparation of the manuscript—assisted in the latter by Hennie Zondervan and Jos Eeckelaer.

INTRODUCTION

FRANK HENY

Semantics often seems to be a particularly unproductive quagmire to those working outside the area, producing little of obvious value outside its own vague borders, guided by dogma rather than theory, and marked by an increasing balkanization within those borders, the devotees of each "approach" having little to say to each other, let alone to the outside world. Although it has to be admitted that such characterizations do have an element of truth, recent years have undoubtedly seen important changes. The papers in this volume, for instance, which originated at a small conference unconstrained by theme beyond being devoted to "the semantics of natural language" and deliberately set up to attract proponents of different approaches, nevertheless exhibit a surprising number of shared assumptions, recurring topics, mutually relevant observations, and even converging results. Above all, there is real evidence of progress in the detailed analyses of particular problem areas. To be sure, there is still no really widely accepted theory of meaning which determines in detail the direction of research and the significance of results. That obviously lies in the future. However it does seem possible to guess with some confidence at what that theory might look like—at least insofar as it is intended to be a theory of meaning for natural language.

SYNTAX AND SEMANTICS
VOLUME 10

ISBN: 0-12-613510-X

Most of these papers, for example, assume that some sort of truth-conditional analysis will play a role in the representation of meaning, and those dealing with specific analytical problems tend, further, to employ the apparatus of model theory. (Indeed, the reader unfamiliar with the application of model theory to semantic problems will make very heavy weather of several of the papers; it would be out of place, even if possible, to provide an adequate introduction to this aspect of the field here, and no attempt will be made, beyond pointing out that suitable background reading and further references will be found in, for example, Partee, 1975; Thomason, 1974; and Cresswell, 1973.) Second, in addition to the assumption that the techniques of logical analysis must be incorporated in some way into a theory of meaning, there is an almost equally universal recognition of the fact that the presently available logical apparatus will need considerable extension, perhaps radical modification and/or incorporation into a wider framework if an adequate account is to be given of the role, in meaning, of the attitudes, assumptions, desires, etc. of speaker, hearer, and others, or of such essentially institutional, socially defined, and rule-governed behavior as questioning, commanding, promising, etc. A third very general tendency can be seen, too: a serious attempt to account in detail for the facts of natural language— not merely isolated facts, but those apparently rather superficial regularities which emerge when one tries to deal with the structure of a natural language as a whole. In several instances the analytical problems have been defined, at least in part, by the recent work in transformational grammar and it is probably the implicit acceptance of the goal of the transformational linguist, to define the notion "possible human language," that binds the contributions to this volume together. Although not all the contributors are linguists, the outlook of the papers is certainly "linguistic" rather than "logical."

Were it not for this last fact, there would be little point in remarking on the others, for it is the acceptance of the overall goal that turns what would otherwise be an option into an obligation, and gives to the quest for a theory of meaning a special status. The existence of pragmatic questions, even the recognition that these might be approached systematically, is nothing new. Similarly, logicians have long been aware of the waywardness of actual human language, of the difference between *I don't like everyone* and *I don't like anyone,* or the problem of reconciling the meaning of *Every man who owns a donkey beats it* with the perfectly reasonable assumption that *every* is something like a universal quantifier, *a* something like an existential, and *it* comparable to a variable bound by *a*. What seems new in the field, and typically linguistic, is the attempt to deal with all these problems at once, precisely and in an integrated, mutually illuminating way, which depends as much (rather than as little) as possible on the structural properties that specifically characterize human language as a whole. Insofar as a theory of meaning might be emerging from this work, then, it is a **linguistic** theory, and might be expected to form part of and interact nontrivially with other aspects of a general linguistic theory. Obviously the resulting theory may well be of interest to logicians, and probably should be of significance to philosophers of language. Equally obviously, the results of logical analysis are, and should be, playing a central role in the development of the theory—but the goals of the logician are not the same as those of the linguist and this is leading to developments which, if anything, are most obvious in those chapters which make most vital use of logical apparatus.

To be more specific, the logician may perfectly well find it an interesting fact worthy of some thought that *every, any,* and *it* interact as they do in the example just touched on. And he even needs to relate his results in some way to ordinary language so as to check the inferences. But he is under no obligation to deal in detail with the vagaries of these words if there are others that will serve his purposes, and he certainly has no need to try to *explain* what is going on. As logician, his goal is to investigate the structure of inference. The system he proposes, however closely or distantly related it be to the sentences of natural language, is justified solely by its ability to throw light on the nature of inference by capturing its essential properties. For the linguist, however, the primary goal is much broader. He is trying to discover the overall structure of language. Any hypothesis about the structure of some part will naturally have to be judged according to how well it can be integrated into that whole. Hence, if the linguist adopts the hypothesis that structures analogous to those of standard first order logic are, or seem to be, a subcomponent of natural language, he cannot gloss over the relationship between *any, every, a,* and *it* and those structures. In fact it is at precisely that point where his real task begins.

It does not seem to be a matter of chance, in fact, that it is precisely to this problem that five of these papers, are devoted, or to be more specific, that these five papers can be readily interpreted as investigating hypotheses about the relationship between pronouns (and hence by implication the quantifiers of natural language too) and variables, and other aspects of standard logical form. Nor does it seem accidental that the tendency is very obviously toward trying to account in a systematic way for features of pronominal meaning which may well result from the essentially pragmatic and social nature of human language.

This can best be illustrated directly, by considering the five papers in more detail. It will be useful to adopt the fiction that these are all investigations of the hypothesis that pronouns possess just the same semantic structure as the bound variables of logic. They are not, of course, since for some time it has been clear that that is much too simple. The idea has been around at least since Quine 1960 and a number of counterexamples have been known for almost as long, but it is obviously the simplest way of relating pronouns to some specific constituent of logical structure and will, hopefully, clarify the issues. The hypothesis can be made more specific by reference to an example like *Some man danced and he sang*. Limiting quantification to the set of men, we can identify such a sentence more or less uncontroversially with the logical structure $\exists x(Dx \,\&\, Sx)$. The pronoun *he* in the example corresponds then in an intuitively clear sense with the final occurrence of x, and generalizing from this we might suppose that whenever a pronoun occurs it will be possible to identify it in like manner with some particular occurrence of a bound variable in the corresponding logical formula. Now, if in determining correspondence, the mapping onto logic is to be done compositionally, beginning with minimal logical units, there is obviously no simple way in which pronouns can be identified in one step with **bound** variables. The quantifier that binds the relevant variable can, in principle, be any distance away, just as a pronoun can be, in principle, any distance away from its antecedent. So in the first instance a pronoun will presumably have to be identified with a **free** variable, with the proviso, for example, that no free variables occur in any formulas corresponding to the sentences of natural language. We will assume therefore that some such constraint will form a component of the hypothesis that pronouns correspond to bound variables, an assumption which will be questioned in due course. First however, let us examine some of the difficulties that arise in maintaining the hypothesis itself.

1. We have already bumped up against one minor problem. Even if there is some way of mapping all pronouns onto bound variables, not all occurrences of bound variables can be related to pronouns. In the sentence about the man who danced and sang, for example, the occurrence of x in the argument place of D(*anced*) corresponds to nothing at all by way of a constituent in the surface form of that sentence. Most nearly, and in part, it corresponds to the antecedent of the pronoun, *some man*. Such variables occur even in formulas corresponding to sentences containing no pronoun at all, such as *Some man danced,* and we shall refer to them as "essentially nonpronominal variables" from time to time. At the very least there is a descriptive problem here: Some way will have to be found of mapping pronouns onto just the "right" variables. Whether this constitutes a deeper problem for the hypothesis that pronouns correspond to bound variables depends on the exact nature of that hypothesis, and a great many other factors, too. For the hypothesis to remain interesting there obviously does not have to be a one-to-one mapping of the elements of natural language onto those of logic; too close a correspondence brings with it the risk of triviality, in fact. Yet where there is a mismatch, such as here, it would obviously be nice if we could show that the lack of correspondence followed, as an automatic consequence, from some other aspect of the theory.

2. There is a second problem which may or may not simply turn out to be an extension of the first, again depending on details of the hypothesis. Other than the ordinary definite pronouns such as *it* there are word classes which sometimes show morphological relationships with that class, but not always,

and which are at least good candidates for identifying with variables; for example both relative pronouns and reflexives, the latter as in *Some man hurt himself,* $\exists x(Hxx)$. Looked at descriptively this simply amounts to the problem of determining which classes of words will correspond to which occurrences of variables. Again, however, there is a deeper problem lurking below the surface: If in fact more than one word class corresponds to bound variables, is there some principled way of explaining this fact?

3. A third problem arises in connection with the fact that the original sentence may be used, perhaps only when accompanied by pointing or said with special emphasis, to state not that one man sang and danced but that one sang and another danced. Pronouns in many environments possess this "ambiguity," being interpreted either as "anaphoric" to some other phrase in the linguistic context or "deictic," when they pick out, as it were, some object or event which may not be expressed (mentioned, represented) in the linguistic context. Only the first usage seems to be even a moderately good candidate for identification as a bound variable; the deictic usage has no obvious relationship to quantifiers at all. Assuming that **some** uses of pronouns can be identified with bound variables, it will apparently be necessary to provide a separate account of the meaning of these others. It is worth adding the observation that this is not a matter of accounting for a separate morphological class of deictic pronouns, but a general fact about the meaning of all pronouns: They appear to be used in two different ways.

4. This "ambiguous" interpretation gives rise to yet another problem. There are constraints which have to be put on the possibility of establishing an anaphoric relationship between a pronoun and some other constituent. One of these appears to be purely structural, the simplest (but not, perhaps the most accurate) statement of which can be extracted from the following:

> ***John*** *saw the man that* ***he*** *knew.*
> *The man that* ***John*** *knew saw* ***him.***
> *The man that* ***he*** *knew saw* ***John.***
> *He saw the man that John knew.*

In the last case, *he* must be interpreted deictically to refer to someone other than John; in the others it may but need not be so interpreted. This appears to result from a restriction on the structural relationship that may hold between an anaphorically interpreted pronoun and its antecedent, which for English may be very roughly stated thus: A pronoun may not have a noun phrase to its right as an antecedent unless it is in a clause "lower" in structure than that antecedent. Stated this way, it is a constraint on "backward" anaphora (but see Wasow's paper in this volume for a much more accurate account and much discussion).

5. There are other constraints on anaphora which are much more obviously semantic in nature, such as that illustrated by the following:

> ***John*** *went home when* ***he*** *had finished the test.*
> ***John*** *went home.* ***He*** *had finished the test.*
> ***Every competitor*** *went home when* ***he*** *had finished the test.*
> *Every competitor went home. He had finished the test.*

The pronoun *he* seems to be anaphoric to *John* even in the second example, when it occurs in a separate main clause; this does not seem to be possible in the fourth sentence, however, when the potential antecedent is the obviously quantified noun phrase *every competitor,* though *he* seems to be "bound" by that phrase when it appears within a structure dominated by one main clause, as in the third sentence. When necessary this will be referred to as the problem of the special constraints on anaphora to quantifiers.

The situation is actually quite complicated, much more so than is obvious from these cases; some additional relevant data will be found in Bartsch's paper, which has reference to German, and in Postal 1972. The main problem which arises is however quite plain; it is the same for these cases and for those falling under the previous heading: Assuming that certain instances of pronouns, namely those that are "anaphoric," are to be identified with the bound variables of logical form, it will still be necessary to limit—in two apparently separate ways—the relationships that can hold between the pronoun and the constituent which is its antecedent (that is, which in some way corresponds to the element binding its corresponding variable in logical form). As with the previous problems, this can be looked at from two points of view. Descriptively, one must at least define the relevant environments, and this goal appears to be more or less in sight. However, unless a natural way can be found of accounting for all these mismatches, the question must sooner or later arise whether there is any explanatory value in maintaining that **some** pronouns correspond to **some** bound variables. We will return to this point.

6. The next problem does not at first glance seem to have anything at all to do with pronouns. It has already been briefly mentioned in passing. Given the assumption, under which we have been working, that the first step in exploring a plausible working hypothesis about the semantic structure of English is to relate English sentences quite precisely to formulas of logic, then such a mapping will be required for the quantifiers as well as the pronouns. The quantifier *every* seems likely to correspond to the universal $\forall$ and *a*, roughly, to the existential $\exists$. But combine these hypotheses and it would seem that the essential structure associated with

$$\textit{Every man} \begin{Bmatrix} \textit{if he} \\ \textit{who} \end{Bmatrix} \textit{owns a donkey beats it}$$

could be represented as $\forall x(\exists y(Oxy) \supset Bxy)$. But that is wrong. If, as is usual, the existential quantifier is limited to having narrower scope than $\supset$, it will not bind the second *y*. (If it were permitted to have wider scope it would of course possess too weak a meaning.) If we simply reverse the order of the quantifiers, thus permitting ∃ to bind the second *y*, the meaning is still wrong, corresponding now to the sentence *There is a donkey such that* . . . . One way to approximate the intended meaning is to permit *a* to correspond in such a case to a universal quantifier—but not in for example, *John has seen a man fall down the stairs.* Logicians might well consider such a solution uncontroversial, and to the extent that it is accepted, pronoun interpretation is not involved. Consequently it is worth remarking right away that for such similar examples as *Every man who owns a donkey buys carrots* or even *Every man who owns a donkey beats that donkey* there is no need for special treatment of the word *a*; it could be permitted to correspond in these, too, to an existential quantifier with narrower scope than ⊃. Only the decision that *it* corresponds to a variable bound by the counterpart of *a* leads to a scope problem with the quantifier.

Rather more complex but essentially similar examples exist, involving crossing anaphoric pronouns, such as *Every man who wants* ***it*** *will get a/the prize* ***he*** *deserves.* There are in fact a number of additional problems connected with these, but this is not the place to attempt to distinguish them, even in outline. Suffice it to say that some seem to be of essentially the same sort as that just discussed, where it seems necessary to complicate the process of mapping quantified noun phrases onto logical form if pronouns are interpreted simply as bound variables.

7. The last problem that will be mentioned here concerns a case where there is just no natural way at all of continuing to maintain such an account of pronouns; it arises in connection with examples like *The man who gave his paycheck to his wife is wiser than the one who gave* ***it*** *to his mistress.* This appears to be a fairly standard use of the pronoun *it*, which may appropriately be considered anaphoric in some sense, since it does seem to relate to the phrase *his paycheck* in the first clause. Yet there is no obvious way of capturing the relevant relationship by means of a bound variable; nor is there any obvious candidate for a quantifier to bind such a pronoun.

Before turning to the first five papers in order to show briefly how they may be viewed as reacting to these problems, it is worth summarizing the situation. There are a number of different ways in which the claim that pronouns correspond to bound variables does not obviously fit known facts about pronouns. Any attempt to incorporate such a claim into linguistic theory is obliged to consider these instances of mismatch seriously and at least provide a descriptively adequate account of how the hypothesis relates to the facts in question. First, there are variables which never correspond to pronouns at all, and natural language phrases of various sorts other than "ordinary" pronouns which may just as well as those pronouns be supposed to correspond to variables. Second, most instances of ordinary pronouns can be interpreted deictically—which appears to have nothing to do with bound variables. Third, there are various constraints on anaphora which independently serve to limit the possible cases of binding. And finally, even among the apparently anaphoric pronouns there are two kinds of case where to interpret the pronoun as a bound variable brings problems, either because it complicates the analysis of quantifiers or because it simply does not seem feasible at all.

Bartsch, extending her own previous analysis of pronouns, concentrates on the special constraints on anaphora to quantifiers and on the donkey problem, dealing also with aspects of the contraints on backward anaphora —which appear to differ for German and English. So, while she clearly interprets relative pronouns and ordinary pronouns differently (the former as essentially noun-forming operators with open sentences as input), she does not discuss this aspect of her approach. Nor does she deal with reflexives. Limiting our attention therefore to the account of definite pronouns, we find that these are mapped ultimately onto two distinct kinds of constituent in the logic: names as well as bound variables and this principally to deal with anaphora to quantifiers and with the donkey sentences.

The entire strategy works as follows. First, a quantified noun phrase may be interpreted as a suitable matrix incorporating the lambda operator and predicate variables and one free individual variable repeated in the argument places of the predicates. The matrix for *ein* ('a') is $\lambda Q \lambda P(Q(\nu) \ \& \ P(\nu))$. An individual variable is assigned to ν and is "blocked," which is to say that it is available only for a pronoun in the same sentence to "hook onto." Proper names are at this same initial stage mapped onto constants. Certain quantified noun phrases may also optionally be so treated, namely those which contain quantifiers like English *some,* which may be used to "refer" to particular individuals. *Every* may not be so interpreted; nor in general may noun phrases in negatives, etc.

Pronouns, meanwhile, are interpreted simply as $\lambda PP(\nu)$, in conjunction with a search procedure which seeks to replace the metavariable ν by either a name or a blocked variable occurring in the existing context. The interpretation of the pronoun is thus incorporated into the existing structure as a result of the fact that ν in $\lambda PP(\nu)$ takes as a value a recurring variable or name of the logic language. The final stage of interpretation is at the level of the highest clause in a sentence. Any formulas containing blocked variables are generalized, the matrices being subject to quantification according to a further set of specific instructions. It is at this stage that the donkey problem is handled, for in appropriate contexts, set out in the paper, the matrix assigned to *ein* ('a') is quantified by a universal quantifier with wide scope. Beyond the level of material dominated by one main clause a pronoun cannot, of course, "hook onto" a blocked variable, for after generalization all blocking is undone. However, names are fixed throughout a discourse, so any quantifiers interpreted by names can, like proper names, have pronouns in later main clauses take the values of their associated names, becoming anaphorically related to them. In this way, the special constraints are not merely dealt with in isolation but are related to the possibility of giving certain quantifiers a "specific" interpretation.

Several points are worthy of comment. First, notice that the system as described does not adhere to the principle of strict compositionality, or to put it another way, the meaning of a composite whole is not simply a function of the meanings of its parts. The two senses of, for example, *Martha spoke to Mary when she arrived* will be built up from precisely the same components and will differ crucially in respect to which name is substituted for ν. This will depend in turn upon whether, for example, "discourse coherence" is destroyed by one or the other substitution—for example, whether *Mary has just arrived,* or *Martha has just arrived* precedes the sentence in question. Thus the search procedures form a central part of the "meaning," in some sense, of pronouns.

Second, notice that the paycheck interpretation is dealt with by a special, additional, frankly ad hoc rule carrying *it* into *his paycheck,* from where the procedures applied in the normal way yield appropriate results. Nor is any specific account given of the deictic interpretation of pronouns, a point to which we shall return shortly since it seems that an account **could** be given along lines suggested by other papers in the volume. (This possibility is in fact discussed by Bartsch in her footnote 8.)

Returning to the original hypothesis, we may now ask how far in fact Bartsch identifies pronouns with bound variables. The answer is of course not simple. She maps pronouns directly onto an intermediate stage, where they are represented by a formula incorporating a free metavariable. But this variable simply stands for either names or "blocked" variables. In addition, the interpretation crucially makes use of search procedures, replacement devices, and generalization, which all appear to make necessary contributions to the solution of the donkey problem, the explanation of the special constraints and (although we have not considered this in detail) contraints on backward anaphora, as well as contributing crucially to meaning. Thus, there really is no simple relationship at all between pronouns and any constituent type in the logic language.

We return later to aspects of Bartsch's solution; let us now consider that of Cooper. He begins explicitly with a different hypothesis, namely, that pronouns correspond to free variables and argues that this, combined with normal assumptions about binding, permits no natural solution of the donkey problem, among others. (In addition he adduces certain facts about relative clauses in Hittite and other languages.) On the basis of these observations he, like Bartsch, proposes to assign two quite distinct interpretations to pronouns. Unlike Bartsch, he does not propose names for the second interpretation but definite descriptions containing free variables over properties or relations. He argues that these provide a natural account of both the donkey and the paycheck sentences.

In identifying pronouns with free rather than bound variables, Cooper simply follows Montague (1970). As already noted, the initial mapping will in any case be onto variables that are not bound. The need to ensure that this is not the final outcome results from the fact that a sentence containing a free variable is generally, and in fact very naturally, identified with its universal generalization. Clearly that will not do. In the simplest case, for example, we do not want to claim that *It is red* (or even *They are red*) means the same as *Everything is red*. However, Montague suggested that it was natural to interpret a free variable relative to a context of use as picking out that particular object which was assigned to it in that context. Of course, the free variable would not "mean" that individual, nor would a pronoun identified with it. Relative to other contexts of use, they could pick out other objects. But in this way, Montague, and following him, Cooper, suggests that the deictic interpretation of pronouns is very naturally accounted for, without in any way interfering with the other.

It is necessary to draw attention to the fact that this proposal makes crucial reference to the assumption that the logical forms (which up to now have formed the end-point of the analysis of meaning in this account) will themselves be further interpreted. Hitherto we have not made any attempt to raise the fundamental question why it might seem reasonable (even as a working hypothesis) to suppose that logical structures might play any role in a theory of meaning for natural language. The question is complex and we cannot do it justice in this introduction; nevertheless without some discussion the significance of these proposals must remain totally obscure.

The incorporation of logical form in a linguistic theory of meaning is in the first place a hypothesis about the **structure** of meaning. Without taking the matter any deeper it might be compared with the supposition by some linguists that a grammar of English including the rule S $\rightarrow$ NP VP forms a part of linguistic theory, to be justified by reference to the fact (if it is one) that numerous other structural features of the language can be best accounted for if there is a constituent VP available for other rules to make use of. What general, comparable utility might such structures as logical forms incorporating bound and free variables possess? Obviously these specific aspects of logical form were introduced because they permitted construction of a more perspicuous, transparent, inferential calculus than hitherto. Aspects of propositional structure which are obscured in natural language but which appear to play a constant role in inference are represented in constant form, while the superficial similarity of *John ran* and *Some man ran*, irrelevant to such issues, is allowed to fall away. But to represent those aspects of the structure of an utterance which are relevant to the derivation of inferences from it, which is to say to the derivation of its consequences, is to represent aspects of its structure which are vital to meaning, provided we assume, as we have done, that there is a close relationship between the meaning of an utterance and the conditions under which it would be true. For the consequences of (or inferences from) an utterance represent one way of analyzing its truth conditions. (Set forth, of course, in turn, in that same language specifically designed to make transparent **their** possible consequences.) Thus, one possible way of providing a general justification for the introduction of such logical structures into a theory of natural languages, and hence paving the way, as it were, for the hypothesis we have been assuming all along (in which pronouns are structurally identified with variables), is to suppose or show that it is in this way that the truth conditional aspects of meaning are to be captured. There may be other reasons, of course, for supposing structures of this same form to play a role, but that is beside the point, since it is not at issue here.

What is at issue, and has in fact just become crucial, is the fact that there are other ways of defining truth conditions, one of these being the use of model theory, in which the crucial features of states of affairs under which utterances are true are represented directly in the relatively sparse language of set theory. Two sides of the same coin, of course, but they may throw different features into relief. In this case, they do, or at least that is the claim. For whereas one certainly would not want to say that a free variable (or pronoun) **means** this or that individual, corresponding to the fact that the inferences one would want to derive from an utterance containing a free variable necessarily treat the variable as arbitrarily satisfied, nevertheless that necessary arbitrariness is introduced **beyond** any particular context of use, relative to which some individual is "picked out" or assigned to it. At any rate this seems to be the essence of the proposal.

But while Cooper adopts it, he points out that it will not deal adequately with the donkey and paycheck sentences. In its simplest version, the alternative analysis of pronouns added by Cooper amounts to interpreting them as (his [5]):

$$\{\lambda \boldsymbol{K}\, \exists x[\forall y[[P_i](y) \equiv y = x] \wedge \boldsymbol{K}(x)]\text{: } i \text{ is a natural number}\}$$

Essentially, P_i being a free variable over properties, this amounts to a definite description, the content of which is undefined. And just as free variables over individuals will, under the interpretation outlined above, pick out a particular individual in a given context of use, so the intention is that this will pick out a particular, and of course appropriate property in a given context of use. Sometimes it will correspond directly to the interpretation of some other part of the sentence itself. Then the pronoun is anaphoric. But if this is not the case, then it is deictic. Thus there is absolutely no difference between the interpretations assigned to the two different usages.

At least that is the general rule; and corresponding to it, Cooper makes no attempt along the lines of Bartsch, for example, to impose explicit conditions so as to constrain anaphoric interpretation of such "demonstrative" pronouns (or indeed of the bound and free variable interpretations either) in order to deal systematically with the special constraints, or backward anaphora, etc. There is one instance, where he does force anaphora. When a pronoun appears as head of the Hittite relative clause construction, which provides the initial justification for introducing this analysis, the pronoun becomes subject to the application of a rule which he schematically sets out as his example (29), which in effect forces "binding" of the pronoun by limiting the interpretation of the variable to the property expressed by the relative clause. Some such mechanism might well be developed to handle the relevant facts such as those discussed by Bartsch and others in this volume.

There are, in addition, cases where, he argues, the special constraint on anaphora to quantifiers is automatically handled without additional apparatus. For example, in *Every competitor left. He had finished,* the uniqueness clause in the demonstrative interpretation of *he* prevents it from picking out any individual on the basis of the preceding sentence. This is in contrast with such superficially parallel cases beginning *John* . . . or *Some man* . . . , and also with *Every competitor left when he had finished,* where the standard bound variable approach (operating as for Bartsch only within a single sentence structure) will permit anaphora. In fact it is clear that Cooper's use of a demonstrative interpretation is playing at this point a very similar role to Bartsch's search procedure and mapping onto an arbitrary name assigned to some other element. The details of their accounts are otherwise so different that it would not be profitable to attempt a comparison here, though it may be worth conjecturing whether Bartsch's system could be extended without discomfort so as to permit a pronoun to pick up a "new" name directly as the value for v and thus represent the deictic usage. (See, in fact, her Footnote 8.)

Whatever the difficulties encountered in comparing these systems in detail, one thing is clear. Their analyses of the donkey sentence are necessarily totally different. Actually, Cooper could permit a solution quite comparable to Bartsch's. Though he does not discuss it in the paper in this volume, he obviously needs and has elsewhere proposed a method of place holding quite comparable to Bartsch's two-stage interpretation of noun phrases like *a donkey.* Otherwise there would be no way of representing the ambiguity of *Every man thinks that he loves some woman,* where the quantifier in the lower sentence can be assigned wider scope. One device proposed in Cooper 1975 is called "NP storage." It would permit him to delay interpreting *a* in the donkey sentence. However, extending a proposal of Rodman (1976) he has proposed to block this from operating into/out of relative clauses. He justifies this at length (see Cooper, 1977). Accepting that, he must obviously deal with the donkey a different way. This he does using an extended version of the demonstrative pronoun interpretation, which permits any structure composed solely of bracketed variables to interpret the property denoting expression. This in turn permits him to include the paycheck sentence—which actually turns out on this account to be rather simpler than the one about the donkey.

Thus Cooper, like Bartsch, ends up with two distinct accounts of the interpretation of pronouns. Under neither of his analyses do they correspond to **bound** variables; they are either free variables (which sometimes crucially, **remain** free) or hidden definite descriptions. (It may be worth remarking that at least in his detailed account of Hittite (Cooper, 1975), such hidden property-denoting expressions were assigned not merely to pronouns but to definite noun phrases in general; hence pronouns regarded in this second way contain, or are, a kind of lowest common denominator of definite noun phrases and have no special interpretation of their own.)

It is perhaps in rejecting this twofold analysis that Hausser, in the next paper, differs most from Cooper, for he, too, makes crucial use of a "hidden" definite description—though in addition the mode of interpretation is quite different. Hausser provides for just a single property variable in the interpretation; thus at least superficially the structure might seem to resemble more the first version of Cooper's analysis (given above) than the later one which he needed to deal with the donkey structures. Yet while Cooper interprets his variables in a completely orthodox way, Hausser makes it clear that his variable is not to be interpreted but rather **replaced** (somewhat after the fashion of Bartsch's ν) with, in Hausser's case, actual linguistic material, even in the case of deictic interpretation. For Hausser, a context is given in purely linguistic terms.

This is crucial, it turns out, for his analysis of the paycheck cases, since there is in such instances a **double** application of "reconstruction": first the replacement of the variable in the interpretation of *it* by *his paycheck,* and then the replacement of the relevant variable in the interpretation of *his*. The only special aspect of this—beyond the notion of "reconstruction" itself—is the double application, which might be said to follow automatically because of the existence of a suitable variable after the first application.

Hausser pays special attention, unlike Cooper, to the problems of backward anaphora, etc., endeavoring to build these conditions specifically into the reconstructive process; likewise he attempts to specify precisely all relevant contextual factors. However, he does not deal explicitly with the interpretation of such sentences as *Every man sat down when he came in,* which were those that motivated Cooper's retention of the original variable analysis. It is, of course, interesting to speculate on how far Hausser's process of reconstruction provides a natural account of the difference between the possibilities of anaphora within a single sentence structure (that is, material lying within the bounds of a single main clause) and those that exist when one goes beyond those bounds—differences that both Cooper and Bartsch emphasize, with "real" binding, in their accounts, occurring only within the confines of a single sentence structure. It is worth mentioning in this connection the distinction between "sentence grammar" and "discourse phenomena" introduced in recent transformational work (for example, in Chomsky, 1978). This distinction is specifically applied, within the framework of that grammatical approach, to various problems (other problems) relating to pronouns by Williams (1977), Hankamer and Sag (1976), and others. But such questions lie beyond the scope of this introduction.

We have come full circle—to another unitary hypothesis. Notice one very significant difference between the one we set out with and Hausser's. According to this latter account, there is no one-to-one relationship between the class of natural language pronouns and some class of elements in the logical structures which serve as the representations of semantic structure. The pronouns are mapped onto quite complex bits of the logic. Moreover, as far as can be seen, no other class of words will ever relate simply to the variables, for Hausser gives explicit accounts of the reflexives, which might have done so but do not, and it seems unlikely that relative pronouns are to be interpreted other than as operators on sentences in some way comparable to Bartsch's explicit account. But if there is no word class that is mapped directly onto the **variables,** then there is no obvious reason for thinking that the quantifiers of natural language should, in turn, be closely related to the **quantifiers** of standard logic. In particular, with no variables, the quantifiers would not need to possess the power to bind an indeterminate number of variables indeterminately far away. For binding (over indeterminate distances) becomes the job of the pronoun instead. Hence, to the extent that a proposal along the lines suggested by Hausser were shown to be justified, it would lead naturally, or at least permit readily, radical changes in the structures associated with quantifiers, reducing still further the similarity between the proposal and standard logical form.

And, curiously, just such a proposal is made in the chapter by Potts. The binding function of pronouns is there separated entirely from that of quantifiers. Potts does not attempt to apply his theory to as wide a range of problems as we have been considering, and there is no point in speculating on how it could be applied. The significance of the chapter is in any case other than that. Potts emphasizes above all that a semantic analysis possesses a structure and shows how pronouns may plausibly and parsimoniously be analyzed not as variables but as argument-reducing operators on verbs, possessing thus a binding function distinct from, and in some ways parallel to, that which he assigns to quantifiers. Deictic pronouns are not dealt with; both relative pronouns and reflexives are interpreted essentially like ordinary definite pronouns.

In setting up this analysis he shows that proofs are perfectly possible in such a system, but emphasizes that a certain transparency of structure possessed by standard logic in regard to this goal may have to be sacrificed in order to represent most perspicuously other aspects of structure. (It should not, incidentally, be too surprising if it should turn out that the semantic structure of natural language is somewhat opaque from the viewpoint of an ideal system of inference; that may after all be thought of as one of the factors underlying the need to construct for the logician's purposes an explicit, regimented formal system, with variables etc.) Potts's semantics is not model theoretic—for reasons he has given elsewhere. It would be pointless to describe it in detail here; the main points are simple and have been summarized, and the technicalities of the analysis are dealt with in the chapter itself.

The last paper in this section is quite different. Wasow explicitly refrains from tackling semantic questions as such. He traces in detail the attempts of transformational grammarians to deal with the constraints on backward anaphora and related matters. These have in fact formed the focus of work on pronouns within transformational grammar, and have resulted in most of the discoveries which underlie attempts, such as those discussed above, to constrain the interpretation of anaphoric pronouns to appropriate environments. For the most part these transformational analyses have, as Wasow points out, been implicitly semantic.

The first, in effect, denied pronouns any separate semantic status, deriving *John sat down when he came in* from *John sat down when John came in*. By limiting appropriately the environments in which the rule turning repeated *John* (etc.) into a pronoun, the constraints were handled automatically along with the semantics. But of course the account foundered hopelessly once quantified noun phrases were tackled in earnest.

The second was a proposal to derive all sentences from underlying structures based on logical form, containing variables, etc. Again, all that was needed to deal with backward anaphora was a set of appropriate restrictions on the transformations lowering sometimes a "full noun phrase" (the superficial antecedent) into the position occupied by a variable, and sometimes a pronoun. These were intended to block the undesirable cases of lowering. However, as Wasow himself showed in earlier work (continued in this paper), this proposal seems unable to avoid generating such forms as **Some students knew that there were they/them outside* and has other, similar, unsatisfactory consequences. What is interesting about Wasow's conclusion, from the point of view of this summary, is the fact (which he himself points out), that it does **not** call in question the possibility of identifying pronouns with bound variables. That, from the point of view of transformational grammar, remains. Wasow's arguments, then, show only that a particular **way of mapping** pronouns onto bound variables is not consistent with other aspects of the transformational account of language structure. As we have seen, there may well be reason to doubt whether the interpretation of natural language pronouns is ever the same as that of a variable in first order logic; but Wasow's arguments are not relevant to that more general point.

Before turning to the papers of the second section, let us briefly relate those of the first to the three characteristics of recent work in semantics noted at the beginning of this introduction: the assumption that the meaning of an expression is intimately related to the conditions under which it could be true; the attempt to integrate such a view with the essentially pragmatic factors that play a crucial role in meaning; and the recognition, at the same time, that a theory of meaning for natural language cannot be developed in isolation from the increasingly documented regularities peculiar to natural language. All three are obviously to be found in these papers. In the first place, the implicit or explicit starting point of all of these is the possibility that pronouns might be identifiable semantically with bound variables—and the bound variable exists only in the logical forms within which it is defined, which in turn are designed specifically to exhibit truth conditions as these relate to inference. Second, as we have seen, the papers of Bartsch, Cooper, and Hausser all explore alternative hypotheses, in all of which the role of context is strongly emphasized in one way or another. Moreover that of Potts, although certainly not concerned with context, explores the possibility of developing a semantics relatively distantly (if at all) related to or influenced by the specific characteristics of a system of inference. Finally, it scarcely needs to be pointed out that all these works are seriously trying to explain the peculiar characteristics of the pronoun class of a particular natural language, and not merely some abstract formal system.

We have discussed the papers on pronouns in some detail; partly because of the complexity of the issues involved, partly because they are so closely interrelated that it seemed worth trying to trace the various threads that run through them, and partly because they seemed peculiarly centrally situated in relation to the issues referred to in the previous paragraph. If the introduction to the other papers is considerably less detailed, this is not because of any lesser value attached to them! Further lengthy discussion of the general issues would be otiose, and of isolated details pointless.

The second section contains just two papers. The first, by Ballweg, builds up a lexical entry for the verb *kill.* First he defines an operator of gradual becoming, CHANGE. He then defines a conditional operator and on the basis of that a causal operator. This latter is a refinement of that of Lewis; its chief advantage is that certain cases of regularly time-related events which are not themselves causally related but, for example, possess a common cause are correctly excluded. The lexical entry for *kill* is based on the two operators CHANGE and CAUSE using the lambda operator to create a two-place predicate. Ballweg shows briefly how the CHANGE operator might figure in the entries for members of certain verb classes; he does not do this for other verbs in the case of the other operators, but that this could be done is implicit. With the reawakening of interest in the lexicon among linguists, it seems clear that such studies are a much needed prerequisite to an understanding of language structure below the level of the word, in particular, a prerequisite to serious morphology.

Schwarz's paper is very different in approach from any discussed so far. In fact it is the only one in this volume adopting a Gricean approach to semantic problems. He modifies an analysis of presuppositions, including those of the definite determiner, proposed originally by Wilson, showing first that her account suffers from certain deficiencies and then endeavoring to rectify them. She has argued that presupposition is essentially pragmatic, resulting from the application of Grice's maxims to the potential interpretations of certain sentences, selecting as the preferred interpretation of a negation, that interpretation which incorporates an entailment of the corresponding positive sentence. Schwarz argues that such utterances are in fact characterized by the fact that while to utter the relevant entailment of such a sentence, S, when not-S is conversationally in order, would in turn be out of order; there is no way of conveying the information that not-S holds on the basis of a state of affairs in which that entailment also holds, without simply uttering not-S—which in turn makes the entailment in question an implicature of not-S in addition to its being an entailment of S.

Such studies as this, seeking to limit the domain of a particular part of semantic theory (in this case seeking to remove certain aspects of meaning from the domain of inference) are extremely relevant to the attempt to integrate a truth-conditional account of meaning into a wider framework. In particular, it is obvious that any account of the semantics of pronouns which makes use of the notion of definite description will, in the end, need to make a choice between a many-valued logic (Hausser's choice) and a more conservative evaluation. As the context of utterance and speaker–hearer relations are dealt with more seriously, the issue will increase in importance.

The final section of the book includes five metatheoretical studies, the first two most closely related to the earlier papers. In the opening section of his paper, Kasher directly tackles the general problem just mentioned: How to define in a motivated way the limits of relevance of a theory or component of a theory. In so doing, he emphasizes strongly the very different goals of the logician and of the linguist, which lead to different demands being placed on, for example, a theory of meaning, even when the same subject matter appears to be in focus. In the paper as a whole, Kasher is most concerned with defining the task of the linguist who accepts the need for a precise theory of meaning. He attempts to set principled limits on an acceptable linguistic theory of meaning, on the basis of the fact that such a theory must characterize an essentially learnable language, and at the same time on the basis of relationships between language and other human characteristics.

Partee's paper is directed to almost the same problem: the discovery of limitations which can be placed on representations of grammatical systems—including the semantics—on the basis of considerations of learnability. Within the framework of Universal Grammar set out in Montague 1970, she proposes an additional constraint, namely, that the input to any rule be itself a possible final output, that is "grammatical sentence." Obviously this would limit the abstractness of possible analyses considerably and she argues that it is plausible on the basis of considerations having to do with learnability. Furthermore, it has the merit of fitting many facts—yet not being trivially or obviously true. In fact much of her paper is taken up in considering possible counterexamples. One of the most interesting and difficult has to do with pronouns, as analyzed by Cooper in the paper appearing in this volume. Partee is thus very directly concerned with the problem of adapting a logician's system for incorporation into an explanatory, empirical theory of language; we have suggested that this is, either directly or indirectly, the main burden of much current work in linguistic semantics.

The three remaining papers all deal with alternatives to a theory based solely on an analysis of truth conditions. Harnish tackles the general problem of relating that part of Searle's theory which **is** related directly to truth conditions to the essentially non-truth-conditional part having to do with speech acts proper, and of relating these in turn to an explicit transformational syntax. He points out that it is by no means obvious that such an integration can be carried out—at least for the versions that he considers. It has often been assumed that there would be some relatively simple way of relating the rules determining regularities of use to those determining regularities of structure. The fact that this is probably not the case for the two well-developed theories he discusses is therefore of considerable potential significance, expecially if it can be generalized to other theories. It is interesting to note that the status of referring expressions, and hence of pronouns (discussed in some detail above), occupies a central and somewhat ambiguous role with regard to use and structure.

Hiż provides an alternative account of the nature of propositional meaning itself. Based not on the notion of truth, but on that of consequence alone, the paper thus gives a purely language-internal account of meaning, which nevertheless exhibits many similarities to a truth-conditional account. The system is axiomatized.

Lieb argues, on general grounds, that any adequate semantic theory of natural language must possess certain properties. In particular he points, like Kasher and in line with the general trend observed in the more detailed analytical papers discussed above, to the relationship which may be expected to hold between such a semantic system and other human institutions, defining basic (morphological) meaning in psychological terms and syntactic meaning as meaning-as-use. To illustrate his conception of syntactic meaning he analyzes certain aspects of the English definite determiner.

REFERENCES

Chomsky, N. (1978). On Wh-movement. In P. Culicover, A. Akmajian, and T. Wasow (Eds.), *Formal Syntax*. New York: Academic Press.

Cooper, R. (1975). Montague's Semantic Theory and Transformational Syntax. Unpublished doctoral dissertation, University of Massachusetts, Amherst.

Cooper, R. (1977). Variable Binding and Relative Clauses. In *Proceedings of the Formal Semantics Workshop held in Bad Homburg*.

Cresswell, M. J. (1973). *Logics and Languages*. London: Methuen.

Hankamer, J., and Sag, I. (1976). Deep and surface anaphora. *Linguistic Inquiry* **7,** 391–428.

Montague, R. (1970). Universal grammar. In R. Thomason (Ed.), *Formal Philosophy: Selected Papers of Richard Montague.* New Haven: Yale University Press.

Partee, B. (1975). Montague grammar and transformational grammar. *Linguistic Inquiry* **6,** 203–300.

Postal, P. (1972). Some further limitations on interpretive theories of anaphora. *Linguistic Inquiry* **3,** 349–370.

Quine, W. V. O. (1960). *Word and Object.* Cambridge, Massachusetts: MIT Press.

Rodman, R. (1976). Scope phenomena, 'movement transformations', and relative clauses. In B. Partee (Ed.) *Montague Grammar.* New York: Academic Press.

Thomason, R. (1974). Introduction. In R. Thomason (Ed.), *Formal Philosophy: Selected Papers of Richard Montague.* New Haven: Yale University Press.

Williams, E. (1977). Discourse and logical form. *Linguistic Inquiry* **8,** 101–139.

PART I

THE ANALYSIS OF PRONOUNS

THE SYNTAX AND SEMANTICS OF SUBORDINATE CLAUSE CONSTRUCTIONS AND PRONOMINAL COREFERENCE

RENATE BARTSCH

1. SKETCH OF THE GRAMMATICAL FRAMEWORK

One prominent goal of linguistic theory is to explicate the relationship between linguistic forms and the results of interpreting them. This explication of the interpretation of linguistic forms of a language L is achieved by formulating what is called "a grammar of L." The first step in achieving that explication for certain linguistic forms is to construct a grammatical fragment that takes into account those forms. I first want to sketch the grammatical framework in which the fragment considered here is located.

I assume that categorial syntax is a means of representing the roles that the constituents of an expression play with regard to other constituents in determining the semantic value of a sentence. The semantic role of an expression is the role it plays in the determination of the semantic value of an expression of which it is a constituent. The role in which an expression is used is indicated by its position within a bigger expression, and especially within a sentence, by its morphological marking, and often also by its se-

SYNTAX AND SEMANTICS
VOLUME 10

ISBN: 0-12-613510-X

mantic content. A single lexical item can be used in different roles; for example, the German adjective *seltsam* 'mysterious' in (1)–(4), where it is used as an adnominal, a manner adverbial, a sentence adverbial, and an adsentential, respectively:

(1) *Frans ist ein seltsamer Mann.*
'Frans is a mysterious man.'

(2) *Frans soll seltsam singen.*
'Frans, presumably, sings mysteriously.'

(3) *Ein Mann will seltsamerweise singen.*
'A man, mysteriously, wants to sing.'

(4) *Seltsamerweise will ein Mann singen.*
'Mysteriously, a man wants to sing.'

In (1)–(4), we receive different positional and morphological information, which is represented by the categorial forms (5)–(8):

(5) $(\textit{seltsam}'_{\mathrm{AN}}(\textit{Mann}_{\mathrm{N}})_{\mathrm{N}})$

(6) $(\textit{seltsam}_{\mathrm{Av,m}}(\textit{sing-}_{\mathrm{V}})_{\mathrm{V}})$

(7) $(\textit{seltsam}_{\mathrm{AV,s}}(\textit{sing-}_{\mathrm{V}})_{\mathrm{V,s}})$

(8) $(\textit{seltsam}_{\mathrm{AS}}(\textit{Ein Mann will singen}_{\mathrm{S}})_{\mathrm{S}})$

These categorial structures form a conventional means used in grammatical theory to express the information about the semantic roles of expressions that we get from the surface structure. This information is called "categorial information," or "categorial structure." It is written down in such a way that function expressions are in front of argument expressions, and subcategorial information is marked beside the categorial information. Both kinds of information play a role in the interpretation of an expression. Special information from a surface expression, such as positional and morphological information, together with general grammatical knowledge, such as lexical and syntactic knowledge with regard to language L, permit a speaker to recognize the categorial and subcategorial structure of the expression.

Besides this syntactical–lexical structure, we need to find out the pattern of pronominal coreference in order to interpret an expression. Very often, coreference cannot be established unambiguously from mere surface information. Even though there are surface restrictions placed on coreference patterns by morphological agreement in number and gender, these are not nearly sufficient to prevent ambiguity. Positional distance between noun phrase and pronoun, and intervening noun phrases with the same number and gender markings do not consistently narrow down the possibility of a multiplicity of coreference patterns for a single expression, because semantic information and knowledge of the world can intervene and make a structurally improbable coreference pattern highly probable. Instead of trying to formulate coreference patterns through grammatical information received only from surface structure information, I would ultimately prefer to propose strategies for interpreting pronouns which are guided by semantic information from the lexical items involved, as well as context, knowledge of the world, and surface structure information. It hardly seems to be possible to establish coreference patterns on the level of logical representation of the text, because at that stage no notion of context is available that includes information like "grammatical gender," "distance in the text," or "main clause/subordinate clause" distinction.[1] In the present paper, I can only formulate a very imperfect interpretation procedure for pronouns that is explicitly based on grammatical information received from the surface structure, and that only implicitly takes into account semantic information in the notion of "text coherence."[2] I assume that anaphoric and kataphoric relationships within sentences (especially in clause constructions, that is, sentences consisting of clauses) receive preference over such relationships extending over the sentence boundary, as long as the coreference patterns established under this assumption do not destroy text coherence.

The interpretation of expressions is achieved by translating categorial and subcategorial information (that is, categorial structure) and information about coreference patterns into an intensional logic with expressions of different types, as well as some sortal distinctions in the realm of basic entity expressions, such as individuals, events and circumstances, actions, activities, etc. This translation is, as far as coreference is concerned, context-dependent. This means that coreference is taken into account in the process of translating into the logic language. The logic language serves as a means for giving a partial description of the worlds that are compatible with the information built up in the course of the interpretation of a text.

In the fragment presented in this paper, I will concentrate on formalizing the relationship between categorial structures and their interpretation through the logic language chosen, which is assumed to be interpreted in the usual way. I will pay less attention to the formulation of the relationship between surface information and the categorial structures, and will only mention that this relationship can be made explicit by general rules of serialization of the hierarchical categorial structures, by language specific rules of

[1] This makes it difficult to see how Ballmer's approach (1975 and 1976) can work. He uses pronouns in his "logical forms" on an intermediate level of representation between natural language and predicate logic. But on that level a lot of information that is necessary for the interpretation of pronouns is already lost, and I do not see that it is somehow incorporated into his "logical forms."

[2] This is a very problematic concept; but some good work on this notion has been done in recent studies in the theory of texts, as, for example, in van Dijk 1977.

serialization, by lexicalization rules, by morphological rules, and by morphophonemic rules.[3]

Besides context-dependent translation into the logic language, another deviation from current developments in Montague grammar should be pointed out: I assume that syntactic ambiguity in surface structures is restricted to a minimum. Thus, I would only assume a surface structure ambiguity if it is caused by morphological ambiguity, for example, *Flying planes can be dangerous*. But I would not assume a syntactic ambiguity corresponding to the *de re* and *de dicto* readings of sentences, and neither would I assume it for sentences like *Every man loves a woman,* where, in Montague grammar, two structurally different orders of the quantifier expressions are assumed resulting in two different meanings and, furthermore, several syntactically different derivations are possible without different results in meaning. Restricting syntactic ambiguity to morphological ambiguity simplifies the set of syntactical rules and makes the parsing of linguistic expressions a finite endeavor. It rules out all the unwanted syntactic ambiguities created in Montague grammar, which can generate an infinite set of syntactic structures for every sentence. The *de re/de dicto* distinction is taken care of by providing two translations for verbs that are intensional in place i, namely, an extensional reading and an intensional reading in i, and correspondingly, two translations of terms, differing in types for the application to a place marked as intensional or extensional, respectively. The second reading of sentences like *Every man loves a woman,* in which there is a woman such that every man loves her, will be inferred from a context that yields one specific reading for *a woman*. The general stategy is to try to make do with one reading, namely, the weaker one, and then derive the other reading from the weaker one together with the contextual information.

Since I do not want to overload this paper with details not directly relevant to the topic of subordinate clauses and pronominal reference, I will keep the fragment for the construction of simple sentences as small as possible, only quoting two rules of categorial syntax.[4]

(S1) If β' is a V^n (n-place verb) with the set of n term-places K, $i \in K$, and if α' is a T (term), then $(\alpha', i)(\beta')$ is a V^{n-1} with the set of term-places $K - \{i\}$.

For this we write $(\alpha'_{T\langle i\rangle}(\beta'_{V^n})_{V^{n-1}})$.

[3] Compare Bartsch and Vennemann 1972: In German, most hierachies are serialized according to A(B)——AB, and not as BA. Thus, the subordinate clause word order, also used in all infinitival constructions, of . . . *der Junge dem Vater den Ball gibt* is a serialization of (*der Junge*(*dem Vater*(*den Ball*(*gibt*)))), and . . . *der Junge den Ball dem Vater gibt* is a serialization of (*der Junge*(*den Ball*(*dem Vater*(*gibt*)))). In constructing the grammar, I assume the preferred type of natural serialization for German, if that is possible. In this way the number of specific rules of serialization is minimized.

[4] Others can be found in Bartsch 1976a.

(S2) If α' is a D (determiner) and β' is an N (noun), then $\alpha'(\beta')$ is a T.

The corresponding translation rules are

(T1) If α'' is the translation of α' as a T, and $\lambda x_j \cdot\cdot\cdot x_m \beta''(x_j, \ldots, x_m)$, with n places, the translation of β' as a V^n, then the translation of $(\alpha', i)(\beta')$ is

$$\lambda x_j \cdot\cdot\cdot x_i' \; 'x_i \cdot\cdot\cdot x_m(\alpha''(\lambda x_i(\beta''(x_j, \ldots, x_m)))), \quad \text{with } x_i'$$

as the variable that precedes x_i and $'x_i$ as the variable that follows x_i.[5]

(T2) If α'' is the translation of α' as a D, and β'' is the translation of β' as an N, then $\alpha''(\beta'')$ is the translation of $\alpha'(\beta')$.

It should be noted that in **T2** the translation has to be what later in the paper will be introduced as the "instantiating translation."

2. PRONOMINAL COREFERENCE

2.1. Relative Clause Constructions

The grammar of German should be able to account for the interpretation of sentences like (9), the English counterpart of which in Montague's fragment (1974) cannot be interpreted in its most natural reading with a coreferential reading between *it* and *a car*.

(9) *Jeder Mann, der* ***ein Auto*** *kauft, versichert* ***es.***
'Every man who buys **a car** insures **it.**'

Similarly, the well-known Bach–Peters sentences are not interpretable in Montague's fragment (1974), for example (10) and (11).

[5] Note that this rule corresponds to **S14/15** in Montague 1974, which is the basis for the construction of *de re* readings. In the present framework, we get *de dicto* readings in the following way: In the lexicon, we have $\lambda x_1 \cdot\cdot\cdot x_n \beta''(x_1, \cdot\cdot\cdot, x_n)$ as the translation of n-place verbs that are extensional in all places. For an n-place verb that can be intensional in its i-th place, we have $\lambda x_1 \cdot\cdot\cdot \mathbf{P}_i \cdot\cdot\cdot x_n \beta''_{*i}(x_1, \cdot\cdot\cdot, \mathbf{P}_i, \cdot\cdot\cdot, x_n)$ as the intensional translation where $\mathbf{P}$ is a variable for expressions that denote intensions of translations of terms. Thus, certain verbs are lexically ambiguous relative to whether they are translated extensionally or intensionally in one of their places. (One could, of course, just assume the intensional reading in the lexicon and introduce an "extensionalization" operation that depends on context information.) Correspondingly, a term α' can be translated not only by the extensional reading $\alpha'' = \lambda P \cdot\cdot\cdot$, but also by an intensional one, $\alpha''_* = \lambda \mathbf{P}\mathbf{P}(\hat{}\lambda P \cdot\cdot\cdot)$, where $\mathbf{P}$ is a variable for predicates of intensions of term translations, that is, a variable for predicates of intensions of sets of predicates of individuals. With this, $(\alpha'_{T\langle i\rangle}(\beta'_{V^n})_{V^{n-1}})$ translates into $\lambda x_j \cdot\cdot\cdot x_i' \; 'x_i \cdot\cdot\cdot x_m(\alpha''_*(\lambda(\lambda \mathbf{P}_i (\beta''_{*i}(x_j, \cdot\cdot\cdot, \mathbf{P}_i, \cdot\cdot\cdot, x_m))))$ n-places $x_j \ldots x_m$, in its *de dicto* reading. The extensional reading is the one described by Rule **T1**.

(10) ***The man/a man/every man*** *who finds* **a woman** *who loves* ***him*** *dates* **her.**

(11) ***The man/a man/every man*** *who finds* **her** *dates* **the woman** *who loves* ***him.***

On the other hand, we also want to take into account the reading with non-coreferentiality within the sentence, which we get in certain contexts, as, for example, in (12) and (13).

(12) *John knows* ***a nice girl.*** *Every man who loves every woman that pays attention to him dates* ***her.***

(13) ***Kathrin*** *is* ***a well-known psychologist.*** *A man who listens to* ***her*** *dates a woman who loves him.*

Here, it is text coherence that forces us to look for the coreferential term for *her* in the previous sentences. With the usual reading of Bach–Peters sentences, the sequence of two sentences in (12) and (13) would not be a text.

Bach–Peters sentences have been treated in detail in Bartsch (1976b). Now, I want to extend the interpretation procedure presented there in such a way that it covers pronominal coreference in other kinds of subordinate clause constructions and in texts.

2.2. Subject Clause and Object Clause Constructions

In the following examples, the cross (+) marks sentences and texts that are deviant with a coreferential reading for the underlined terms, or in other words, that do not allow for coreferentiality between the bold face terms. We make two distinctions. First, whether the subordinate clause is preposed or not, and second, whether the pronoun in question is in the subordinate clause or not.

A. SUBORDINATE CLAUSE FIRST

(i) *Full term in subordinate clause, pronoun in superordinate clause*
Subject clause:

(14) a. *?Dass* $\left\{\begin{array}{l}\textbf{\textit{John}}\\ \textbf{\textit{der Mann}}\end{array}\right\}$ *die Stelle bekam, freute* ***ihn.***

b. *+Dass* $\left\{\begin{array}{l}\textbf{\textit{ein Mann}}\\ \textbf{\textit{jeder Mann}}\\ \textbf{\textit{kein Mann}}\end{array}\right\}$ *die Stelle bekam, freute* ***ihn.***

'That **John/the man/a man/every man/no man** got appointed pleased **him.**'

Object clause:

(15) a. ?*Dass* {***John*** / ***der Mann***} *die Stelle bekam, erwartete* ***er.***
lit. 'That **John/the man** got appointed **he** expected.'

b. +*Dass* {***ein Mann*** / ***jeder Mann*** / ***kein Mann***} *die Stelle bekam, erwartete* ***er.***

c. +*Dass sie* ***John*** *betrogen hatte, verzieh* ***er*** *Maria nie.*
lit. 'That she had deceived **John he** never forgave Maria.'

But: *Dass* ***Maria*** *ihn betrogen hatte, verzieh John* ***ihr*** *nie.*
lit. 'That **Maria** had deceived him John never forgave **her.**'

(ii) *Pronoun in subordinate clause, full term in superordinate clause*
Subject clause:

(16) *Dass* ***er*** *die Stelle bekam, freute* {***John*** / ***den Mann*** / ***einen Mann*** / ***jeden Mann*** / ***keinen Mann***}.

'That **he** got appointed pleased **John/** . . . .'

Object clause:

(17) *Dass* ***er*** *die Stelle bekam, erwartete* {***John*** / ***der Mann*** / ***ein Mann*** / ***jeder Mann*** / ***kein Mann***}.

lit. 'That **he** got appointed expected **John/** . . . .'

B. Main (Superordinate) Clause First

(i) *Full term in subordinate clause, pronoun in superordinate clause*
Subject clause:

(18) +*Es freute* ***ihn,*** *dass* {***John*** / ***der Mann*** / ***ein Mann*** / ***jeder Mann*** / ***kein Mann***} *die Stelle bekam.*
'It pleased **him** that **John/** . . . got appointed.'

Object clause:

(19) +***Er*** *erwartete, dass* {***John*** / ***der Mann*** / ***ein Mann*** / ***jeder Mann*** / ***kein Mann***} *die Stelle bekam.*

'**He** expected that **John/** . . . got appointed.'

(ii) *Pronoun in subordinate clause, full term in superordinate clause*
Subject clause:

(20) {***John*** / ***Den Mann*** / ***Einen Mann*** / ***Jeden Mann*** / ***Keinen Mann***} *freute es, dass* **er** *die Stelle bekam.*

'It pleased **John/** . . . that **he** got appointed.'

Object clause:

(21) {***John*** / ***Der Mann*** / ***Ein Mann*** / ***Jeder Mann*** / ***Kein Mann***} *erwartete, dass* **er** *die Stelle bekam.*

'**John/** . . . expected that **he** got appointed.'

According to these data (which reflect my linguistic intuition as a speaker of German), subject clauses and object clauses may not contain a full term that is coreferential with a pronoun in the main (superordinate) clause. The pronoun has to appear instead in the subordinate clause and the full term in the main (superordinate) clause, regardless of clause position. The same does not hold for other subordinate clause constructions. (Note that the grammar to be developed in the present paper will render the sentences marked questionable as grammatical with the coreferential reading, because my intuition with regard to these sentences is not very clear, and there are people who find them acceptable with the coreferential reading.)

2.3 Conditional Clause Constructions

A. Subordinate Clause First

(i) *Full term in subordinate clause, pronoun in superordinate clause*

(22) a. *Wenn* {***John*** / ***der Mann***} *gewinnt, ist* **er** *glücklich.*

'If **John/the man** wins **he** is happy.'

b. *Wenn* ***ein Mann*** *gewinnt, ist* ***er*** *glücklich.*
'If **a man** wins **he** is happy.'

c. +*Wenn* $\left\{\begin{matrix}\textbf{\textit{jeder Mann}}\\ \textbf{\textit{kein Mann}}\end{matrix}\right\}$ *gewinnt, ist* ***er*** *glücklich.*
'If **every man/no man** wins **he** is happy.'

(ii) *Pronoun in subordinate clause, full term in superordinate clause*

(23) *Wenn* ***er*** *gewinnt, ist* $\left\{\begin{matrix}\textbf{\textit{John}}\\ \textbf{\textit{der Mann}}\\ \textbf{\textit{ein Mann}}\\ \textbf{\textit{jeder Mann}}\\ \textbf{\textit{kein Mann}}\end{matrix}\right\}$ *glücklich.*
'If **he** wins **John/** . . . is happy.'

B. Main (Superordinate) Clause First

(i) *Full term in subordinate clause, pronoun in superordinate clause*

(24) +***Er*** *ist glücklich, wenn* $\left\{\begin{matrix}\textbf{\textit{John}}\\ \textbf{\textit{der Mann}}\\ \textbf{\textit{ein Mann}}\\ \textbf{\textit{jeder Mann}}\\ \textbf{\textit{kein Mann}}\end{matrix}\right\}$ *gewinnt.*
'**He** is happy if **John**/ . . . wins.'

(ii) *Pronoun in subordinate clause, full term in superordinate clause*

(25) $\left\{\begin{matrix}\textbf{\textit{John}}\\ \textbf{\textit{Der Mann}}\\ \textbf{\textit{Ein Mann}}\\ \textbf{\textit{Jeder Mann}}\\ \textbf{\textit{Kein Mann}}\end{matrix}\right\}$ *ist glücklich, wenn* ***er*** *gewinnt.*
'**John/** . . . is happy if **he** wins.'

2.4. Adverbial Clause Constructions

A. Subordinate Clause First

(i) *Full term in subordinate clause, pronoun in superordinate clause*

(26) a. $\left\{\begin{matrix}\textit{Obgleich}\\ \textit{Weil}\\ \textit{Nachdem}\end{matrix}\right\}\left\{\begin{matrix}\textbf{\textit{John}}\\ \textbf{\textit{der Mann}}\\ \textbf{\textit{ein Mann}}\end{matrix}\right\}$ *gewonnen hat, ist* ***er*** *glücklich.*
'Although/because/after **John/the man/a man** has won, **he** is happy.'

b. + {*Obgleich* / *Weil* / *Nachdem*} {***jeder Mann*** / ***kein Mann***} *gewonnen hat, ist* ***er*** *glücklich.*
'Although/because/after **every man/no man** has won, **he** is happy.'

(ii) *Pronoun in subordinate clause, full term in superordinate clause*

(27) {*Obgleich* / *Weil* / *Nachdem*} ***er*** *gewonnen hat, ist* {***John*** / ***der Mann*** / ***ein Mann*** / ***jeder Mann*** / ***kein Mann***} *glücklich.*
'Although/because/after **he** has won **John/** . . . is happy.'

B. Main (Superordinate) Clause First

(i) *Full term in subordinate clause, pronoun in superordinate clause*

(28) +***Er*** *ist glücklich,* {*obgleich* / *weil* / *nachdem*} {***John*** / ***der Mann*** / ***ein Mann*** / ***jeder Mann*** / ***kein Mann***} *gewonnen hat.*
'**He** is happy, although/ . . . **John/**. . . has won.'

(ii) *Pronoun in subordinate clause, full term in superordinate clause*

(29) {***John*** / ***Der Mann*** / ***Ein Mann*** / ***Jeder Mann*** / ***Kein Mann***} *ist glücklich,* {*obgleich* / *weil* / *nachdem*} ***er*** *gewonnen hat.*
'**John/** . . . is happy, although/ . . . **he** has won.'

If we compare nominalized adverbial constructions in German with the above data, we realize that they behave like subordinate clause constructions with respect to possible coreferentiality, except that the nominalized form given in (30) corresponding to (26) is impossible with coreferentiality:

(30) + {*Trotz* / *Wegen* / *Nach*} {***Johns*** / ***des Mannes*** / ***eines Mannes*** / ***jedes Mannes*** / ***keines Mannes***} *Sieg(es) ist* ***er*** *glücklich.*
'In spite of/because of/ after **John's/** . . . victory **he** is happy.'

In English, unlike German, (30) parallels (26).

(31) $\left\{\begin{matrix}\left\{\begin{matrix}\textit{Trotz}\\ \textit{Wegen}\end{matrix}\right\} & \textit{\textbf{seines} Sieges}\\ \textit{Nach} & \textit{\textbf{seinem} Sieg}\end{matrix}\right\}$ *ist* $\left\{\begin{matrix}\textbf{\textit{John}}\\ \textbf{\textit{der Mann}}\\ \textbf{\textit{ein Mann}}\\ \textbf{\textit{jeder Mann}}\\ \textbf{\textit{kein Mann}}\end{matrix}\right\}$ *glücklich.*

'In spite of/ . . . **his** victory **John/** . . . is happy.'

(32) +***Er*** *ist glücklich* $\left\{\begin{matrix}\textit{trotz}\\ \textit{wegen}\\ \textit{nach}\end{matrix}\right\}\left\{\begin{matrix}\textbf{\textit{Johns}}\\ \textbf{\textit{des Mannes}}\\ \textbf{\textit{eines Mannes}}\\ \textbf{\textit{jedes Mannes}}\\ \textbf{\textit{keines Mannes}}\end{matrix}\right\}$ *Sieg.*

'**He** is happy in spite of/ . . . **John's/** . . . victory.'

(33) $\left\{\begin{matrix}\textbf{\textit{John}}\\ \textbf{\textit{Der Mann}}\\ \textbf{\textit{Ein Mann}}\\ \textbf{\textit{Jeder Mann}}\\ \textbf{\textit{Kein Mann}}\end{matrix}\right\}$ *ist glücklich wegen **seines** Sieges.*

'**John/** . . . is happy because of **his** victory.'

If it were not for the acceptable readings of the examples (22) and (26), we could conclude (A):

(A) Full terms in subordinate clauses cannot be coreferential with pronouns in superordinate clauses.

Note that the exceptions to this statement are sentences in which the full term can be interpreted as denoting an individual; that is, in the semantic representation we can use an individual constant rather than a quantitifier expression in the translation of these terms. Thus, we could replace (A) by (B):

(B) Full terms with a quantificational analysis in subordinate clauses cannot be coreferential with pronouns in superordinate clauses. This means that quantifiers in the semantic representation of subordinate clauses cannot bind variables in superordinate clauses.

Statement (B) also fits nicely with the pronominal patterns in the texts in (34), which are to be compared with (26). Instead of a subordinate clause, we have in (34) a preceding sentence with a full term and a following sentence with a pronoun.

(34) a. $\left\{\begin{matrix}\textbf{\textit{John}}\\ \textbf{\textit{Der Mann}}\\ \textbf{\textit{Ein Mann}}\end{matrix}\right\}$ *hat gewonnen. Darum ist **er** glücklich.*

'**John/** . . . has won. Therefore **he** is happy.'

But: b. +{***Jeder Schüler dieser Klasse*** / ***Kein Schüler dieser Klasse***}
hat gewonnen. Darum ist ***er*** *glücklich.*
'**Every student of this class/no student of this class**
has won. Therefore **he** is happy.'

But even (B) does not seem to be quite correct. If we look at the indefinite term in (22b) *Wenn* ***ein Mann*** *gewinnt, ist er glücklich,* it seems that ***ein Mann*** 'a man' has a universal reading, and thus is not represented by an individual constant on the entity level. (We also find this universal reading in relative clause constructions, for example, *A man who buys a car insures it,* which can be interpreted as "Every man who buys a car insures it.")[6] But we can also interpret the indefinite term in (22b) by assuming a reading that may be called a "stereotype reading." There seems to be a difference between the stereotype reading and the universal reading with respect to pronominalization: In the universal quantifier reading, we have no coreferentiality between the phrase *jeder Teilnehmer* 'every participant' and the pronoun *er* 'he' in text (35).

(35) +***Jeder Teilnehmer*** *kam zur Eröffnungssitzung zu spät.* ***Er*** *wurde durch Schneefall aufgehalten.*
'**Every participant** came late to the opening session. **He** got delayed by a snowstorm.'

But in (36), where we have a stereotype reading, the phrase *jeder Seefahrer* 'every sailor' admits coreferentiality with the pronoun *er.*

(36) *Früher litt* ***jeder Seefahrer*** *an Vitaminmangel.* ***Er*** *musste Sauerkraut essen.*
'In former times, **every sailor** suffered from vitamin deficiency. **He** had to eat sauerkraut.'

Here, it seems that the term *every sailor* refers to the stereotype sailor (of former times), and so does the pronoun. The same reference to the stereotype sailor is established by *the sailor/a sailor* in (37).

(37) *Früher litt* ***der Seefahrer/ein Seefahrer*** *an Vitaminmangel.* ***Er*** *musste Sauerkraut essen.*
'In former times, **the sailor/a sailor** suffered from vitamin deficiency. **He** had to eat sauerkraut.'

The same holds for the stereotype reading of *kein arktischer Penguin* 'no arctic penguin' in (38).

[6] The treatment of relative clauses in Bartsch 1976b takes care of this, as well as the universal reading of *If a man buys a car he insures it.*

(38) ***Kein arktischer Penguin*** *überlebt die Gefangenschaft.* ***Er*** *stirbt binnen drei Jahren.*
'**No arctic penguin** survives captivity. **He** dies within three years.'

The logical structure of stereotype sentences seems to be the following: The stereotype α^* is the set of properties that entities of the kind α normally (or mostly) have.[7] The stereotype α^* *has the property* β means that the property β belongs to α^*: $\alpha^*(\beta)$; that is, $(\lambda P(\forall x(\alpha(x) \rightarrow \textit{normally}\ (\,^\wedge(P(x))))))(\beta)$. The negation, *no* α^* *has* β, that is, $(\textit{no}\ \alpha^*)(\beta)$ is $\alpha^*(\lambda x(-\beta(x)))$, and thus $(\textit{no}\ \alpha^*) = \lambda Q\alpha^*(\lambda x(-Q(x)))$. The pronoun *er* in (38) has as a coreferent α^*. The first sentence of (38) is thus equivalent with the stereotype readings of *Der arktische Penguin überlebt die Gefangenschaft nicht,* or *Ein arktischer Penguin überlebt die Gefangenschaft nicht.*

We can maintain our postulate (B) for texts also, if we assume a stereotype reading, which is not a quantifier reading, in those cases where coreference between a full nonspecific term in the preceding sentence and a pronoun with gender–number agreement in the following sentence is possible.

3. THE TRANSLATION PROCEDURE FOR TERMS

3.1. Outline of the Translation Procedure

In the following, the interpretation procedure for full terms and for pronouns from Bartsch (1976b will be repeated and extended. Proper names are translated following Montague 1974; for example, *Hans'* by $\lambda PP(h)$. The semantic representations of subordinate clause constructions will be suggested in the next section, and finally, the interpretation procedure for pronouns will be connected with the procedure by which subordinate clause constructions are translated into their semantic representations expressed in the logic language.

To account for the linguistic data just presented, pronouns must be translated in such a way that coreferentiality is established on the level of semantic representation in the logic language either by a single contextually introduced constant or by a single variable bound by a single quantifier. The first is possible within texts, that is, domains extending beyond the sentence boundary as well as within clause constructions, or, within the sentence boundary. Binding of the one variable by a single quantifier is assumed to take place only within the sentence boundary. The contextually introduced constants are brought in by the translation of proper names or by a proce-

[7] For social stereotypes this analysis is given by Quasthoff (1973). For other stereotypes, it is implicit in Putnam's use of the concept "stereotype" in Putnam 1975.

dure called "referent introduction,"[8] which is always possible for the specific interpretation of terms and for the stereotype interpretation of terms. For proper names and definite and indefinite terms in the specific reading, constants of the entity type are introduced in the process of translation of the terms into the logic language. For indefinite terms with an unspecific reading, referent introduction on the entity level takes place if the term is not within the scope of a negation operator preceded by a term of the form *every . . .* or *no . . . .* Within the scope of an intensional verb, there is referent introduction for the specific reading of terms on the entity level, and there is referent introduction for unspecific readings of terms on an ideational level. If a term is interpreted as unspecific with regard to the speaker and specific with regard to the subject of the intensional verb, we get the "unspecific" translation (namely, translation by a quantifier expression), and referent introduction. If we have texts like (39), we assume entity referents.

(39) *John wants to have* ***a car. It*** *is red.*

We can interpret this text in such a way that the speaker has a specific car in mind. This is the specific reading, where we use a constant in the translation of *a car.* We can also interpret the text to mean that the speaker does not know which car, but only knows that John has a specific car in mind. This is the unspecific reading for which we use a quantifier translation, and in addition introduce a new constant on the entity level for use in the translation of pronouns that might appear in the following text and that are coreferential with *a car.* But in texts like (40), we assume ideational referents, that is, referents in the worlds characterized by the intensional verb in question.

(40) *John wants to have* ***a car. It*** *must be red.*

The modal verb *must* indicates that the speaker makes the requirement (maybe taking over a requirement John makes) that the property RED belongs to the set of properties of the car John has in all the worlds that are partially described by the fulfillment of John's wishes.

For terms of the form *every . . .* , *no . . .* , and for *a/an . . .* under the above-mentioned conditions, there is no referent introduction on the entity level, though there is something that at least within the sentence boundary can fulfill the same purpose with regard to the translation of pronouns that are coreferential with those terms: the use of blocked variables. The translation with blocked variables occurs in two phases: The first step is the

[8] This term was introduced by Ballmer (1972), and is based on Karttunen's concept of "discourse referents" in Karttunen 1971b. The interpretation of deictic pronouns relies on referents being introduced by deixis. In this way the presented procedure can be extended to deictic pronouns as suggested by Heny in the introduction to this volume. Furthermore, referent introduction for deictic pronouns in general can be added as a special case of referent introduction for specific terms in **RI**, Section 3.3, as done in Bartsch 1976a.

instantiating translation and the second step is the generalization of this translation. In the instantiating translation the term is translated in such a way that in the translation a variable is blocked. This means that it cannot be used in the translation of other phrases of the sentence except for pronouns that are coreferential with the term in question. After the instantiating translation of a sentence is completed, the step of generalization takes place, in which the quantitifers are added to the so far open formula.[9] Only when the instantiating translation has become generalized (with respect to all quantificationally translated terms) do we obtain the translation of the sentence.

3.2. The Two-Stage Translation of Full Terms

This is the translation strategy in sentence interpretation:

IN: In translating the terms *ein'* . . . ('a/an . . .') and *d-'* . . . ('the . . .') in their unspecific readings, and *jed-'* . . . ('every . . .') and *kein-'* . . . ('no . . .') introduce and block a new variable ν as indicated in the following provisional translations:

Instantiating translations

ein-': $\lambda Q \lambda P(Q(\nu) \ \& \ P(\nu))$
d-': $\lambda Q \lambda P(Q(\nu) \ \& \ (\forall y)(Q(y) \rightarrow \nu = y) \ \& \ P(\nu))$
jed-': $\lambda Q \lambda P(Q(\nu) \rightarrow P(\nu))$
kein-': $\lambda Q \lambda P(Q(\nu) \rightarrow -P(\nu))$

After having completed the sentence translation with instantiating term translations, universal and existential generalization takes place according to the following procedure:

Generalized translation

(i) *Order of quantifier application*

The hierarchical order of quantifiers reflects the linear order of terms in the sentence to be translated; that is, the quantifier with the broadest scope corresponds with the leftmost term and the

[9] The methods of instantiating and generalizing are adopted from well-known strategies in logical deduction, namely, existential and universal instantiation and existential and universal generalization, for example, in Quine 1956 (and further editions). It is not surprising that a procedure based on this method has some similarity to Hintikka's language games with respect to quantifiers, as used in Edmondson 1976; Hintikka's language games (1973) are very similar to procedures used in logical deduction, as, for example, in Quine 1956 (and further editions) and in the method of semantic and deductive tableaux developed by Beth (1969 and elsewhere). Hintikka's own approach to the problem of Bach–Peters sentences (cf. Hintikka and Saarinen, 1975) takes up the problem from a different angle, namely, from a generative semantic point of view. As a whole, it is more suggestive than explicit because the relationship between logical forms and natural language sentences is not explicated.

quantifier with the narrowest scope corresponds with the rightmost.[10]
(ii) *Choice of quantifiers*
(iia) If a term A is part of a relative clause with a governing term of the form *jed-* . . . or *kein-* . . . (generally: If a term A is preceded by an uneven number of head terms of the form *jed-* . . . or *kein-* . . . in relative clause embeddings)[11], then if A = *ein-* . . . or *d-* . . . , the generalization is achieved by binding the variable introduced with instantiating translation of A by the universal quantifier ∀; if A = *jed-* . . . or *kein-* . . . then the generalization is achieved by the existential quantifier ∃.
(iib) Otherwise, we achieve generalization with regard to terms of the form *ein-* . . . and *d-* . . . by the existential qualtifier ∃, and with regard to terms of the form *jed-* . . . and *kein-* . . . by the universal quantifier ∀.
(iii) *Cancelling of variable blockade*
After the generalization is completed, variable blockade is cancelled. With this, the translation of the sentence is completed.

This procedure of instantiating and generalizing translations can be used as an obligatory one, that is, in the translation of every sentence. But it also does not hurt to use it obligatorily only when coreference between pronouns and noun phrases (in their unspecific readings) has been assigned by Rule **IP** (page 42). If one chooses the latter option, one can in all cases use the generalized translations of noun phrases to begin with, as in Montague 1974 or in Bartsch 1976a.

[10] The restriction on the order of quantifiers, according to (i), is a kind of reversal of the "global constraints" of Lakoff 1971 for the derivation of noun phrases with quantifiers. These constraints are not only reversed, but also sharpened, because they hold for passive as well as active sentences. According to my intuition with respect to German, sentences like *Jeder Mann liebt eine Frau* 'Every man loves a woman' are not ambiguous between "For every man there is a woman who he loves" and "There is a woman who every man loves." Only the first interpretation is the one derivable following the syntax of the sentence in question. The second reading would be expressed by the passive sentence *Eine Frau wird von jedem Mann geliebt* or by the active sentence with reversed word order *Eine Frau liebt jeder Mann*. The second reading, the stronger one, can be derived from the first reading in certain contexts, where information is provided to the effect that the respective woman is the same for every man. The other possibility is that there are indications for a specific reading of *eine Frau*, as in *Jeder Mann liebt eine Frau. Sie ist Maria* 'Every man loves a woman. She is Mary.'

[11] The reversal of quantifiers according to (iia) is necessary because under this condition the translation of A lies within the antecedent clause of a conditional construction. Since the scope of the generalization exceeds the antecedent clause, the quantifier has to be reversed. This restriction is based on the fact that the formula $(\forall x)(\ldots x \ldots) \rightarrow p$ is not equivalent with $(\forall x)(\ldots x \ldots \rightarrow p)$, but rather with $(\exists x)(\ldots x \ldots \rightarrow p)$, and $(\exists x)(\ldots x \ldots) \rightarrow p$ is not equivalent with $(\exists x)(\ldots x \ldots \rightarrow p)$, but rather with $(\forall x)(\ldots x \ldots \rightarrow p)$

3.3. Referent Introduction

In the preceding section, we have seen that terms of the form *jed-* . . . and *kein-* . . . , quantifying over entities cannot be coreferential with a pronoun in the following text or in the following superordinate clause, though they can be coreferential with a pronoun in the stereotype reading, where the pronoun does not refer to an entity but to a stereotype. The same is true for indefinite articles under the scope of negation or certain other conditions that prevent referent introduction on the entity level.[12] This can be seen in the following examples:

(41) *It is not the case that* ***an arctic penguin*** *dies in captivity.* ***He*** *can even raise young.*

But:

(42) a. +*It is not the case that* ***an arctic penguin*** *walked into the bedroom.* ***He*** *was below in the swimming pool all night.*

b. ***An arctic penguin*** *walked into the bedroom.* ***He*** *woke me up.*

To provide for coreferentiality through constants, we now have to formulate the procedure of referent introduction. The indefinite article can be understood as specific or as unspecific (*a*) with regard to the information possessed by the speaker, (*b*) with regard to the intensional verb in question and the information possessed by the subject of that verb, (*c*) with regard to the modal verb in question and the contextual information presupposed for its use,[13] or (*d*) with regard to a future tense marker. In order to formulate referent introduction, we need to define the concept of a *relevant universe of discourse:*

(i) If there is a verb, intensional in place *i*, to which a term of the form *ein-* . . . is applied in *i* in its intensional reading, the universe of discourse that is relevant for referent introduction with regard to this term is the universe of those worlds that are specified by the intensional verb and its subject.

(ii) If the term *ein-* . . . in its intensional reading is applied to a place of a verb marked for modality (for example, a modal verb + infinitive) or future tense, or if the term is in the scope of a modal or tense-indicating adsentential, the relevant universe of discourse with regard to this term is the set of worlds or points of time specified by the modal or future indicator involved.

(iii) For all other unspecific readings and for all specific readings the relevant universe of discourse with regard to the term *ein-* . . . is the basic uni-

[12] Cf. Karttunen 1971b and Edmondson 1976.

[13] For the relevance of the information possessed by the speaker to the semantics of modal verbs cf. Groenendijk and Stokhof 1975.

verse of discourse that is assumed to be included in the common speaker–hearer knowledge, including information from the previous text.

Referent Introduction (**RI**):
Unspecific indefinite article:

> If *ein-* . . . is not in the scope of a negation or preceded by a term of the form *jed-* . . .[14] or *kein-* . . . , then, after the translation (according to procedure **IN**) that specifies the variables P and Q, add to the relevant universe of discourse a new individual name with the specifications of P and Q.

Specific indefinite article:

> *ein-'* translates into $\lambda Q \lambda P(Q(v) \; \& \; P(v))$, with v being a newly introduced constant (cf. Bartsch, 1976a).

Specific definite article:

> *d-'* translates into $\lambda Q \lambda P(Q(v) \; \& \; P(v))$, with v determined according to the following reference instruction.

Reference instruction for $d\text{-}\alpha'$ (that is, $Q = \alpha''$):

(i) Take the last individual name introduced or used with the translation of the last sentence such that its referent has the property α'' (that is, the predicate specifying Q). If there is no such name, apply (ii).

(ii) Take the second last individual name introduced or used with the translation of the second last sentence such that its referent has the property α''. If there is no such name, apply (iii).

(iii) Check whether it holds in the information contained in the previous text and in presupposed information of the speaker–hearer that $(\exists x)(\alpha''(x) \; \&$

[14] Texts like the following can be handled, if interpreted as stereotypical sentences.

> *Jeder Mann heiratet eine Frau. Sie kocht für ihn. Er unterhält sie.*
> 'Every man marries a woman. She cooks for him. He supports her.'

The pronouns refer to stereotypes, as do the definite descriptions in *Der Mann unterhält die Frau, die er geheiratet hat. Die Frau kocht für den Mann, der sie geheiratet hat.* We do not assume a semantic representation with *jeder Mann* analyzed by the normal universal quantifier which then somehow gets extended beyond the sentence boundary. That this should not be done can be seen from the following text, which is deviant with the boldface terms as coreferential.

> +***Jeder Junge*** *in dieser Klasse besitzt* **einen Atlas.** *Gestern hat* **er ihn** *nicht mitgebracht.*
> 'Every boy in this class owns an atlas. Yesterday, he didn't bring it to school'.

This deviance can be explained if we assume that the scope of a quantifier does not extend beyond the sentence boundary; that is, sentence translation is generalized before the following sentence is interpreted. Since this text refers to a particular situation, the stereotypical reading is not possible, and the whole text is semantically deviant. It would not be deviant if *er* and *ihn* could refer deictically to individuals in the speech situation, or if the text were preceded by a sentence like *Hans hat einen besonders schönen Atlas* 'Hans has an atlas that is especially good,' where *er* could refer to Hans, and *ihn* to his atlas.

$(\forall y)(\alpha''(y) \rightarrow x = y))$. If that is the case, introduce a new name ν, with $\alpha''(\nu)$, unless there already exists some individual name, μ, possibly introduced, such that $\alpha''(\mu)$, and use the introduced or existing name.

(iv) Check whether in the presupposed speaker–hearer information that does not belong to the previous text information, it holds that $(\exists x)(\alpha''(x)$ & $(\forall y)(\alpha''(y) \rightarrow x = y))$. If that is the case, introduce a new name ν, with $\alpha''(\nu)$, unless there already exists some individual name, μ, possibly introduced, such that $\alpha''(\mu)$, and use the introduced or existing name.

There exists the following restriction of use for definite articles. *d-*α', 'the α'', can only be used if $(\exists x)(\alpha''(x)$ & $(\forall y)(\alpha''(y) \rightarrow x = y))$ holds in one of the realms of information described in the strategy (i)–(iv) above. If *d-*α' is used, this restriction gives reason to assume $(\exists x)(\alpha''(x)$ & $(\forall y)(\alpha''(y) \rightarrow x = y))$ for one of the realms of information described in the strategy (i)–(iv).

Stereotype reading:

The stereotype reading of *ein-*α', *d-*α', *jed-*α' can be obtained by translating these terms into the constant α^* on the stereotype level, with $\alpha^* = \lambda P(\forall x)(\alpha''(x) \rightarrow normally({}^\wedge P(x)))$, and *kein-*$\alpha'$ is translated into $\lambda Q\alpha^*(\lambda x(-Q(x)))$, with α^* as the introduced constant.

3.4. The Interpretation of Pronouns

We are now in a position to formulate a strategy for the translation of pronouns—except for pronouns of laziness.[15] The translation of nonreflexive

[15] Unlike Partee 1975 (p. 233) I do not treat *her* in *Every man such that he loves* ***a woman*** *loses* ***her*** as a pronoun of laziness. It is a true instance of coreferentiality and gets translated in my approach by a single bound variable common for the translation of *a woman* and *her*. Thus, in my framework this example can be handled without any problem (compare with sentence (9)) and therefore does not need to be treated as an instance of phrase substitution, as Partee (1975) suggests. But in the sentence *The man who gave* **his paycheck** *to his wife was wiser than the man who gave* ***it*** *to his mistress* from Karttunen 1969, the *it* is a pronoun of laziness. In this case, Partee's position of looking at *it* as a substitution instance must be taken, as, for example, is done in Bennett 1971. Pronouns of laziness are not instances of coreferentiality within sentences or texts for which my procedure is designed. For noncoreferential use of pronouns, that is, lazy use of pronouns, my procedure is only applicable after *it* has been replaced by *his paycheck*. In other words, the phrase *his paycheck* has to be translated twice; first in the context where it stands and second, in the context where *it* stands. Then the interpretation runs according to my procedure **IP** (see additional remarks about possessive pronouns, p. 42) by translating *his* by means of the constant or blocked variable used for the translation of the directly preceding phrase *the man* (**IP**, (ia)). Thus, the pronoun *his* in the phrase *his paycheck* gets translated by the constant or blocked variable used in the translation of the first phrase *the man;* and *it* gets translated by a translation of the phrase *his paycheck* translated in the immediate context of the second phrase *the man,* that is, by use of the constant or variable blocked with the translation of the second phrase *the man*. In the Karttunen example, we get different constants or variables in the translation of the pronoun and its assigned preceding noun phrase *his paycheck,* which we therefore should not call its preceding "coreferential" noun phrase. Note that for the interpretation of deictic pronouns (see Footnote 8) the reference instruction has to be extended to constants introduced by deictic reference.

pronouns *er* 'he,' *sie* 'she,' *es* 'it' (and morphologically related forms of their paradigms) proceeds according to the following procedure:

IP:

Reference instruction:

(i) (a) Search within the preceding superordinate or subordinate clause of the clause construction for the nearest term with gender–number agreement with the pronoun in question (such that the term and the pronoun do not bind places of the same verb). If there is no such term, apply (b).

(b) Search for the nearest following term with gender–number agreement in the following superordinate clause, if there is one within the clause construction (which consists of the subordinate clause and the superordinate clause). If there is no such term, apply (c).

(c) Search for the nearest preceding term with gender–number agreement in the preceding text.

(ii) If a term has been found according to (i) then translate the pronoun by $\lambda PP(\nu)$ with ν being the variable introduced and blocked in the course of translation of the term, or with ν being the constant used or introduced with the translation of the term. If the constant is a stereotype constant, the translation of the pronoun is just this constant. If there is no blocked variable or constant connected with the term found in the search according to (i), then follow procedure (i) for the next term with gender–number agreement.

(iii) If the interpretation achieved through the translation according to (i) and (ii) is very improbable or destroys text coherence, then follow procedures (i) and (ii) for the next nearest term with gender–number agreement.

The translation of possessive pronouns follows the same principles, except that the possessive relationship has to be included. If we have, for example, *sein Pferd* 'his horse' with *Hans* as coreferential term found according to procedure **IP**, *sein-'* is translated by $\lambda Q\lambda P(Q(\nu)\ \&\ \text{POSS}(h,\nu)\ \&\ P(\nu))$, and *sein Pferd'*, accordingly, by $\lambda P(Pferd''(\nu)\ \&\ \text{POSS}(h,\nu)\ \&\ P(\nu))$.

In all those cases where coreference between a term and a pronoun is not captured by binding a single variable by a single quantifier, we assume that it is captured by a single constant introduced with the translation of the preceding coreferential term. Binding variables by quantifiers can be done only when the quantifiers are based on full terms in superordinate clauses, but not if they are based on full terms in subordinate clauses or preceding sentences. In these cases, coreference can be expressed only by a single constant and thus it can only take place if referent introduction is possible. This means that full terms in subordinate clauses and in preceding sentences have to be translated fully, that is, with generalization, before the translation of the subordinate clause is connected with the translation of the superordinate clause. But in subordinate clauses, pronouns can be translated through coreference to the superordinate clause. To get the syntactic structure of the scope of quantifiers right, we have to incorporate into the verb phrase of the superordinate clause the adverbial and conditional subordinate clauses containing pronouns that are coreferential with quantificational terms within the superordinate clause. For example:

(43) *Wenn* ***er*** *krank ist, nimmt* ***jeder Engländer*** *Aspirin.*
'If **he** is sick **every Englishman** takes aspirin.'

The subordinate clause has to be incorporated into the P of $\lambda P(Englishman(x) \rightarrow P(x))$. Therefore, the categorial structure of these subordinate clauses has to be such that they are treated as a kind of adverbial. If sentences (15b), (22c) and (26b) were acceptable, we could treat the preceding subordinate clause as an adsentential applied to the main clause, rather than as an adverbial applied to the verb phrase of the main clause.

The preposing of the subordinate clause is a means of making its content thematic. If, on the other hand, the superordinate clause is first, then its content is thematic. I shall **not** express the pragmatic notion "thematic" by means of categorial structure in such a way that the preposed subordinate clause would be treated as an adsentential to the main clause, and the postposed subordinate clause as an adverbial. To do this would conflict with the semantic relationships involved in the coreference patterns. These require us to treat the subordinate clause as a part of the verb phrase of the superordinate clause. How this is done is the topic of the next section.

4. SUBORDINATE CLAUSE CONSTRUCTIONS

4.1. Relative Clause Constructions

I shall be brief about relative clause constructions because they are treated in detail in Bartsch 1976b, where other literature about this topic is reviewed also. There, I considered whether we should assume two syntactic constructions for relative clauses; one for restrictive and one for nonrestrictive clauses. It was decided against this and a single syntactic construction was assumed for all relative clauses, namely, the term-modifier construction, although from a mathematical point of view this made the translation rules for the restrictive relative clauses less elegant.

Relative clauses are built like sentences, with the restriction that the term which is applied to the verb phrase last is a relative pronoun or contains a relative pronoun. In the categorial syntax, this "relative term" does not have gender and number markings, but gets them when the relative sentence is bound to a governing noun or term and thus becomes a relative clause agreeing in gender and number with the governing noun or term. Morphologically, the relative term will be realized as a relative pronoun with the case marking that characterizes its function in the sentence, and with the gender–number marking of the governing noun or term. (An alternative would be a copying transformation followed by a context-free morphologization rule.) The morphological realization of the relative pronoun is, in this version, context dependent, because it is dependent on the governing noun or term. The relative clause in German is serialized in the subordinate clause word order, that is, the hierarchy of the terms filling in the case places of the verb is simply linearized from right to left: $(\alpha'(\beta')) \rightarrow \alpha\beta$. Semantically, the relative pronoun is taken as an open place which is assigned to the variable or constant used in the translation of the governing term or noun. This means that the relative pronoun has a double role:

(i) It stands for an open term-place in the relative clause and this role is indicated by the case marking.
(ii) It has an assignment relationship with the governing noun or term and this role is indicated by number and gender agreement with this noun or term.

The translation of the relative pronoun into the logic language is split up according to these two roles.

(i′) We bind the open place with the lambda operator thus obtaining from the open sentence a predicate, which, according to (ii′), is integrated into the translation of the governing noun or term.
(ii′) We introduce relative operators RELN and RELT, which relate the predicate constructed according to (i′) to the governing noun or term. Thus, the translation of a relative clause β' is $\text{RELN}(\lambda x\beta''(x))$ or $\text{RELT}(\lambda x\beta''(x))$.

In our categorial syntax, we have the following syntactical and semantical rules:

(i) *Adnominal relative clauses* (*restrictive relative clauses*):

(S3) If α' is a noun and β' a relative sentence, then $\beta'(\alpha')$ is a noun, (and β' in this role is called a "relative clause"). We write $(\beta'_{\text{RN}}(\alpha'_{\text{N}})_{\text{N}})$.

(T3) If α'' and $\text{RELN}(\lambda x\beta''(x))$ are the translations of α' and β' from **S3**, then $(\text{RELN}(\lambda x\beta''(x)))(\alpha'')$ is the translation of $\beta'(\alpha')$, with $\text{RELN}(\lambda x\beta''(x)) = \lambda P\lambda x(P(x) \;\&\; \beta''(x))$.

According to **T3**, we have $(\text{RELN}(\lambda x\beta''(x)))(\alpha'') = \lambda x(\alpha''(x) \;\&\; \beta''(x))$.

Example: *Mann, der trinkt*
'man who drinks'

This phrase has the categorial form $((d\text{-}'_{\text{T}\langle 1\rangle}(trink'_{\text{V1}})_{\text{RN}})(Mann'_{\text{N}})_{\text{N}})$. The translation of the relative clause is RELN($\lambda x\, trink''(x)$), which applied to *Mann''* results in $\lambda x(Mann''(x)\ \&\ trink''(x))$.

(ii) *Adterminal relative clauses* (*nonrestrictive relative clauses*)

(S4) If α' is a term and β' a relative sentence, then $\beta'(\alpha')$ is a term. We write $(\beta'_{\mathrm{RT}}(\alpha'_{\mathrm{T}})_{\mathrm{T}})$.

(T4) If α'' is the translation of the term α' and RELT($\lambda x \beta''(x)$) is the translation of the relative clause β' from **S4**, then (RELT($\lambda x \beta''(x)$))(α'') is the translation of $\beta'(\alpha')$, and, for all terms α' with $\alpha'' = \lambda P(\ldots P(v) \ldots)$, we have (RELT($\lambda x \beta''(x)$))($\lambda P(\ldots P(v) \ldots)$) $= \lambda P(\ldots \beta''(v)\ \&\ P(v) \ldots)$.

Example: *Hans, der trinkt*
'Hans, who drinks'

This phrase has the categorial form $((d\text{-}'_{\mathrm{T}\langle 1\rangle}(trink'_{\mathrm{V}1})_{\mathrm{RT}})(Hans'_{\mathrm{T}})_{\mathrm{T}})$. *Hans'* has the translation $\lambda PP(h)$, and the translation of the relative clause is RELT($\lambda x\, trink''(x)$). According to **T4**, the translation of the phrase is $\lambda P(trink''(h)\ \&\ P(h))$.

Example: *Auf der Bank sitzt ein Mann. Der Mann, der schläft, . . .*
'There is a man sitting on the bench. The man, who is asleep, ...'

Let us assume that we have introduced the constant a to denote the man on the bench in the course of the translation of the first sentence of this text. Then the translation of *der Mann* in the second sentence is $\lambda P(Mann''(a)\ \&\ P(a))$. The translation of *der Mann, der schläft* is $\lambda P(Mann''(a)\ \&\ schlaf''(a)\ \&\ P(a))$.

4.1.1. Examples of Bach–Peters Sentences

(i) *Adnominal relative clauses*

(44) *Jeder Mann, der eine Frau, die ihn beachtet, liebt, beschenkt sie.*[16]
'Every man who loves a woman who pays attention to him gives her presents.'

[16] Other Bach–Peters sentences are treated in Bartsch 1976b. From the examples presented there, I have selected those cases which were not successfully dealt with in Bennett 1971, where he only develops a procedure for Bach–Peters sentences of the classic form with definite noun phrases. He himself realizes that the weakness of his solution is that it cannot be used to treat other forms, for example, those with *every* . . . .

Instantiating Translation

jeder: $\lambda Q \lambda P(Q(x) \rightarrow P(x))$; x blocked! **IN**
eine: $\lambda Q \lambda P(Q(y) \& P(y))$; y blocked! **IN**
ihn: $\lambda P P(x)$ **IP**(ia)(ii)
sie: $\lambda P P(y)$ **IP**(ia)(ii)
beachtet: $\lambda vz\ beachtet''(v, z)$ Lexicon
ihn beachtet: $\lambda v((\lambda P P(x))(\lambda z\ beachtet''(v, z))) = \lambda v\ beachtet''(v, x)$ **T1**
die ihn beachtet: $\mathrm{RELN}(\lambda v\ beachtet''(v, x))$ **T3**
Frau: $\lambda u\ Frau''(u)$ Lexicon
Frau, die ihn beachtet: $\lambda u(Frau''(u) \& beachtet''(u, x))$ **T3**
eine Frau, die ihn beachtet: $\lambda P(Frau''(y) \& beachtet''(y, x) \& P(y))$ **T2**
liebt: $\lambda vz\ liebt''(v, z)$ Lexicon
eine Frau, die ihn beachtet, liebt: $\lambda v(Frau''(y) \& beachtet''(y, x) \& liebt''(v, y))$ **T1**
Mann, der eine Frau, die ihn beachtet, liebt: $\lambda u(Mann''(u) \& Frau''(y) \& beachtet''(y, x) \& liebt''(u, y))$ **T3**
jeder Mann, der eine Frau, die ihn beachtet, liebt: $\lambda P(Mann''(x) \& Frau''(y) \& beachtet''(y, x) \& liebt''(x, y) \rightarrow P(x))$ **T2**
beschenkt: $\lambda vz\ beschenkt''(v, z)$ Lexicon
beschenkt sie: $\lambda v\ beschenkt''(v, y)$ **T1**
jeder Mann, der eine Frau, die ihn beachtet, liebt, beschenkt sie: $Mann''(x) \& Frau''(y) \& beachtet''(y, x) \& liebt''(x, y) \rightarrow beschenkt''(x, y)$ **T1**

This is the end of the provisional, instantiating, translation. The complete translation is obtained by generalization. The hierarchy of the quantifiers results from the linear order of the terms. This means that we first generalize with respect to *eine Frau,* the innermost term, and then with respect to *jeder Mann,* the outermost term.

Generalizing Translation

With respect to *eine Frau:*
$(\forall y)(Mann''(x) \& Frau''(y) \& beachtet''(y, x) \& liebt''(x, y) \rightarrow beschenkt''(x, y))$ **IN**

With respect to *jeder Mann:*
$(\forall x)(\forall y)(Mann''(x) \& Frau''(y) \& beachtet''(y, x) \& liebt''(x, y) \rightarrow beschenkt''(x, y))$ **IN**

This is the complete translation of the above sentence. In the following example, the pronoun precedes the coreferential full term.

(45) *Ein Mann, der sie liebt, beschenkt eine Frau, die ihn beachtet.*
'A man who loves her gives presents to a woman who pays attention to him.'

We only refer to the important steps of the translation that leads to the non-universal interpretation. The term *eine Frau,* which is coreferential with the pronoun *sie,* is located in the superordinate clause of the clause construction.

Instantiating Translation

ein:	$\lambda Q \lambda P(Q(x) \ \& \ P(x))$; x blocked.	**IN**
eine:	$\lambda Q \lambda P(Q(y) \ \& \ P(y))$; y blocked.	**IN**
sie:	$\lambda P P(y)$	**IP**(ia)(ii)
er:	$\lambda P P(x)$	**IP**(ia)(ii)

and further, according to the translation for example (44).

Generalizing Translation

$$(\exists x)(\exists y)(Mann''(x) \ \& \ Frau''(y) \ \& \ beachtet''(y, x) \ \& \ liebt''(x, y) \ \& \ beschenkt''(x, y))$$

Compare the following text with example (45).

(46) *Brigitte ist eine Psychologin. Ein Mann, der auf sie hört, beschenkt die Frau, die ihn liebt.*
'Brigitte is a psychologist. A man who listens to her gives presents to the woman who loves him.'

Here, the pronoun *sie* has to be coreferential with *Brigitte,* and this in the universal as well as in the nonuniversal reading, otherwise text coherence would not be achieved. Therefore, translate *sie* by $\lambda P P(b)$, where b is the constant used in the translation of *Brigitte*.

(ii) *Adterminal relative clauses*

In the next example, we have two nonrestrictive relative clauses, which are bound to a governing term according to **S4/T4.**

(47) *Gestern sah ich Inge mit einem Mann. Der Mann, der Inge, die ihn beachtet, liebt, beschenkt sie.*
'Yesterday I saw Inge with a man. The man, who loves Inge, who pays attention to him, gives her presents.'

Since, in this example, we treat both relative clauses as nonrestrictive, the instantiating translation coincides with the generalizing translation. This is the case because we do not have variables, but rather the constants *i* for *Inge* and a for *the man,* introduced in the course of the translation of *ein Mann* in the previous sentence.

Translations

der Mann: $\lambda P(Mann''(a)\ \&\ P(a))$; *ihn:* $\lambda PP(a)$ **RI, IP**
Inge: $\lambda PP(i)$; *sie:* $\lambda PP(i)$ **RI, IP**
Inge, die ihn beachtet: $(\text{RELT}(\lambda x\ beachtet''(x, a)))(\lambda PP(i)) =$
$\lambda P(beachtet''(i, a)\ \&\ P(i))$ **T4**
der Mann, der Inge, die ihn beachtet, liebt:
$(\text{RELT}(\lambda x(beachtet''(i, a)\ \&\ liebt''(x, i))))$
$(\lambda P(Mann''(a)\ \&\ P(a))) = \lambda P(Mann''(a)\ \&$
$beachtet''(i, a)\ \&\ liebt''(a, i)\ \&\ P(a))$ **T4**
beschenkt sie: $\lambda x\ beschenkt''(x, i)$ **T1**
der Mann, der Inge, die ihn beachtet, liebt, beschenkt sie:
$Mann''(a)\ \&\ beachtet''(i, a)\ \&\ liebt''(a, i)\ \&\ beschenkt''(a, i)$ **T1**

If our text continued with *Sie schätzt das* 'she appreciates that,' we would have to extend our fragment to include pronouns that are coreferentially related to whole sentences. To achieve that we would need an additional referent introduction rule for events and circumstances, which can be described by whole sentences and directly referred to by nominalizations of those sentences (cf. Ullmer-Ehrich, 1977).

4.2. Subject Clauses and Object Clauses[17]

For subject clause constructions and object clause constructions I assume a semantics along the lines of Hintikka 1969. The translation into the logic for a two-place verb α that expresses a propositional attitude of x towards the content of a sentence β comes down to $\alpha''({}^\wedge\beta'', x)$ and $\alpha''(x, {}^\wedge\beta'')$, for object and subject clauses, respectively. $\alpha''(x)$ can be treated as an operator on ${}^\wedge\beta''$ such that $(\alpha''(x))({}^\wedge\beta'')$ is true if and only if β'' is true in all worlds characterized by $\alpha''(x)$, that is, for example by x's beliefs, or expectations, or by what x has forgiven somebody, etc., if α is *believe, expect,* and *forgive,* respectively. Another view of the matter is to assign properties to senses of sentences, such as "to be believed by John," etc. We assume the following syntactical and semantical rules:

[17] In German, preposed subject and object clauses behave differently from other preposed subordinate clauses in that they do not admit nonquantificational full terms that are coreferential with a pronoun in the superordinate clause (cf. example (14) from Section 2), although this does not happen completely consistently. If the pronoun is the subject term of the superordinate clause, coreferentiality with a term in the preceding object clause is not possible, but it *is* possible if the pronoun is an object term. I do not see how I could explain this situation by means of the categorial syntax and the translation rules for these constructions, and therefore choose to ignore that problem, which in English does not arise anyhow, and about which intuitions in German are not very secure.

S5: If β' is a sentence then $dass(\beta')$ is a term.

Note that subject clauses and object clauses are treated as a special kind of term. The grammatical formative *dass* 'that' is an operator that makes this kind of term out of a sentence.

T5: If β'' is the translation of β' then $\lambda \mathbf{P}\mathbf{P}({}^\wedge\beta'')$ is the translation of *dass* (β').

The variable **P** is a variable over predicates of senses of sentences. These predicates, such as *glaubt* 'believes', *erwartet* 'expects', are translated into $\lambda x\mathbf{S}\ glaubt''(x, \mathbf{S})$, $\lambda x\mathbf{S}\ erwartet''(x, \mathbf{S})$, with **S** as a variable over senses of sentences. If, for example, *dass* β is *dass Peter ankommt* 'that Peter arrives', we receive for *glaubt, dass Peter ankommt* 'believes that Peter arrives': $\lambda x((\lambda \mathbf{P}\mathbf{P}({}^\wedge Peter\ ankommt''))(\lambda \mathbf{S}\ glaubt''(x, \mathbf{S}))$. This is $\lambda x\ glaubt''(x, {}^\wedge Peter\ ankommt'')$. Note that before the translation of the subordinate clause is integrated into the translation of the superordinate clause, it has to be completed. This means that it has to be a generalized translation and not an instantiating translation with regard to full terms that are quantificationally translated. A generalization that captures the pronouns translated by blocked variables (from the translation of the coreferential term in the superordinate clause) is only possible after integration of the translation of the subordinate clause into the translation of the superordinate clause. Thus, translations of subordinate clauses with pronouns that are coreferential with quantificationally analyzed terms in the superordinate clause are always open in the blocked variables, before they are integrated.

Examples

(48) *Jeder Mann glaubt, dass er ankommt.*
'Every man believes that he will arrive.'

This sentence has the following categorial structure:

$$(jeder\ Mann'_{T\langle 1\rangle}((dass(er\ ankommt')_{T\langle 2\rangle})(glaubt'_{V2})_{V1})S)$$

The translation proceeds roughly as follows:

jeder Mann: $\lambda P(Mann''(x) \rightarrow P(x))$; x blocked!	**IN, T2**
er: $\lambda PP(x)$	**IP**
er ankommt: $ankommt''(x)$	**T1**
dass er ankommt: $\lambda \mathbf{P}\mathbf{P}({}^\wedge(ankommt''(x)))$	**T5**
glaubt: $\lambda x\mathbf{S}\ glaubt''(x, \gamma)$	Lexicon
glaubt, dass er ankommt: $\lambda y\ glaubt''(y, {}^\wedge(ankommt''(y))$	**T1**
jeder Mann glaubt, dass er ankommt: $Mann''(x) \rightarrow glaubt''(x, {}^\wedge(ankommt''(x)))$	**T1**

Generalization

$(\forall x)(Mann''(x) \rightarrow glaubt''(x, {}^\wedge(ankommt''(x))))$ **IN**

If this reading would destroy context coherence, we could translate *er* by some constant received from the preceding text, or by a blocked variable from another superordinate clause, as, for example, from *jeder Junge* 'every boy' in *Jeder Junge glaubt, dass jeder Mann glaubt, dass er ankommt* 'Every boy believes that every man believes that he will arrive'.

If we take the categorial structure of *Dass er ankommt, glaubt jeder Mann,* $(dass\ er\ ankommt'_{T\langle 2\rangle}(jeder\ Mann'_{T\langle 1\rangle}(glaubt'_{V2})_{V1})S)$ we get the same result of translation, and thus the same interpretation.

(49) *Dass John ankommt, glaubt er.*
lit. 'That John will arrive, he believes.'

According to the search procedure for pronouns, *er* can be taken to be coreferential with *John*. There is nothing in our procedure that prevents us from getting the reading that we also get for *Dass er ankommt, glaubt John,* lit. 'That he will arrive, John believes,' or for *John glaubt, dass er ankommt* 'John believes that he arrives,' namely, $glaubt''(j, \hat{}(ankommt''(j))$. But our procedure prevents us from getting a reading with coreferentiality for

+*Dass* ***jeder Mann*** *ankommt, glaubt* ***er***.

The search procedure for pronouns assigns *jeder Mann* as a coreferential term to *er*, but the translation procedure rules out this reading, because the translation of the subordinate clause is generalized before it is integrated into the translation of the main clause. We thus get:

$$glaubt''(x, \hat{}((\forall x)(Mann''(x) \rightarrow ankommt''(x)))$$

The *x* corresponding to the *er* in the main clause remains unbound. It could as well be any other variable. Of course, we can treat *er* as coreferential with another term in the context, and then *er* would be translated by a constant, or by a variable blocked in the translation of a term in a superordinate clause, as, for example, in *Dass dass jeder Mann ankommt,* ***ihn*** *beruhigt, hofft* ***John*** *sehr* lit. 'That that every man will arrive will calm *him* down *John* believes very much.'[18]

4.3. Adverbial Clause Constructions

Conditional clauses have the same coreference patterns as the so-called adverbial clauses, constructed with conjunctions *weil* 'because', *obgleich* 'although', *während* 'while', *nachdem* 'after', etc., with one exception: The preceding subordinate conditional clause may contain a term of the form *ein-* . . . 'a/an . . .' that is coreferential with a pronoun in the following superordinate clause, even in a nonspecific reading. As has been shown in Sec-

[18] Note that the unacceptability of some of the examples in the last paragraph (in English) has nothing to do with pronominalization, but with the restriction on word order in English, which, unlike German, does not permit the object (here, the object clause) to precede the verb.

tion 2.4, we get instead a universal reading. We can avoid making an exception for such clauses by assuming that they are not truly universal sentences, but rather stereotypical sentences. That is to say, the sentence *If a man wins a car he is happy*, for example, is not analyzed directly as "For every man it holds that if he wins a car then he is happy," but rather in the sense that it belongs to the stereotype of a man that he is happy if he wins a car. Thus, the antecedent of the conditional sentence is integrated into the property that is predicated of the stereotypical man. This means, if *a man** is the set of properties defining the stereotype, then the complex property $\hat{}\lambda x(\mathit{wins\ a\ car}''(x) \rightarrow \mathit{happy}''(x))$ belongs to it. This way we also can interpret sentences like *Because a man must die he is unhappy*. The complex property "to be unhappy because one must die" belongs to the stereotype *a man**.

As has already been pointed out, we will integrate all adverbial clauses, including conditional clauses, into the verb phrase of the superordinate clause. This ensures the proper binding of blocked variables by quantifiers. It does not create difficulties with adverbial or conditional clauses that are not open, that is, that do not contain any blocked variables. In such cases, we will obtain by our translation rules formulas like $(\forall x)(\alpha(x) \rightarrow (p \rightarrow \beta(x)))$, $(\exists x)(\alpha(x)\ \&\ (p \rightarrow \beta(x)))$, and $p \rightarrow \beta(a)$, for sentences such as *If it is cold every man is unhappy*, *If it is cold a man is unhappy* (in the nonuniversal reading), and *If it is cold John is unhappy*, respectively. Our translation rules, which are designed to treat coreferentiality, work on categorial structures where conditional and adverbial clauses are incorporated into the verb phrase.

To represent the conditional, causal, concessive, and temporal relationships between events or circumstances, we assume an operation of event nominalization that is applied to sentences and results in expressions that denote events or circumstances described by these sentences (cf. Reichenbach, 1947; Davidson, 1966; Bartsch, 1972, 1976c; and Ullmer-Ehrich, 1977). This operation will be called EV. If the sentence to which it is applied contains an open variable x, $\mathrm{EV}(p_x)$ is a function that assigns to every instance of x an event or circumstance of the kind described in p. For example, EV(*x eats*) assigns to John the event that John eats, or of John's eating. We have the following semantic rules that explicate the truth functional aspect of these connectives, which are here interpreted as relationships between events or circumstances.

It necessarily holds that:

$$\mathrm{COND}(\mathrm{EV}(p),\mathrm{EV}(q)) \rightarrow (p \rightarrow q)$$
$$\mathrm{CAUS}(\mathrm{EV}(p),\mathrm{EV}(q)) \rightarrow (p\ \&\ q)$$
$$\mathrm{CONC}(\mathrm{EV}(p),\mathrm{EV}(q)) \rightarrow (p\ \&\ q)$$
$$\mathrm{TEMP}(\mathrm{EV}(p),\mathrm{EV}(q)) \rightarrow (p\ \&\ q)$$

The relations COND, CAUS, CONC, and TEMP and the operator EV are such that variables within their scope can be bound by quantifiers from outside, in other words, we can quantify in.

When, in the following discussion, we speak about adverbial clauses, it should be understood that conditional clauses are included. We have the following syntactic and semantic rules.

The Formation of Adverbial Clauses

(S6) If β' is a sentence and α' a clause connective, then $\alpha'(\beta')$ is an adverbial clause. (α': *wenn* 'if', *weil* 'because', *obgleich* 'although', *nachdem* 'after', etc.)

(T6) If β'' is the translation of β' and α'' the translation of α' from **S6**, then $\alpha''(\beta'')$ is the translation of $\alpha'(\beta')$.

The translation of the adverbial clause $\alpha''(\beta'')$ is a special kind of function that is described in **T7** by the effect it has on the verb to which it will be applied according to **S7**.

(S7) If α' is an adverbial clause and β' is an n-place verb then $\alpha'(\beta')$ is an n-place verb.

(T7) If α'', with $\alpha'' = \gamma''(\delta'')$, is the translation of the adverbial clause α' and $\lambda x_1 \cdots x_n \beta''(x_1, \cdots, x_n)$ is the translation of the n-place verb β' from **S7**, then the translation of $\alpha'(\beta')$ is:

$$\begin{aligned}\alpha''(\beta'') &= (\gamma''(\delta''))(\lambda x_1 \cdots x_n \beta''(x_1, \cdots, x_n)) \\ &= \lambda x_1 \cdots x_n(G(\mathrm{EV}(\delta''), \mathrm{EV}(\beta''(x_1, \cdots, x_n))),\end{aligned}$$

with G as the relation between events or circumstances that corresponds to the clause connective γ.

Examples

(50) *Weil er betrunken ist, leitet Hans jeden Gast hinaus.*
'Because he is drunk John dismisses every guest.'

This sentence is three-ways ambiguous. It may be that every guest is drunk, or that Hans is drunk, or, perhaps, that the barkeeper is drunk, and therefore Hans has to dismiss the guests. Our search procedure for pronouns will first give coreferentiality between *Hans* and *er*; then, if that is improbable or destroys text coherence, coreferentiality between *jeden Gast* and *er* will be tried; and if that is improbable or destroys text coherence, then coreferentiality between *er* and a term in the previous text will be established, as, for example, in *Peter is not a very reliable barkeeper. Because he is drunk John has dismissed every guest.*

(i) Coreference between *Hans* and *er*:
We assume for sentence (50) Categorial Structure A:

A: $(Hans'_{T\langle 1\rangle}(jeden\ Gast'_{T\langle 2\rangle}((Weil'_{C}(er\ betrunken\ ist'_{S})_{AC})$
$(hinausleitet'_{V2})_{V2})_{V1})_{S})$

According to this structure, we get the following translation:

Instantiating Translation

Hans: $\lambda PP(h)$	Lexicon
jeden Gast: $\lambda P(Gast''(x) \rightarrow P(x))$; x blocked!	**IN**, **T2**
er: $\lambda PP(h)$	**IP**(ib)

Translation of the subordinate clause:

Er betrunken ist: $betrunken''(h)$ **T1**

weil er betrunken ist, hinausleitet:
$\lambda zy(\mathrm{CAUS}(\mathrm{EV}(betrunken''(h)),\mathrm{EV}(hinausleitet''(z,y))))$ **T7**

jeden Gast, weil er betrunken ist, hinausleitet: $\lambda z(Gast''(x) \rightarrow$
$\mathrm{CAUS}(\mathrm{EV}(betrunken''(h)),\mathrm{EV}(hinausleitet''(z,x))))$ **T1**

Hans jeden Gast, weil er betrunken ist, hinausleitet:
$Gast''(x) \rightarrow$
$\mathrm{CAUS}(\mathrm{EV}(betrunken''(h)),\mathrm{EV}(hinausleitet''(h,x)))$ **T1**

Generalization

$$(\forall x)(Gast''(x) \rightarrow \mathrm{CAUS}(\mathrm{EV}(betrunken''(h)),\mathrm{EV}(hinausleitet''(h,x))))$$

This is the final translation.

We could also proceed by assuming another categorial structure, namely *B*.

B: $(Hans'_{\mathrm{T}\langle 1\rangle}((Weil'_{\mathrm{C}}(er\ betrunken\ ist'_{\mathrm{S}})_{\mathrm{AC}})(jeden\ Gast'_{\mathrm{T}\langle 2\rangle}(hinausleitet'_{\mathrm{V2}})_{\mathrm{V1}})_{\mathrm{V1}})_{\mathrm{S}})$

The translation would be equivalent to that given for Categorial Structure *A*.

(ii) Coreference between *jeden Gast* and *er*:

In this case, we are not free to choose between Categorial Structures *A* and *B* as given in (i), because we have to ensure that the blocked variable used in the translation of the pronoun is at the right place within the scope of the universal quantifier used for the generalization with respect to *jeden Gast*. This means that the adverbial clause has to be within the scope of *jeden Gast,* and therefore we can only use Categorial Structure *A*, but not *B*. Another possibility is Structure *C*, which we have not considered so far:

C: $(Jeden\ Gast'_{\mathrm{T}\langle 2\rangle}(Hans'_{\mathrm{T}\langle 1\rangle}((weil'_{\mathrm{C}}(er\ betrunken\ ist'_{\mathrm{S}})_{\mathrm{AC}})(hinausleitet'_{\mathrm{V2}})_{\mathrm{V2}})_{\mathrm{V1}})_{\mathrm{S}})$

Also, Structure *D*, the last one to be considered, is a possible choice.

D: $(Jeden\ Gast'_{\mathrm{T}\langle 2\rangle}((weil'_{\mathrm{C}}(er\ betrunken\ ist'_{\mathrm{S}})_{\mathrm{AC}})(Hans'_{\mathrm{T}\langle 1\rangle}(hinausleitet'_{\mathrm{V2}})_{\mathrm{V1}})_{\mathrm{V1}})_{\mathrm{S}})$

Note that from these categorial structures, we get different word orders according to natural serialization in the German subordinate clause word order:

A: . . . , *(obgleich) Hans jeden Gast, weil er betrunken ist, hinausleitet*
B: . . . , *(obgleich) Hans, weil er betrunken ist, jeden Gast hinausleitet*
C: . . . , *(obgleich) jeden Gast Hans, weil er betrunken ist, hinausleitet*
D: . . . , *(obgleich) jeden Gast, weil er betrunken ist, Hans hinausleitet*

Indeed, word order B does not permit a reading with coreferentiality between *er* and *jeden Gast*. Word orders C and D are deviant for German subordinate clauses, but the corresponding word orders for main clauses, where the finite verb takes the second place, are acceptable.

A′: *Hans leitet jeden Gast, weil er betrunken ist, hinaus.*
B′: *Hans leitet, weil er betrunken ist, jeden Gast hinaus.*
C′: *Jeden Gast leitet Hans, weil er betrunken ist, hinaus.*
D′: *Jeden Gast leitet, weil er betrunken ist, Hans hinaus.*

Again, word order B does not admit a reading with coreferentiality between *er* and *jeden Gast*. Word order D does not permit coreferentiality between *er* and *Hans*. A and C permit both possibilities. The "natural" serialization with its modification for main clause word order does not yield the word order with the subordinate clause preposed, nor the word order with the subordinate clause postposed; these are obtained by special serialization rules from the Categorial Structures *A–D*. *Weil er betrunken ist, leitet Hans jeden Gast hinaus* can be derived from *A* and *B*; *Hans leitet jeden Gast hinaus, weil er betrunken ist* can also be derived from *A* and *B*; *Jeden Gast leitet Hans hinaus, weil er betrunken ist* can be derived from Categorial Structures *C* and *D*. Which word order will be derived is dependent on conditions of the context, such as thematic and topic–comment relationships.

If we had based the translation of the sentence *Weil er betrunken ist, leitet Hans jeden Gast hinaus* on Categorial Structure *B* and at the same time assumed the (impossible) coreference pattern *er–jeden Gast*, we would end up with a translation that is not intended. The translation would be such that its interpretation would be: "Everybody is drunk and that causes it to be the case that if he is a guest he gets dismissed by Hans." This means that coreference between a pronoun in a preposed or postposed adverbial clause and a quantificationally translated term in the superordinate clause can only be interpreted correctly if the categorial structure assumed is such that the subordinate adverbial clause is within the scope of the quantificational term. We could prevent unwanted translations by either restricting the search procedure for pronouns so that it takes into account the categorial structure, thus ruling out the coreference pattern *er–jeden Gast* for word order B, or by making the assignment of the "right" categorial structure for preposed and for postposed adverbial clauses dependent on the coreference pattern. The last suggestion seems appropriate, because it keeps our search procedure for pronouns working directly on surface structure information. Thus, we restrict the choice of categorial structures for adverbial clause constructions with preposed or postposed adverbial clause in the following way:

> If there are pronouns in the subordinate clause that are coreferential with terms in the superordinate clause that are quantificationally analyzed, then choose a categorial structure such that these terms are higher in the hierarchy than the subordinate adverbial clause.

This restriction is, of course, fulfilled in all cases where we choose a categorial structure for the adverbial clause construction such that the adverbial clause is applied to the verb first, that is, before any term is applied to the verb of the superordinate clause, as in Structure *A*.

For the translation of sentence (50), with the coreference pattern (ii), we now assume Categorial Structure *A* (*C* and *D* are excluded because of the word order *Hans . . . jeden Gast* in (49), since they result in the word order *Jeden Gast . . . Hans,* if we keep word order relative to the hierarchy of the categorial structure as natural as possible. The translation proceeds as follows.

Instantiating Translation

Hans: $\lambda PP(h)$	Lexicon
jeden Gast: $\lambda P(Gast''(x) \rightarrow P(x))$; x blocked!	**IN, T2**
hinausleitet: $\lambda yz\ hinausleitet''(y, z)$	Lexicon

Translation of the subordinate clause:

er: $\lambda PP(x)$	**IP**(iii, ib)
er betrunken ist: $betrunken''(x)$	**T1**
weil er betrunken ist, hinausleitet: $\lambda yz(\mathrm{CAUS}(\mathrm{EV}(betrunken''(x)),\mathrm{EV}(hinausleitet''(y, z))))$	**T7**
jeden Gast, weil er betrunken ist, hinausleitet: $\lambda y(Gast''(x) \rightarrow \mathrm{CAUS}(\mathrm{EV}(betrunken''(x)),\mathrm{EV}(hinausleitet''(y, x))))$	**T1**
Hans jeden Gast, weil er betrunken ist, hinausleitet: $Gast''(x) \rightarrow \mathrm{CAUS}(\mathrm{EV}(betrunken''(x)),\mathrm{EV}(hinausleitet''(h, x)))$	**T1**

Generalization completes the translation:

$$(\forall x)(Gast''(x) \rightarrow \mathrm{CAUS}(\mathrm{EV}(betrunken''(x)),\mathrm{EV}(hinausleitet''(h, x))))$$

(51) *Weil jeder Gast betrunken ist, leitet Hans ihn hinaus.*
'Because every guest is drunk John dismisses him.'

Here, we have an example that is not interpretable with coreferentiality between *jeder Gast* and *er*. Although the search procedure for pronouns does give us a coreferentiality assignment between *jeder Gast* and *ihn* (as it is designed to do because we want a coreferential reading for *ein Gast, der Gast, Peter,* etc. and *ihn*), the translation procedure leaves the pronoun used in the translation for *ihn* 'him' unbound.

Instantiating Translation

jeder Gast: $\lambda P(Gast''(x) \rightarrow P(x))$; x blocked!
Hans: $\lambda PP(h)$
ihn: $\lambda PP(x)$

Translation of the subordinate clause:

jeder Gast betrunken ist: $(\forall x)(Gast''(x) \rightarrow betrunken''(x))$

Note that the translation of the subordinate clause is completed, that is, generalized with respect to all those of its terms that are generalizable.

weil jeder Gast betrunken ist, hinausleitet: λyz
$\text{CAUS}(\text{EV}((\forall x)(Gast''(x) \rightarrow betrunken''(x))),\text{EV}(hinausleitet''(y,z))$ **T7**
ihn, weil jeder Gast betrunken ist, hinausleitet:
$\lambda y\ \text{CAUS}(\text{EV}((\forall x)(Gast''(x) \rightarrow$
$betrunken''(x))),\text{EV}(hinausleitet''(y,x)))$ **T1**
Hans ihn, weil jeder Gast betrunken ist, hinausleitet:
$\text{CAUS}(\text{EV}((\forall x)(Gast''(x) \rightarrow betrunken''(x))),\text{EV}(hinausleitet''(h,x))$ **T1**

The last x remains dangling. Of course, in a suitable text, the *ihn* 'him' could be coreferential with some other term, as in *Peter regt sich über Betrunkene auf. Weil jeder Gast betrunken ist, leitet Hans ihn hinaus* 'Peter gets upset about drunken people. Because every guest is drunk, Hans dismisses him,' where instead of the dangling x, we would have a constant referring to Peter.

The translation of *Weil jeder Gast betrunken ist, leitet Hans jeden Gast hinaus* 'Because every guest is drunk, John dismissed every guest' is unproblematic:

$$(\forall x)(Gast''(x) \rightarrow \text{CAUS}(\text{EV}((\forall y)(Gast''(y) \rightarrow betrunken''(y)),\text{EV}(hinausleitet''(h,x))))$$

In concluding, it should be pointed out that the search procedure for pronouns, by which coreferentiality is assigned, can be reformulated, if one wants complete ambiguity of pronouns, by assigning alternative coreferential patterns. Instead, I have formulated the search procedure hierarchically, giving a preference to certain readings on structural grounds. Depending on further considerations, it might well turn out to be better to include in the search procedure a reference to categorial structures and not just to allow it to work on the linear order of sentence constituents. This would at least bring about the restriction that quantificationally translated terms that are lower in the hierarchy than the subordinate clause, are ruled out as candidates for coreferentiality with pronouns in the subordinate clause. Here, I have chosen the alternative course, namely, to rule out specific categorial structures if a certain coreferential pattern is established first.[19]

[19] To formulate an adequate procedure for assigning antecedent noun phrases to pronouns in English, one certainly has to take into account the conditions on antecedents formulated in

Since I cannot extend the analysis further here, pronouns that have coreferential relationships with whole sentences or with verb phrases have been left out. To treat them, I would use other sorts of "entities" and, correspondingly, other sorts of constants and variables that refer to these entities, as in Bartsch 1972 and 1976c and in Ullmer-Ehrich 1977. The details of such a treatment have yet to be worked out. The extension for plurals can be achieved in the present framework along the lines of Bartsch 1973, where constants and variables are used that refer to sets or groups of entities (in a nontechnical sense) with more than one member, and where the extensions of plural nouns and plural verbs are the sets of all subsets with more than one member of the extensions of the corresponding singular forms of these nouns and verbs. Integration of this into the present fragment does not seem to pose serious problems. Pronouns of laziness have to be integrated according to Footnote 15 of this chapter. Deictic pronouns can be treated according to Footnotes 8 and 15, in the way hinted at in the introduction of this volume. Further, the whole treatment of pronouns will eventually have to be connected with the treatment of proforms other than pronouns, such as pro-verbs like *tun* 'do,' or pro-adverbs like *so* 'so,' in sentences like *Maria spricht langsam, und Hans spricht auch so* or . . . , *und Hans tut das ebenso* 'Mary talks slowly, and so does Hans.' All this goes much beyond the scope of this chapter.

Finally, I should like to point out that my proposal for the treatment of pronouns seems roughly in accordance with the treatment considered in Wasow 1975 (p. 381), where he considers combining an interpretive approach, in which pronouns are syntactically generated in the base (that is, nontransformationally), with a formulation of conditions on possible antecedents assigned to pronouns, and rules of interpretation. In my model, pronouns appear in the syntax, that is, on the level of surface syntax, as well as on the more abstract level of categorial syntax; and instead of interpretation rules I use translation rules that translate forms of categorial syntax together with information from the reference pattern established according to the procedure **IP**, into constants or bound variables of the logical language chosen for semantic representation.

REFERENCES

Ballmer, Th. (1972). Einführung und kontrolle von diskurswelten. In D. Wunderlich (ED.), *Linguistische Pragmatik.* Frankfurt: Athenäum. Pp. 183–206.

Ballmer, Th. (1975). Patterns of reference. Unpublished paper, Technical University of Berlin.

Ballmer, Th. (1976). Logical language reconstruction and reference. In *Papers from the 12th*

Jackendoff 1973; as well as in recent work by Wasow, which was not available to me at the time this paper was written. Edmondson (1973) has modified Jackendoff's procedure for assigning antecedents for German, and he checks also other transformational work on pronouns with respect to German.

Regional Meeting of the Chicago Linguistic Society (CLS 12). Chicago: Chicago Linguistic Society.

Bartsch, R. (1972). *Adverbialsemantik.* Frankfurt: Athenäum. Revised English translation (1976c) *The Grammar of Adverbials.* Amsterdam: North Holland.

Bartsch, R. (1973). The semantics and syntax of number and numbers. In J. Kimball (Ed.), *Syntax and Semantics,* Vol. II. New York: Seminar Press. Pp. 51–94.

Bartsch, R. (1976a). The role of categorial syntax in grammatical theory. In A. Kasher (Ed.), *Language in Focus: Foundations, Methods, and Systems. Essays in Memory of Y. Bar-Hillel.* Dordrecht: D. Reidel. Pp. 503–540.

Bartsch, R. (1976b). Syntax and semantics of relative clauses. In J. Groenendijk and M. Stokhof (Eds.), *Proceedings of the Amsterdam Colloquium on Montague Grammar and Related Topics, January 1976, Amsterdam Papers in Formal Grammar,* Vol. 1. Centrale Interfaculteit, Universiteit van Amsterdam. Pp. 1–24.

Bartsch, R., and Vennemann, Th. (1972). *Semantic Structures: A Study in the Relation between Semantics and Syntax,* Frankfurt: Athenäum.

Bennett, M. (1971), Accommodating Bach–Peters sentences in Montague's fragment of English. Unpublished manuscript, UCLA.

Beth, E. W. (1969). *Moderne Logica.* Assen: Van Gorcum.

Davidson, D. (1966). The logical form of action sentences, In N. Resher (Ed.), *The Logic of Decision and Action.* Pittsburgh: University of Pittsburgh Press. Pp. 81–95.

van Dijk, T. (1977). *Text of Context.* Longmans, London.

Edmondson, J. A. (1973). The Generative Theories of Pronominalization in Light of Evidence from Modern German. Unpublished doctoral dissertation, UCLA.

Edmondson, J. A. (1976). Semantics, games and anaphoric chains. In J. Groenendijk and M. Stokhof (Eds.), *Proceedings of the Amsterdam Colloquium on Montague Grammar and Related Topics, January 1976, Amsterdam Papers in Formal Grammar,* Vol. I. Centrale Interfaculteit, Universiteit van Amsterdam. Pp. 25–40.

Groenendijk, J., and Stokhof, M. (1975). Modality and conversational information. *Theoretical Linguistics,* **2,** 61–112.

Hintikka, J. (1969). Semantics for propositional attitudes. In Davis and Hockney (Eds.), *Studies in Philosophical Logic.* Dordrecht: D. Reidel. Pp. 12–45.

Hintikka, J. (1973). *Logic, Language Games and Information.* Oxford: Clarendon.

Hintikka, J., and Saarinen, E. (1975). Semantical games and the Bach–Peters paradox. *Theoretical Linguistics,* **2,** 1–20.

Jackendoff, R. (1973). *Semantic Interpretation in Generative Grammar.* Cambridge, Massachusetts: M.I.T. Press.

Karttunen, L. (1969). Pronouns and variables. In Robert Binnick *et al.* (Eds.), *Papers from the Fifth Regional Meeting of the Chicago Linguistic Society.* Chicago: Department of Linguistics, University of Chicago. Pp. 108–116.

Karttunen, L. (1971a). Definite descriptions with crossing coreference. A study of the Bach–Peters paradox. *Foundations of Language,* **7,** 157–182.

Karttunen, L. (1971b). Discourse referents. Indiana University Linguistics Club, Bloomington, Indiana.

Lakoff, G. (1971). On generative semantics. In D. Steinberg and L. Jakobovits (Eds.), *Semantics: An Interdisciplinary Reader in Philosophy, Linguistics and Psychology.* Cambridge: Cambridge University Press, Pp. 232–296.

Montague, R. (1974). The proper treatment of quantification in ordinary English. In R. Thomason (Ed.), *Formal Philosophy: Selected Papers of Richard Montague,* New Haven: Yale University Press. Pp. 247–270.

Partee, B. (1975). Montague grammar and transformational grammar. *Linguistic Inquiry,* **4,** 203–300.

Putnam, H. (1975). The meaning of "meaning". In H. Putnam, *Philosophical Papers,* Vol. II: *Mind, Language and Reality.* Cambridge: Cambridge University Press, Pp. 215–271.

Quasthoff, U. (1973). *Soziales Vorurteil und Kommunikation—Eine Sprachwissenschaftliche Analyse des Stereotyps.* Frankfurt: Athenäum.

Quine, W. V. O. (1956). *Methods of Logic,* New York: Holt.

Reichenbach, H. (1947). *Elements of Symbolic Logic,* Macmillan (Free Press). Paperback edition, Toronto, 1966.

Ullmer-Ehrich, V. (1977). *Zur Syntax und Semantik von Substantivierungen im Deutschen,* Kronberg/Ts: Scriptor.

Wasow, T. (1975). Anaphoric pronouns and bound variables. *Language,* **51,** 368–383.

THE INTERPRETATION OF PRONOUNS[1]

ROBIN COOPER

1. INTRODUCTION

In this paper I want to demonstrate that an enrichment of the semantic treatment of pronouns proposed by Richard Montague allows us to construct an adequate semantics for a nonabstract syntactic analysis of certain examples of Hittite and English. At the same time, this semantic treatment enables us to avoid certain syntactic problems. In this way I want to suggest that a theory of semantics founded on the work of Richard Montague provides a tool which allows us to give a precise semantic treatment for natural language without having to make concessions in the syntax in order to make it interpretable.

I shall consider examples here within the context of a general Montague framework, though with certain modifications, among them being the replacement of Montague's syntax by a transformational syntax. The se-

[1] I am extremely grateful to Lauri Karttunen and Sue Schmerling for fruitful discussion during the preparation of this paper, and to Stanley Peters for his invaluable insights and suggestions which have helped me to straighten out my ideas. Frank Heny made comments on the final version that went far beyond the editorial. Of course, none of these people are responsible for shortcomings in the paper.

SYNTAX AND SEMANTICS
VOLUME 10

ISBN: 0-12-613510-X

mantics is defined on the deep structure. Following a slight revision of Montague's treatment by Thomason (1974), I shall regard noun phrases as denoting sets of characteristics. A characteristic is to be regarded as a function from entities to truth values. Thus we may talk of the characteristic of running or the characteristic of loving Mary. The characteristic of loving Mary, for example, will be a function that assigns truth to those entities which love Mary and falsity to those entities which do not love Mary. Thus, we may also think of the characteristic in terms of the set of entities which love Mary, that is, the set of entities to which the characteristic assigns truth. It is sometimes conceptually more intuitive to talk about sets and sometimes easier to think in terms of characteristics. For example, I shall talk about sets of characteristics rather than characteristics of characteristics or sets of sets.

Letting noun phrases denote sets of characteristics enables us, among other things, to give a unified syntactic and semantic treatment of proper names and quantified NPs. For example, the NP *John* will denote the set of John's characteristics and the NP *a man* will denote the set of characteristics possessed by at least one man. As it is often somewhat complicated to spell out in English what the denotation of a given phrase is, it is useful to represent it by an expression of Montague's logic. It should be made clear that the translation of natural language expressions into the logic is simply a matter of convenience, and is not intended to represent a claim that such a translation should be included in a grammar of natural language. We may consider a grammar to be a device which, among other things, relates expressions and meanings. In a theory like Montague's, a meaning determines the denotation of an expression at any possible world and context of use, and a logic may be used to represent such denotations. However it is not necessary to go through the logic to find the meaning of an expression. In particular, the logic does not correspond to what linguists term semantic representation or logic form. (For some discussion of the role of meanings in Montague's framework see Cooper, 1977.)

Bearing this in mind, we may represent the fact that the NP *John* denotes the set of John's characteristics by translating it into expression (1).

(1) $$\lambda K K(j)$$

Here, K is to be regarded as a variable over characteristics, and j as a constant denoting John. Example (1) is an expression of the logic which denotes the set of John's characteristics. Similarly *a man* is translated as (2).

(2) $$\lambda K \exists u[man'(u) \wedge K(u)]$$

The logical expression (2) denotes the set represented in (3).

(3) {c: c is a characteristic and there is at least one individual which is a man and has c}

On this sort of treatment, natural language pronouns in a sentence are associated with any one of a whole set of logical translations representing sets of characteristics. There will be one translation corresponding to each of denumerably many variables. The set of translations associated with a single pronoun may be represented as in (4). (I am ignoring any treatment of gender here and will do so for the rest of the paper. For a precise suggestion for a nonsyntactic treatment of gender see Cooper 1975.)

(4) $\{\lambda K K(u_i)\text{:}\quad i \text{ is a natural number}\}$

Thus, a pronoun may denote the set of characteristics of the individual represented by u_i, for any natural number i. There are a number of ways in which we could associate these translations with English pronouns. We could have a different deep structure variable (he_0, he_1, etc.) which turns into an appropriate pronoun on its way from deep structure to surface structure. This would correspond to the style of generative semantics and also to Montague's original syntax. Another possibility would be to include pronouns rather than variables in deep structure and provide an indexing procedure prior to translation which would distinguish the various possibilities. Such a system is discussed in Cooper and Parsons 1976. We could also map English pronouns directly into the set of logical expressions as in the system described in Cooper (1975, Ch. 4). The precise way in which we associate the set of logical expressions with an English pronoun is not important to the discussion in this paper.

I will suggest that pronouns should, in addition to the treatment represented by the set (4), sometimes be treated as definite descriptions containing a free variable over properties. The idea that pronouns should be regarded in certain cases as disguised definite descriptions is not new. Bennett (1976) has suggested that the pronouns *he* and *she* should be treated semantically as definite descriptions when they do not represent bound variables and are not anaphorically related to a noun phrase in an earlier sentence in the discourse. As Bennett points out, Quine 1960 also suggested that pronouns might be treated as abbreviations of definite descriptions, and Karttunen 1971 makes a similar suggestion. I wish to propose that, besides the regular translations represented by the set in (4), pronouns should also be allowed to translate as any of the expressions in the set represented in (5).

(5) $\{\lambda K \exists x[\forall y[[{}^{\vee}P_i](y) \equiv y = x] \wedge K(x)]\text{:}\quad i \text{ is a natural number}\}$

P_i, for any natural number i, is a variable over properties, in other words, intensions of characteristics or functions from possible worlds to characteristics. $[{}^{\vee}P_i]$ is the extension of P_i, that is, the characteristic which P_i finds at whatever world we are evaluating.[2]

[2] It should be made clear that this enrichment of our treatment does not in fact increase the number of translations, and therefore also denotations, associated with a pronoun. There were

Translating a pronoun as one of the members of (5) indicates that the denotation of that pronoun is the set of characteristics possessed by the unique entity which has the property P_i. Note that uniqueness here is indicated by making use of Russell's theory of definite descriptions. Although the translations in (5) are more complex than those in (4), this does not mean that the denotations represented in (5) are more complex. In each case, the denotation associated with a pronoun is a set of characteristics, that is, a set of functions which assign truth to some entities and falsity to others. That we need to represent the functions in one case with a longer expression of the logic does not a priori require that the functions themselves are more complex. The functions may be modeled as sets of ordered pairs in either case.

If we were to analyze a sentence such as *he walks* according to the treatment represented in (5), we might obtain the translation (6), where *walk'* represents a constant of intensional logic denoting the characteristic which is denoted by the English word *walk,* that is, the characteristic of walking.

Here and throughout the paper I shall not give detailed accounts of how translations are related to deep structures as this would obscure the main point. The interested reader may work through the rules in the appendix to verify the treatment of most of the sentences.

(6) $$\lambda K \exists x[\forall y[[{}^{\vee}P_0](y) \equiv y = x] \wedge K(x)](walk')$$

This might be roughly paraphrased as "the set of characteristics possessed by the unique entity which has the property P_0 includes the characteristic of walking." This is equivalent to saying that the unique entity which has the property P_0 walks, and the logic represents this by showing that (6) is equivalent to (7).

(7) $$\exists x[\forall y[[{}^{\vee}P_0](y) \equiv y = x] \wedge walk'(x)]$$

The expression in (7) is obtained syntactically from (6) by lambda conversion where we have substituted *walk'* for K and eliminated the lambda operator, λK. Note that (7) contains a free variable, P_0. The value of this variable is determined in accordance with Montague's theory by the context of use.

In formal language theory, free variables have generally been regarded as a useful means to an end, namely, the closed sentence without free variables. It is for closed sentences that the general definition of truth and logical consequence is given. In Montague's work on pragmatics, however, a slightly different light is cast on the free variable. Montague (1974) interprets formal languages with respect not only to possible worlds but also to con-

denumerably many translations before and there are still denumerably many. It is clear that there must be infinitely many possible denotations for a pronoun because it is possible to construct a sentence of indefinite length of the form *he told him and then he told him and then. . . .* Such sentences are quite acceptable if accompanied by suitable pointing, for example, if we are reporting how a message was passed along a chain of people.

texts of use. While one could imagine a number of things being specified in a context, Montague's particular treatment for the purposes of his fragments of English introduces the context of use "in order to permit a treatment . . . of such indexical locutions as demonstratives, first- and second-person singular pronouns, and free variables (which are treated . . . as a kind of demonstrative) [p. 228]." Thus the sentence *he walks* translated as (7) corresponds to different propositions depending on the context of use, which assigns some property to the free variable P_0, such as the property of being the man wearing the red hat or the property of being the man who is the president of the company. Of course, *he walks* may also be translated using the regular translations for pronouns. An example of this is (8a), which is equivalent to (8b).

(8) a. $\lambda K K(u_0)(walk')$
b. $walk'(u_0)$

We may paraphrase (8a) as "the set of characteristics of u_0 includes the characteristic of walking" and (8b) simply as "u_0 walks." The value of the free variable is determined by the context of use. Thus, according to one context of use the sentence may correspond to the same proposition as *John walks* and according to another context of use it may correspond to the same proposition as *Bill walks*, etc. It is easy to see that in many situations it will not make a lot of difference which type of interpretation we choose for the pronoun. Suppose that John is the man wearing the red hat at some possible world. Then, if we evaluate (7) and (8b) relative to a context of use which assigns John to u_0 and the property of being the man wearing the red hat to P_0, we will get the same truth value for these sentences at that possible world.

In the following sections, I should like to point out some examples where the treatment of pronouns as definite descriptions makes an important difference in the propositions expressed by sentences. First, I shall show how it allows for a natural treatment of Hittite relative clauses where, in the crucial cases, the free property variable in the definite description gets bound. Second, I shall show that this treatment of pronouns allows a syntactically elegant account of some problems involving discourse anaphora in English. Finally, I shall suggest that treating pronouns as definite descriptions in this way provides a reasonable, though perhaps not totally convincing, account of a problem raised by Geach (1962), concerning sentences like *every man who owns a donkey beats it*.

2. HITTITE RELATIVE CLAUSES

In Hittite one finds the following sentence:

(9) aššer-ma-kan kuiēš DINGIR.MEŠ n-aš *INA*
remained-but-ptc which god.pl ptc-them to
E ^{D}Mezulla pihhun
house god-M. I-gave
Which gods remained, I gave them to the temple of Mezzulla
'I gave the gods which were left to the temple of Mezzula'
[Raman, 1973, p. 121].

Hittite is one of a fairly small number of languages, including Walbiri and several languages of India, which place relative clauses not next to the head noun, as one usually finds in English, but to the left or right of the main clause. In Hittite, one normally finds relative clauses to the left of the sentence, occasionally to the right, but never embedded in the sentence. Thus, one might suppose that a reasonable deep structure for (9) would be something like that represented in (10).

(10)

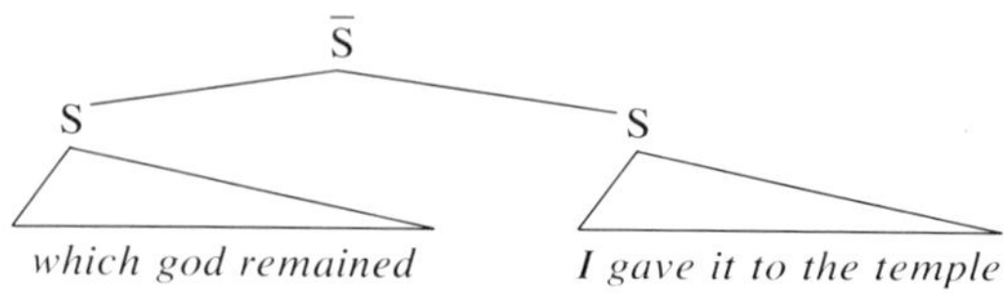

I have simplified a number of things here to facilitate discussion of the main issue. Besides substituting English for Hittite, I have also changed the plural *gods* to a singular in order not to have to consider the semantics of the plural. For a precise treatment of the original sentence within the framework of a fragment of Hittite, see Cooper 1975. There is no syntactic evidence for deriving the sentence from a structure where the relative sentence is embedded. If we needed to have the relative clause embedded next to the head noun, as the normal treatment of relative clauses within a Montague framework would lead us to expect, this would have undesirable consequences for the syntax. The problem, then, is to obtain a reasonable semantics for a structure like (10) where we can naturally associate the denotations of *it* and *which god remained,* even though they are not syntactically adjacent.

Let us first consider the relative clause. The underlying structure of this is given in more detail in (11).

(11)

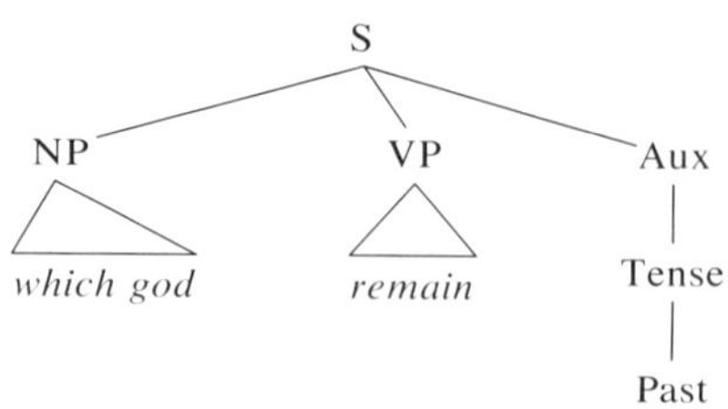

(This is still a much-simplified version of the structure given in Cooper 1975.) In the analysis I am proposing, *wh*-phrases are freely generated in deep structure. Sentences that are overgenerated as a result of this are shown by the semantics to be truth valueless with respect to any model. Since *which god* is an NP, it denotes a set of characteristics—according to the analysis I am proposing, the set of characteristics of the individual represented by z, where z is a particular variable of the logic used only in the translation of *wh*-phrases. Among those characteristics is the characteristic of being a god. Thus, in terms of the logic, we may represent the denotation of this NP as (12).

(12) $$\lambda K[god'(z) \wedge K(z)]$$

The verb phrase *remain* denotes the characteristic of remaining which we represent in the logic as *remain'*. Ignoring the interpretation of tense for the sake of simplicity, we say that the sentence represented in (11) is true just in case the set of characteristics denoted by the subject NP includes the characteristic denoted by the VP. In the logic this is represented as (13).

(13) $$\lambda K[god'(z) \wedge K(z)](remain')$$

Example (13) says that the set of characteristics of z, which includes the characteristic of being a god, also contains the characteristic of remaining. This is the same as saying that z is a god and z remains, and we can obtain an equivalent expression of the logic which represents this more directly. This is given in (14).

(14) $$god'(z) \wedge remain'(z)$$

The interpretation of the main sentence *I gave it to the temple* is a little more complex, and I shall not go through it in detail. The whole sentence may be translated into an expression of which (15) is a slight abbreviation.

(15) $$\lambda KK(i)(give'(\hat{}to\ the\ temple')(\hat{}\lambda KK(u_0)))$$

Here, $\lambda KK(i)$ represents the denotation of the personal pronoun *I* which may be regarded as the set of characteristics of the speaker. The expressions *give'* and *to the temple'* represent the denotations of the corresponding unprimed expressions, and $\lambda KK(u_0)$ represents one of the possible choices of denotations for the pronoun *it* under a theory which does not allow pronouns to correspond to definite descriptions in the way that I am suggesting. The translation in (15) is, by dint of the semantics of the logic and some notational conventions, equivalent to (16).

(16) $$\exists u[\forall v[temple'(v) \equiv u = v] \wedge give(\hat{}to')_*(i, u, u_0)]$$

We can paraphrase (16) as "there is exactly one temple and I give u_0 to it." Now the problem of interpreting the whole sentence *Which god remained I gave it to the temple* reduces to finding a way of combining (14) and (16) to provide us with an appropriate reading. There are not too many hopeful choices. We could put some connective between the two sentences, but no connective we choose will associate the z in (14) with the u_0 in (16) in the way that we want to. For instance, (17) represents what would happen if we conjoined the two sentences.

(17) $$god'(z) \wedge remain'(z) \wedge \exists u[\forall v[temple'(v) \equiv u = v] \wedge give'({}^\wedge to')_*(i, u, u_0)]$$

This says that z is a god and z remains and there is exactly one temple such that I give u_0 to it. There will, of course, be some contexts of use which assign the same entity both to z and to u_0, and such a context would thus establish some connection between z and u_0. However, there will be many contexts of use which assign different entities to z and u_0, and relative to such contexts, (17) would not represent the correct reading for the Hittite sentence at all. We must therefore try a different strategy for putting the two together.

Both sentences contain free variables, and it might therefore seem a reasonable line of attack to do something not with the sentences themselves but with the expressions denoting characteristics corresponding to these sentences, that is, the characteristic of being a z such that z is a god and z remains and the characteristic of being a u_0 such that there is exactly one temple and I give u_0 to it. We can represent these characteristics in the logic simply by prefixing the appropriate lambda operator to the sentence. The resulting expressions are shown in (18).

(18) a. $\lambda z[god'(z) \wedge remain'(z)]$
b. $\lambda u_0[\exists u[\forall v[temple'(v) \equiv u = v] \wedge give'({}^\wedge to')_*(i, u, u_0)]$

We could say that the unique entity which has the characteristic represented in (18a) also has the characteristic represented in (18b). This would be represented in the logic as (19), where the occurrences of (18a) and (18b) have been underlined.

(19) $$\exists x[\forall y[\underline{\lambda z[god'(z) \wedge remain'(z)]}(y) \equiv y = x] \wedge \underline{\lambda u_0[\exists u[\forall v[temple'(v) \equiv u = v] \wedge give'({}^\wedge to'(i, u, u_0)]]}(x)]$$

We could get to this from the translations of the relative and main clauses if we had a translation rule that associated the translations in (20b) with structures of the form (20a).

(20) a.

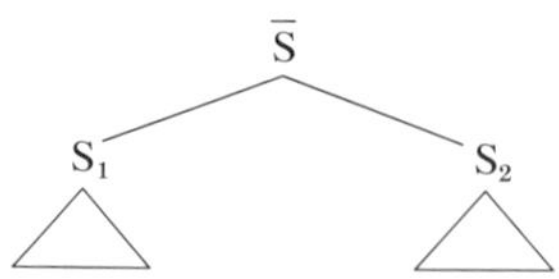

b. $\{\exists x[\forall y[\lambda z[S_1'](y) \equiv y = x] \wedge \lambda u_i[S_2'](x)]$: i is a natural number, S_1' is a translation of S_1, S_2' is a translation of $S_2\}$

Although this solution would work for this particular sentence, it would not fit together with a general account of Hittite relative clauses. If we consider examples such as (21), where there is an explicit quantifier in the head (rather than a pronoun as in [9]), it becomes clear that the quantification should not be introduced while processing the structure (20a). It must be introduced while processing the NP which is the head of the relative clause since, when quantification is overt in the syntax, this is where it resides.[3]

(21)

ŠA	NA$_4$.ḪI.A-ia	kuieš	GUNNLMEŠ	[nu]	kuišša	1	*GÌN*
of	stone-pl-and	which	hearths	ptc	each (one)	1	shekel

which hearths (are) of stones, each (one) (is) one shekel
'Every hearth which is made of stones, costs one shekel' [Held, 1957, p. 24].

It is clear here that it is the *each* in the main sentence that determines universal quantification, and that it would be entirely inappropriate for us to allow the structure (20a) to add the quantification associated with a definite description as would be the case if we adopted the translations (20b). It seems, then, that in sentences where there is a pronoun rather than a quantified NP serving as the head of the relative clause, the semantics of the pronoun should represent the required quantification even if no overt quantifier appears in the syntax.

Let us now consider the alternative of allowing pronouns to translate as definite descriptions in the manner I have suggested. Instead of (15) as the translation of the main sentence we may now get (22).

(22) $$\lambda KK(i)(give'(^\wedge to\ the\ temple'))(^\wedge\lambda K\exists x[\forall y[[^\vee P_0](y) \equiv y = x] \wedge K(x)]))$$

This, through the semantics of the logic and some notational conventions, is equivalent to (23).

(23) $$\exists u[\forall v[temple'(v) \equiv u = v] \wedge \exists x[\forall y[[^\vee P_0](y) \equiv y = x] \wedge give'(^\wedge to')_*(i, u, x)]]$$

The expression in (23) might be paraphrased as "there is exactly one temple and there is exactly one entity with property P_0 and I give that entity to the temple." Now the problem becomes one of combining (14) and (23) in a way which will give the appropriate reading. Because we have translated the pronoun in the main sentence as a definite description with a free property variable, the solution is straightforward. We have a free variable in (23), and we can talk about the set of properties which we could put in for P_0 to make (23) come out true. The way that this set would be represented in the logic can be obtained simply by putting the appropriate lambda operator in front of (23), as shown in (24).

(24) $$\lambda P_0[\exists u[\forall v[temple'(v) \equiv u = v] \wedge \exists x[\forall y[[^\vee P_0](y) \equiv y = x] \wedge give'(^\wedge to')_*(i, u, x)]]]$$

[3] I am grateful to Carleton T. Hodge for suggesting the translation "costs" here in place of "weighs."

Thus, (24) represents the set of properties characterized in (25).

(25) {P: P is a property and there is exactly one temple and exactly one entity having P and I give the entity to the temple}

What we want to say in order for the sentence to come out right is that the property of being the remaining god is in the set represented in (24) and (25). It is a straightforward matter to derive a representation of the property from the open sentence by which we translate the relative clause. We already saw how to represent the characteristic of being the remaining god in (18a). The requisite property is the function that finds the corresponding characteristic in every possible world. This is represented in the logic as (26).

(26) $${}^\wedge\lambda z[god'(z) \wedge remain'(z)]$$

We can now say that the property represented in (26) is in the set represented in (24):

(27) $$\lambda P_0[\exists u[\forall v[temple'(v) \equiv u = v] \wedge \exists x[\forall y[[{}^\vee P_0](y) \equiv y = x] \wedge give'({}^\wedge to')_*(i, u, x)]]]({}^\wedge\lambda z[god'(z) \wedge remain'(z)])$$

The expression in (27) is equivalent to (28), where we have applied lambda conversion twice.

(28) $$\exists u[\forall v[temple'(v) \equiv u = v] \wedge \exists x[\forall y[god'(y) \wedge remain'(y) \equiv y = x] \wedge give'({}^\wedge to')_*(i, u, x)]]$$

We may paraphrase (28) as "there is exactly one temple and exactly one remaining god and I give the god to the temple." This is now a correct representation of the interpretation of the sentence. In order to achieve this we must have a translation rule for the structure represented in (29a) which associates it with the set of translations represented in (29b).

(29) a.

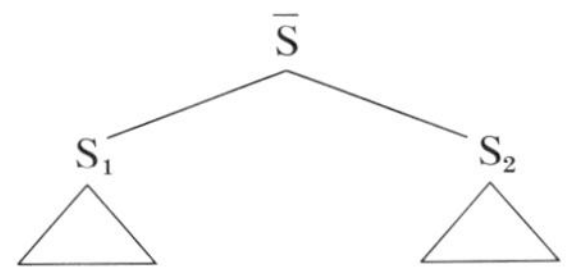

b. $\{\lambda P_i[S_2']({}^\wedge\lambda z[S_1'])$: i is a natural number, S_2' is a translation of S_2, S_1' is a translation of $S_1\}$

The semantics is set up in such a way that $\lambda z[S_1']$ will not denote anything unless z occurs in S_1'. The variable z will only occur in S_1' in case there is a *wh*-phrase in S_1. Thus, for the whole sentence to have a truth value, there must be a *wh*-phrase in S_1. Similarly, $\lambda P_i[S_2']$ will not denote anything unless there is an occurrence of P_i in S_2' and consequently the sentence will not get a truth value unless we interpret a pronoun in S_2 as an appropriate definite description. In this way we can use the semantics to filter out many of the cases of overgeneration caused by allowing free base-generation of pronouns and *wh*-phrases (see Cooper, 1975, Ch. 6).

One of the more interesting aspects of this analysis of Hittite is that the translation (28) which we have assigned to the whole sentence (9) is equivalent to the translation we would give to the English sentence *I gave the god which remained to the temple,* though it may well be the case that lexical items of Hittite correspond to different constants of the logic than English lexical items. However, the two sentences have the same logical structure. If we ignore any differences in the meanings of the lexical items of the two languages, we are in fact assigning the same meaning to the Hittite sentence as we are to the English sentence. Thus, we have made a generalization about the semantics of Hittite and English despite the rather different syntaxes of the two languages. Because we have an explicit semantic theory, we can capture the similarity in the semantics without proposing corresponding similarities in the syntax.

However, the main point I hope to have shown in this section is simply that if we wish to give a semantics for a straightforward syntax of Hittite, we have to include the possibility of interpreting pronouns as definite descriptions of the sort I have proposed. In the next section, I shall show that this interpretation for pronouns allows us to account for seemingly unrelated facts about English.

3. DISCOURSE ANAPHORA

The second use I will put this complex translation of pronouns to is to account for certain problematic cases of discourse anaphora in English. First, I will discuss how I think a reasonable treatment of straightforward discourse anaphora may be accommodated by Montague's notion of context of use, and then I shall discuss the cases which seem to require that pronouns be treated as definite descriptions.

Consider the discourse (30).

(30) *John looked up. He smiled.*

There are several ways we could interpret the pronoun *he*. It might refer to John, and if (30) is the complete discourse, this is the most likely reading. However, if I am pointing at someone other than John at the same time as I say *he*, *he* might well refer to someone other than John. In this second reading, we seem to have a clear case of the demonstrative *he* that Montague had in mind when he introduced his context of use. Thus, we might well consider that a translation of the second sentence of (30) is (31) where I have applied lambda conversion.

(31) $smile'(u_0)$

The expression in (31), of course, contains the free variable u_0, the value of which will be determined by the context of use relative to which (31) is evaluated. The question now arises how we would treat the case of anaphora where *he* is not exactly demonstrative, since it is related to an NP that occurs earlier in the discourse. It seems unlikely that we would want to quantify in *John* across two sentences in order to somehow bind the second pronoun. It is not clear that this is even possible, and if it were it would involve mechanisms that are to my knowledge as yet unformulated. However, we already have a mechanism which allows us to treat *he* as corresponding to a free variable and assign *John* as value to that free variable: namely, the mechanism we just used to get a nonanaphoric reading for (30). In fact, if we translate the second sentence of (30) as (31) and allow the value of u_0 to be determined by any context of use, we would have to introduce some special mechanism in order to *prevent* a context of use from being chosen which would give the anaphoric reading. Thus, it seems that the most straightforward way to treat (30), without adding any new machinery, is to translate the second sentence with a formula containing a free variable and interpret this formula relative to a context of use. Some contexts will assign John to the free variable, others will not. Similar remarks can be made about cases involving certain quantifiers. Consider (32).

(32) *The man looked up. He smiled.*

In order for the first sentence of (32) to be true, there must be exactly one man, in the universe of discourse we may assume. To obtain a nonanaphoric reading for the pronoun we would translate the second sentence as a formula with a free variable, as in (31), and let the context of use decide for us what *he* denotes. Here again, we would have to introduce some special mechanism to stop the context of use from assigning the entity which is the man in question to the free variable. Thus we can obtain the anaphoric reading without adding any extra mechanism to the syntax or the semantics.

Consider (33), which has an indefinite instead of a definite article.

(33) *A man looked up. He smiled.*

In order for the first sentence of (33) to be true, there has to be at least one man who looked up. Again, in order to get a nonanaphoric reading for the pronoun in the second sentence we will treat it as an open formula. But there is nothing in the system to prevent the context of use from assigning one of the men who looked up to the free variable and thus yielding an anaphoric reading. Here again, then, there is no reason for us to introduce a new mechanism for quantifying a noun phrase into a discourse, since Montague's technique for interpreting demonstrative pronouns already does the work for us. Note that in cases where it really would be necessary to bind the pronoun it is not usually possible to have discourse anaphora. This is illustrated by the sentences in (34).

(34) a. *Every man looked up. He smiled.*
b. *No man looked up. He smiled.*
c. *Mary didn't see any man laugh. He smiled.*
d. *Each man looked up. He smiled.*

In these examples, *he* may be assigned some unique individual by the context of use, but no context of use will make it appear that *he* is anaphorically related to the NP in the previous sentence. No matter which single individual is assigned to the variable, it is not possible to convey the idea that every man smiled or that no man smiled, etc. It seems, then, that allowing Montague's context of use to account for discourse anaphora not only prevents us from having to introduce a new mechanism for quantifying into discourses and also possibly block the mechanism for interpreting demonstrative pronouns in certain cases. We also predict that we would not be able to get anaphoric readings in cases like (34), where we cannot in fact get such readings.[4]

There are, however, some examples which intuitively seem to contain instances of anaphoric pronouns but where the pronoun may not be related to an individual variable in this way. Consider the discourses in (35).

(35) a. *This year the president is a Republican. Next year he will be a Democrat.*
b. *The secretary has typed the report himself for years. In the future a typist will often type it for him.*
c. *Mary hasn't yet found the man she will marry. Now that she has a job in Washington she hopes that she will meet him.*

I believe that (35a) can be interpreted in such a way that it does not require that the president (in 1976), Gerald Ford, is going to become a Democrat next year but rather that the next president of the United States will be a Democrat. Similarly, (35b) seems to assert that on several occasions in the past the secretary on that occasion typed the report for that occasion but that on all future occasions there will be a secretary who will type whatever report is involved for whichever secretary is in office at the time. Example (35c) may be taken to mean that there is no time in the past such that Mary found the man she will marry but she hopes that she will meet whoever she will marry, although there is no particular man that she hopes to meet.

Clearly, things will not work out quite right if we translate the second sentences in these discourses with free individual variables as I have been suggesting. A simplified version of a translation of the second sentence of (35a)

[4] Lauri Karttunen has pointed out to me that one can in fact get anaphoric readings for pronouns across sentences associated with NPs containing *every*. The examples, however, all seem to involve cases where one might propose a treatment in terms of an iterative or habitual operator. The analysis proposed in what follows would be able to handle such examples by treating the pronoun as a definite description within the scope of such an operator.

is shown in (36). (I have left out time adverbs since I am not sure how to treat them, but this does not affect the argument.)

(36) $$W[\exists u[democrat'(u) \wedge u = u_0]]$$

Expression (36) is equivalent to (37).

(37) $$W[democrat'(u_0)]$$

This predicates of some particular individual, the individual assigned to the variable u_0 by the context of use, that at some time in the future he will be a Democrat. This will work fine if we wish to intepret the discourse as telling us that Gerald Ford will change parties. We simply choose a context of use which assigns Gerald Ford to u_0. But this does not correspond to the reading where whoever is president next year will be a Democrat. The context of use can only pick out one value for the variable, and there is no way to incorporate the requisite quantification into the interpretation of the variable. In this case it would not help matters if we built into our system some mechanism for quantifying in NPs across sentences. We would not want the discourse to be represented logically as anything like (38).

(38) $$\exists v[\forall u[president'(u) \equiv u = v] \wedge republican'(v) \wedge W[democrat'(v)]]$$

This would require that there is a unique president who is now a Republican and will be a Democrat and thus will not be much of an improvement on having a free individual variable in the second sentence. The *he* in the second sentence seems to be what Geach (1962) would call a pronoun of laziness. That is, it appears to be a pronoun which stands for the form of words *the president*. We do not repeat the words *the president* to avoid repetitious language. It has been suggested by Partee (1973) that there are good reasons for building some such notion into a transformational grammar by making use of a classical pronominalization rule that would turn a full noun phrase into a pronoun. To implement this proposal, we might suggest that the deep structure for the second sentence of (35a) should correspond to the sentence (39).

(39) *The president will be a Democrat.*

A possible translation of this sentence would be (40a), which is equivalent to (40b).

(40) a. $$W[\exists u[\forall v[president'(v) \equiv u = v] \wedge \exists x[democrat'(x) \wedge x = u]]]$$

b. $$W[\exists u[\forall v[president'(v) \equiv u = v] \wedge democrat'(u)]]$$

The expression (40b) seems to be right. It says that there is some time in the future at which the president at that time is a Democrat. Now we have a reasonable semantics, but we are left with a syntactic problem. We have to formulate a pronominalization transformation which will change the NP *the president* in the deep structure of the second sentence (35a) into *he* because of its formal identity with a noun phrase in a different sentence of the discourse. It is not clear how the standard theory of transformations should be extended to encompass such rules, however.

Quite apart from the necessity of introducing transformations that take more than one sentence as their domain, some linguists may object to a transformation that changes a full noun phrase into a pronoun on the grounds that it involves a "drastic change in the form of lexical material [Jackendoff, 1972, p. 109]." There does not seem to be any syntactic motivation for having a full noun phrase underlying the pronoun. For instance, there is no process which crucially makes use of the presence of the full NP and which applies prior to the pronominalization rule. We cannot even make the claim that all pronouns come from full NPs, since it can be argued convincingly that at least those corresponding to bound variables should not. The addition of the pronominalization rule would not actually change the set of sentences generated by the grammar: The language generated by the syntax would be exactly the same whether the pronominalization rule were added to the grammar or not. It seems, then, that the addition of such a pronominalization rule would be a concession we would have to make in the syntax in order to get the semantics to come out right. It is as if we were admitting that there are certain cases of English pronouns that we cannot interpret, and that we therefore decide to derive them from nonpronominal sources for no other reason than to avoid the issue of what the interpretation of the pronoun would be. This would be a legitimate, though perhaps disquieting, move, if it represented the only available way of designing the system in order to get it to work.

However, I want to suggest that such is not the case with regard to the present data. If we allow English pronouns to be interpreted in the same way that we suggested for Hittite pronouns, there is no need for a syntactic rule of discourse pronominalization in order to account for these sentences. A translation of the second sentence in (35a), translating the pronoun as a definite description with a free property variable in it, is given in (41).

(41) $$W[\exists u[\forall v[[{}^{\vee}P_0](v) \equiv u = v] \wedge democrat'(u)]]$$

This, of course, will be evaluated with respect to a context of use. According to some contexts of use, P_0 will denote the property of being a president and relative to such a context (41) will be true under exactly the same conditions as (40b). Other contexts of use will assign a different property to P_0 and we will obtain a nonanaphoric reading for the pronoun. This less usual use of pronouns to represent definite descriptions will be accounted for by means of the context of use in a way similar to that which we suggested for the more straightforward anaphoric pronouns.

Remarks similar to those I have made about (35a) can be made about (35b) and (35c). The second sentence of (35b) under our normal free variable treatment would translate as (42).

(42) $\mathrm{OFT}(\lambda p[W[{}^{\vee}p])({}^{\wedge}\exists u[typist'(u) \wedge for'_{*}(u_0)(type'_{*}(u_1))(u)])$

Although the semantics of *often* is by no means clear, it seems reasonable to suppose that it requires that the sentence within its scope be true more than once and this is all that is necessary to see that (42) does not adequately represent (35b). In evaluating (42) relative to some context of use, we only get to choose one value each for u_0 and u_1. At best we could have a typist typing just one report for the same secretary several times. This is not the most likely reading for the sentence, however. Again, quantifying in the *secretary* or *the report* over the two sentences would not help, since this would also require that there be just one report and one secretary. A solution would be to derive the pronouns from underlying full NPs and thus have a deep structure corresponding to (43a) which could be translated as (43b).

(43) a. *A typist will often type the report for the secretary.*
b. $\mathrm{OFT}(\lambda p[W[{}^{\vee}p]])({}^{\wedge}\exists u[typist'(u) \wedge \exists v[\forall w[secretary'(w) \equiv w = v] \wedge \exists x[\forall y[report'(y) \equiv y = x] \wedge for'_{*}(v)(type'_{*}(x))(u)]]])$

This yields an appropriate reading since it only requires that at a number of times in the future there be a unique secretary and a unique report, but it does not require that it be the same secretary or the same report each time. In order to obtain a surface structure from a deep structure corresponding to (43a), we would have to introduce the sort of pronominalization rule that converts a full NP into a pronoun which I argued against earlier. If, however, we introduce the sort of translation for pronouns that I am proposing, we will be able to obtain the translation in (44) and let the context of use do the work for us.

(44) $\mathrm{OFT}(\lambda p[W[{}^{\vee}p]])({}^{\wedge}\exists u[typist'(u) \wedge \exists v[\forall w[[{}^{\vee}P_0](w) \equiv w = v] \wedge \exists x[\forall y[[{}^{\vee}P_1](y) \equiv y = x] \wedge for'_{*}(v)(type'_{*}(x))(u)]]])$

Finally, we can make similar arguments for (35c), where the relevant pronoun is in an intensional context. If we were to translate the second sentence of (35c) only with variables, we would obtain (45).

(45) $hope'(u_0, {}^{\wedge}W[meet'_{*}(u_0, u_1)])$

Assuming that the context of use assigns Mary to u_0, this would say that Mary hopes that she will meet whoever is assigned by the context to u_1. Let us assume that the context assigns Fred to u_1 and that Fred, unbeknownst to Mary, is the man she is going to marry. If (45) is the translation, then the sentence will come out true since Fred is indeed the man she is going to marry, and she hopes that she will meet the man she is going to marry.

Therefore, it is true that she hopes to meet Fred. But, of course, it is false that she hopes to meet Fred, on the reading we are interested in, because she does not know that he is the man she is going to marry. She stands in the hope relation to quite a different proposition, namely, one that would be expressed in terms of a definite description *the man she will marry*. We could suggest that *him* is really derived from an underlying full NP. But then we would have the same syntactic problems again. However, if we treat pronouns themselves as definite descriptions, we can have the translation (46) and allow the context of use to do the work.

(46) $hope'(u_0, {}^\wedge W[\exists u[\forall v[[{}^\vee P_0](v) \equiv v = u] \wedge meet'_*(u_0, u)])$

I should now like to look at some different examples where it seems that individual variables, whether bound or free, do not adequately represent pronouns of English. The sentence (47) is a well-known example from Karttunen 1969.

(47) *The man who gave his paycheck to his wife was wiser than the man who gave it to his mistress.*

Here, it seems that the pronoun *it*, if it is to be taken with the most likely anaphoric reading, should be a pronoun of laziness standing for the words *his paycheck,* since there is more than one paycheck involved. I will suggest that a slight revision of the treatment I am proposing will allow us to deal with such examples. I shall not consider (47) specifically, partly because I do not wish to become involved in a discussion of the semantics of the comparative. Rather, I shall consider the discourse (48), where I believe there is a similar reading for the pronoun.

(48) *John gave his paycheck to his mistress. Everybody else put it in the bank.*

In (48), we may assume that the NP *everybody else* denotes the set of characteristics possessed by every person other than John, and we may translate it as (49) where John is assigned to u_0 by the context of use.

(49) $\lambda K[\forall u[person'(u) \wedge \sim[u = u_0] \supset K(u)]$

If we were to choose a simple variable translation for *it* in the second sentence of (48), we might obtain a translation (50a), equivalent to (50b).

(50) a. $\lambda K[\forall u[person'(u) \wedge \sim[u = u_0] \supset K(u)]]$
$(put'({}^\wedge in\ the\ bank')({}^\wedge\lambda K K(u_1)))$

b. $\forall u[person'(u) \wedge \sim[u = u_0]$
$\supset put'({}^\wedge in\ the\ bank')_*(u, u_1)]$

This is clearly inadequate if we wish to represent a reading where each person put his own paycheck in the bank, because the context of use can assign only one value to u_1. A solution might be to treat *it* as a pronoun of laziness derived from *his paycheck,* but this would involve a rule of discourse pronominalization that we have argued against. In this case it would not be appropriate to treat the pronoun as a definite description of exactly the kind I am proposing. This would get us the translation (51a) which is equivalent to (51b).

(51) a. $\lambda K[\forall u[person'(u) \wedge \sim[u = u_0] \supset K(u)]](put'(\hat{}in\ the\ bank')(\hat{}\lambda K[\exists x[\forall y[[\check{}P_0](y) \equiv y = x] \wedge K(x)]]))$

b. $\forall u[person'(u) \wedge \sim[u = u_0] \supset \exists x[\forall y[[\check{}P_0](y) \equiv y = x] \wedge put'(\hat{}in\ the\ bank')_*(u, x)]]$

Expression (51b) may be paraphrased as "for any u, if u is a person other than u_0, then there is some unique thing with property P_0 and u puts it in the bank." This is not going to help us much because whatever property we assign to P_0, we are going to be talking about the unique thing which has that property and this will not allow for the possibility that there is a different thing with property P_0 for each person. What we need to represent is the property of being u's paycheck. In order to do this we need to have a bound variable u inside the definite description. Rather than (51b) we need something like (52).

(52) $\forall u[person'(u) \wedge \sim[u = u_0] \supset \exists x[\forall y[[\check{}S_0(u)](y) \equiv y = x] \wedge put'(\hat{}in\ the\ bank')_*(u, x)]]$

S_0 is to be taken as a variable over functions which map entities into properties. Expression (52), then, says "for any u, if u is a person other than u_0, there is some unique thing which bears relation S_0 to u and u puts it in the bank." (I am using the word *relation* loosely here.) According to some contexts of use $S(u)$ will represent the property of being u's paycheck and we will have what is intuitively an anaphoric reading for the pronoun *it*. According to other contexts of use, $S(u)$ will be the property of bearing some other relation to u and we will obtain a nonanaphoric reading. We can bind the u in the definite description in (52) by using a version of Montague's quantifying-in rule to quantify the NP *everybody else* into the sentence *he puts it in the bank.* I will give examples of the application of this rule in the next section.

In order to allow such translations for the pronoun, we may propose that in addition to the normal variable translations, pronouns may be associated with any one of the translations in the set characterized in (53).

(53) $\{\lambda K\exists x[\forall y[[\check{}\Pi](y) \equiv y = x] \wedge K(x)]:$ Π is a property-denoting expression containing only free variables and parentheses$\}$

Note that the difference between this set and the set of definite description translations we presented in example (5) is that instead of having a free-property variable, we only require that there be some property-denoting expression which consists entirely of free variables and parentheses. The use of the schematic letter Π does not require us to make necessary use of the syntax of intensional logic in order to characterize an interpretation for English. We could equally well associate English pronouns directly with a set of meanings.[5]

This proposal allows us to have as many free individual variables as we might need in the translation of the pronoun so that they may get bound by quantifiers and still leave at least one free variable whose value is determined by the context of use. It is not clear to me that we actually need all the possibilities, since it is hard to process examples that would need even two bound variables in the representation of a pronoun. Thus, (54) seems somewhat marginal on the intended reading.

(54) *The new ruling requires that the semester report which professors write about their students be nonconfidential. However, each professor we interviewed told us that not every student would want to see* ***it****.*

The pronoun *it* in (54) should be paraphrasable by *his report about him,* where *his* and *him* would have to correspond to bound variables in the scope of *each* and *every,* respectively. However, it should be noted that even with all the possibilities we have allowed ourselves with (53), we have still not increased the cardinality of the set of denotations that may be associated with a pronoun. There would still be denumerably many possibilities just as there were in the original treatment of pronouns.

4. GEACH'S DONKEY

The third problem I wish to address is one that was raised by Geach, 1962, involving sentences like (55).

[5] The set of all meanings associated with a pronoun may be characterized as follows, using the notations of Universal Grammar (Montague, 1974) except that we regard the context of use, j, as a sequence of possible denotations rather than an assignment to variables:
the union of

> $\{m$: $m \in M_{\langle\langle e,t\rangle,t\rangle,E,I,J}$ and for some natural number n and for all $i, j (i \in I, j \in J)$, $m(i,j) = k$ such that for any $x \in D_{\langle e,t\rangle,E,I}$, $k(x) = 1$ if and only if $x(j_{\langle n,e\rangle}) = 1$, where $j_{\langle n,e\rangle}$ is the nth member of j of type $e\}$

and

> $\{m$: $m \in M_{\langle\langle e,t\rangle,t\rangle,E,I,J}$ and for some nonnull sequence of ordered pairs of natural numbers and types $\langle\langle\alpha_0,\sigma_0\rangle,\langle\alpha_1,\sigma_1\rangle, \cdots, \langle\alpha_{n-1},\sigma_{n-1}\rangle\rangle$, such that for all $i, j (i \in I, j \in J)$, there is some $l \in D_{\langle s,\langle e,t\rangle\rangle,E,I}$ such that l is the result of $f(j_{\langle\alpha_0,\sigma_0\rangle}, j_{\langle\alpha_1,\sigma_1\rangle}, \cdots, j_{\langle\alpha_{n-1},\sigma_{n-1}\rangle})$ (where $j_{\langle\alpha_r,\sigma_r\rangle}$ is the α_rth member of j of type σ_r and f is an n-place composition of functional application or, in case $n = 1$, the identity function) and $m(i,j) = k$ such that if $x \in D_{\langle e,t\rangle,E,I}$, $k(x) = 1$ if and only if there is some $y \in D_{e,E,I}$ such that for all $z \in D_{e,E,I}$, $l(i)(z) = 1$ if and only if $z = y$ and $x(y) = 1\}$

(55) *Every man who owns a donkey beats it.*

The difficulty with this sentence can be illustrated in terms of a framework in which scope is represented syntactically in the deep structure, as in the first definition of the fragment given in Cooper and Parsons 1976. It provides us with a perspicuous way of discussing the problem, although I do not believe that it is in fact necessary to represent quantifier scope at all in a language which provides an adequate base for a model theoretic interpretation. (See Cooper, 1975, for discussion.) I will also use deep structure variables to represent pronouns, though, as I stated in the introduction to this paper, I believe that pronouns themselves can be interpreted directly and that it is not necessary to represent variables in the syntax.

In order to bind the pronoun *it* with the NP *a donkey*, it is necessary to quantify in *a donkey* into the sentence containing *it*. To do this we would need a deep structure approximately represented by the tree in (56).

(56)

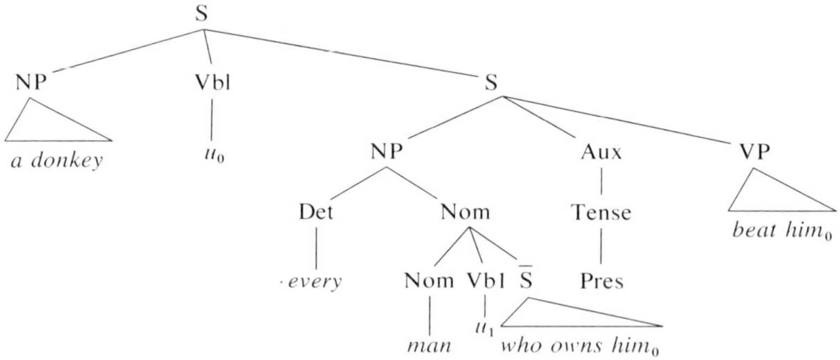

However, this yields an unlikely reading for (55), where the existential quantifier associated with *a* has wider scope than the universal quantifier associated with *every*. We would obtain a translation which is equivalent to (57).

(57) $\exists u[donkey'(u) \wedge \forall v[[man'(v) \wedge own'_*(v, u)] \supset beat'_*(v, u)]]$

If, instead, we put *a donkey* within the scope of *every* in order to obtain the intuitive reading, we cannot bind the pronoun in the VP as is shown by the tree and the translation in (58).

(58) a.

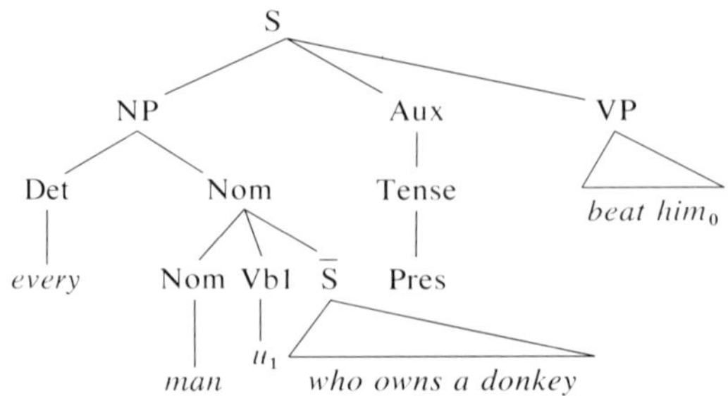

b. $\forall v[man'(v) \wedge \exists u[donkey'(u) \wedge own'_*(v,u)] \supset beat'_*(v,u_0)]$

One way out of this dilemma, which appears to correspond to Geach's solution (see Geach, 1962, paragraph 79), is to regard the indefinite article in *a donkey* as corresponding to a universal quantifier. We could reconstruct this in the sort of framework we are considering here by interpreting a structure such as (56), except that we would choose a different variant of the indefinite article. We might represent this as a^* if we care about keeping the syntax unambiguous. This would provide us with the translation (59).

(59) $\forall u[donkey'(u) \supset \forall v[man'(v) \wedge own'_*(v,u) \supset beat'_*(v,u)]]$

There are three things about this solution we might object to. First, it is not clear that (59) represents the correct reading, at least for some speakers. The expression in (59) requires that every man beats every donkey that he owns. Although the sentence certainly seems to suggest this, it is not clear that the semantics should require that each man beat all his donkeys. The sentence (60) casts some suspicion on treating the indefinite here as if it were a universal.

(60) *Every man who has a daughter thinks she is the most beautiful girl in the world.*

It does not seem to me, at least, that (60) commits any father of more than one daughter to the contradictory belief that each of his daughters is the most beautiful girl in the world.

The second objection that might be raised is that, given the way our systems of model-theoretic interpretation have worked so far, if we allow the indefinite article to be associated with a universal at all, then we have to allow for the possibility that it is so interpreted wherever it occurs. This means that besides the usual reading of (61a) there will be an additional reading which it has in common with (61b).

(61) a. *A donkey died.*
b. *Every donkey died.*

This does not accord with our intuitions.

The third objection, raised by Parsons 1972, is that such a treatment would not generalize to sentences of a slightly greater complexity. Consider the sentence (62), which is Parsons' example.

(62) *Every man who owns a donkey that he likes beats it.*

If we give this an analysis similar to the tree in (56) in order to bind the pronoun *it*, we have to quantify in the NP *a donkey that he likes* as shown in (63).

(63)

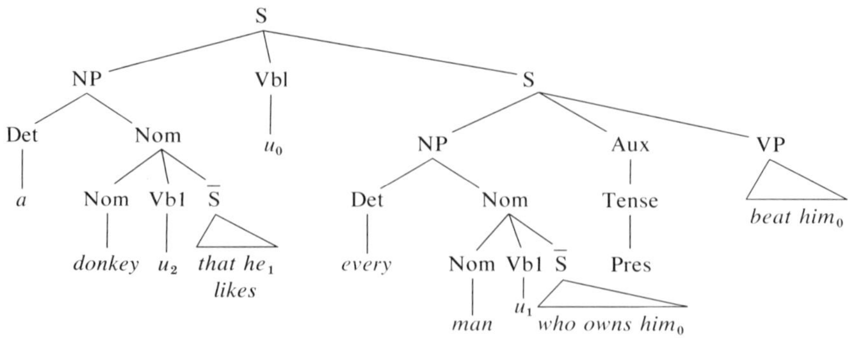

But now the *he* in *a donkey that he likes* does not get bound by the universal associated with *every*. Thus, whichever way we analyze this sentence, either the *it* or the *he* remains unbound.

This dilemma has led some researchers to propose that the *it* should not correspond to a variable at all, but should rather be interpreted as if it were an abbreviation for a full noun phrase, that is, it should be a sort of pronoun of laziness. The most obvious solution in strictly transformational terms would be to make the NP underlying *it* formally identical with the NP *a donkey,* so that the deep structure for (55) would be something like (64).

(64)

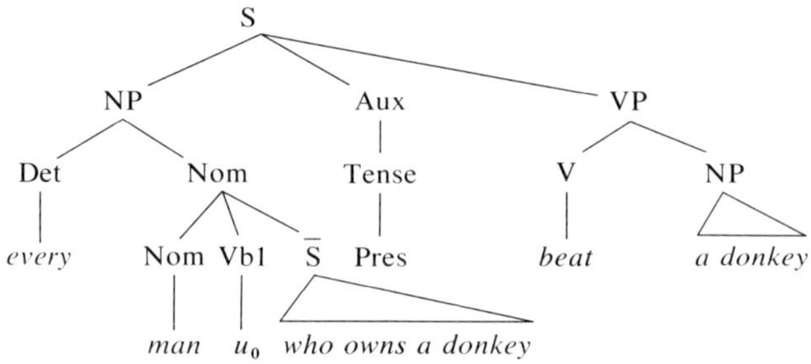

In (64), the rightmost occurrence of *a donkey* could be changed to *it* by a pronominalization rule. However, such a proposal is semantically inadequate. The translation of (64) would be (65).

(65) $$\forall u[man'(u) \wedge \exists v[donkey'(v) \wedge own'_*(u, v)] \supset \exists x[donkey'(x) \wedge beat'_*(u, x)]]$$

Clearly, (65) does not say the same thing as the original sentence, since it allows for the possibility that every man beats a donkey other than the one he owns, whereas the original sentence requires at least that every man beat a donkey that he owns, if not all of them. What we need underlying the pronoun is something like *a donkey that he owns* or *the donkey that he owns*. In order to bind the *he* in this underlying representation of the pronoun, we have to quantify in the whole NP *every man who owns a donkey*. For this we will need a structure something like (66).

(66)

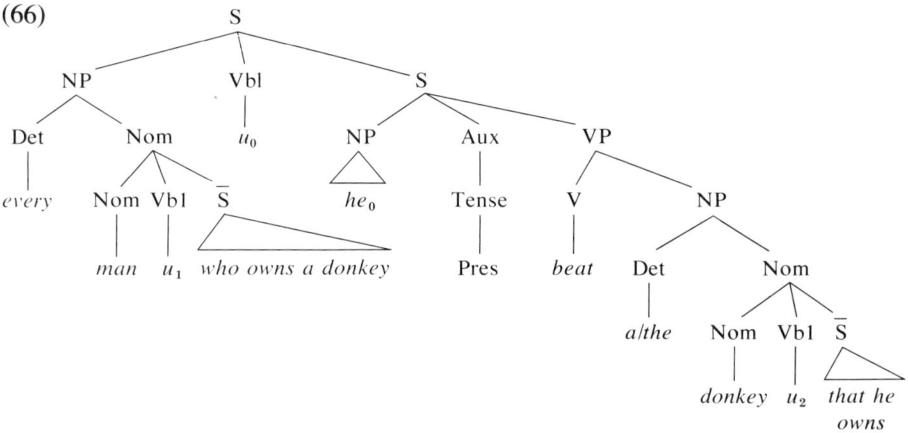

Trees like (66) would certainly provide us with more reasonable readings for these sentences. But now we have a syntactic problem. The NP *the/a donkey that he owns* must somehow turn into *it* by a rule of pronominalization. However the NP is not formally identical to anything else in the tree, and there is no way that I am aware of to encode the relationship that *the donkey that he owns* bears to *man who owns a donkey* into the structural description of a transformation, given the usual theory of transformations. We appear, then, to have a conflict between syntax and semantics. The pronoun of laziness treatment that would work in the syntax does not give the correct reading, and the pronoun of laziness treatment which would provide at least a reasonable semantics does not provide a workable syntax. However, if we allow complex translations for pronouns of the sort that I have been suggesting, then we may provide a reasonable semantics and avoid the syntactic problem of the pronoun of laziness.

In order to treat this sentence, we have a deep structure like (67) which is similar to (66) except that we have replaced *the donkey that he owns* with it^0, which we will use as a representation for a pronoun with the same denotation as the pronoun used in example (48).

(67)

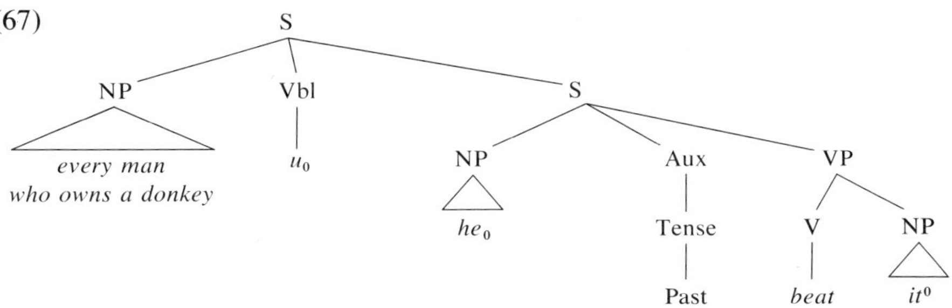

Here, it^0 would have a translation with two free variables rather than one. One of the free variables, u_0, will actually get bound by the NP *every man*

who owns a donkey, and the other will be a variable over relations whose value will be determined by the context of use. (I am again using the word *relation* loosely.) The translation of *it*^{0} will be (68), which is the same as that used in (52).

(68) $$\lambda K \exists x[\forall y[[{}^{\vee}S(u_0)](y) \equiv y = x] \wedge K(x)]$$

Translation (68) indicates that the variable *it*^{0} denotes the set of characteristics of the unique entity which stands in the relation S to u_0. After some reductions, the translation of the lower sentence will be (69).

(69) $$\exists x[\forall y[[{}^{\vee}S(u_0)](y) \equiv y = x] \wedge beat'_*(u_0, x)]$$

This may be paraphrased approximately as "there is some unique individual which bears the relation S to u_0 and u_0 beats that individual." We can now quantify in the translation of the NP *every man who owns a donkey* and we get (70a) which reduces by lambda conversion to (70b).

(70) a. $\lambda K \forall u[man'(u) \wedge \exists v[donkey'(v) \wedge own'_*(u, v)] \supset K(u)]$
$(\lambda u_0[\exists x[\forall y[[{}^{\vee}S(u_0)](y) \equiv y = x] \wedge beat'_*(u_0, x)]])$

b. $\forall u[man'(u) \wedge \exists v[donkey'(v) \wedge own'_*(u, v)]$
$\supset \exists x[\forall y[[{}^{\vee}S(u)](y) \equiv y = x] \wedge beat'_*(u, x)]]$

Expression (70b) might be paraphrased approximately as "for any individual, if it is a man and there is a donkey which he owns, then there is some unique individual that bears relation S to him and he beats that individual." Whatever the relation S is, of course, will be determined by the context of use. According to some contexts of use $S(u)$ will represent the property of being the donkey which u owns and thus will correspond to an anaphoric reading of *it*^{0}. According to other contexts of use, it will represent something different and may correspond to nonanaphoric readings of *it*^{0}.

There is one slight semantic advantage to this treatment over a corresponding one where the pronoun of laziness is represented as a definite description in the syntax, which would in this case be *the donkey that he owns*. If we were to paraphrase the pronoun in this way, then the sentence would require that every relevant man own only one donkey and would be false if one of the men owned two donkeys. However, it is not necessary that $S(u)$ be determined by the context of use as the property of being the only donkey owned by u—it could well be some other property which uniquely picks out one of u's donkeys. This would get us an apparent anaphoric reading without requiring that u own just one donkey. Neither will the sentence have to come out false if some of the men beat more than one or even all of their donkeys, as long as the context of use is such that $S(u)$ will pick out a particular one of u's donkeys for each u. The sentence will in no way require that every man beat all his donkeys, but as we saw with sentence (60), this would probably not be correct. The minimum requirement for this sentence to be true, under what we would intuitively consider to be an anaphoric reading, is that for each man who owns a donkey there is some donkey which he owns and beats and that the context of use is set up so that the unique entity which stands in the relation S to each man is a donkey of his that he beats. Of course, we have moved the problem of determining precisely what proposition is expressed by a given utterance of this sentence out of the semantics and into the area of pragmatics. I do not have a clear idea how we should set up a pragmatic theory to deal with this, but it does seem likely to me that much of the difficulty we have in determining exactly what such sentences mean might be explained by such a theory. To determine precisely what the value of S is would require a good deal more background of common knowledge between speaker and hearer than is normally present when this sentence is uttered. We would have to have in mind a single relation that would uniquely determine an appropriate donkey for each man. For a good number of utterances of this sentence, the context will be defective, and this might well explain why we have such difficulty deciding exactly what it means. The only case where it seems to be quite clear what the sentence means is where we assume that all of the relevant men each own exactly one donkey. Then the sentence itself provides us with enough information to determine a reasonable value for S. It will be such that $S(u)$ represents the property of being u's donkey.

Although I have no precise proposal to offer for how the pragmatic aspect of the analysis would work, it seems that we have a semantic analysis which is very similar to a pronoun of laziness treatment but does not have the syntactic problem of having to change an appropriate NP into a pronoun by a transformation. In addition, it utilizes a treatment of pronouns similar to the one we needed to account for Hittite relative clauses and cases of English discourse anaphora.

5. CONCLUSION

In this paper I have tried to give evidence for a precise adjustment to the treatment of natural language pronouns in the framework of Montague's theory. I hope, however, that I have illustrated two points which I think are of importance for a general theory of language. First, an explicit model-theoretic semantic theory does not necessarily, as we might at first believe, require us to develop abstract syntactic structures for which there is only semantic motivation. It is possible to interpret Hittite relative clauses without positing underlying structures where the relative clause is embedded to the head, and it is possible to interpret examples where it seems that a pronoun of laziness is required without having to propose rules of pronominalization.

The second point is that examining natural language in the light of an explicit semantic theory enables us to consider generalizations which would otherwise not be apparent. Despite the obvious syntactic disparity of Hittite and English relative clauses we can show them to have a basically similar semantics, and the semantic treatment of pronouns as definite descriptions allows us to draw parallels between seemingly unrelated problems of English and Hittite.

APPENDIX

The following is a definition of a fragment containing a number of the English examples in the text. (For a definition of a fragment containing the Hittite examples see Cooper, 1975, Ch. 6.) For the sake of clarity, I have followed the style and conventions of Cooper, 1975, (Ch. 2), because this employs a syntactic representation of scope and variables. I believe, however, that the fragment could also be defined employing the conventions of Chapter 4, where this is not the case. As in the text of the present paper, I shall ignore gender entirely. Thus, for example, the sentence *she hopes that she will meet him* will be represented in the fragment by he_0 *hopes that* he_0 *will meet* $him_{[0,\langle s,\langle e,t\rangle\rangle]}$. For a suggestion of how this might be improved see Cooper, 1975, (Ch. 3).

SYNTAX

Phrase Structure Rules

$$(1)\quad S \rightarrow \left\{\begin{array}{lll} NP & Vbl & S \\ NP & Aux & VP \end{array}\right\} \quad \begin{array}{l} a. \\ b. \end{array}$$

$$(2)\quad VP \rightarrow \left\{\begin{array}{llll} VP & PP & & \\ V_i & & & \\ V_t & NP & & \\ V_S & S & & \\ V_{t,PP} & & NP & PP \end{array}\right\} \quad \begin{array}{l} a. \\ b. \\ c. \\ d. \\ e. \end{array}$$

$$(3)\quad NP \rightarrow \left\{\begin{array}{ll} Det & Nom \\ N_{prop} & \end{array}\right\} \quad \begin{array}{l} a. \\ b. \end{array}$$

$$(4)\quad Nom \rightarrow \left\{\begin{array}{lll} N_{com} & & \\ Nom & Vbl & \bar{S} \end{array}\right\} \quad \begin{array}{l} a. \\ b. \end{array}$$

$$(5)\quad \bar{S} \rightarrow COMP \quad S$$

$$(6)\quad PP \rightarrow P \quad NP$$

$$(7)\quad Aux \rightarrow \left\{\begin{array}{l} Adv_{Aux} \\ \varnothing \end{array}\right\} Tns \quad \begin{array}{l} a. \\ b. \end{array}$$

$$(8)\quad Adv_{Aux} \rightarrow \textit{often}$$

(9) $\mathrm{Vbl} \rightarrow u_0, u_1, \ldots$

(10) $\mathrm{V_i} \rightarrow$ *walk, smile, look up, die*

(11) $\mathrm{V_t} \rightarrow$ *be, type, meet, own, beat, like*

(12) $\mathrm{V_S} \rightarrow$ *hope-that*

(13) $\mathrm{V_{t,PP}} \rightarrow$ *put*

(14) Det $\rightarrow$ *the, a, every, no, each*

(15) $\mathrm{N_{com}} \rightarrow$ *man, Democrat, typist, bank, donkey*

(16) $$\mathrm{N_{prop}} \rightarrow \left\{\begin{array}{l} \textit{John, who, everybody else}_i \text{ for any natural number } i \\ he_i \text{ for any natural number } i \\ he[i_0, \tau_0], \ldots, [i_{n-1}, \tau_{n-1}] \text{ for any } n > 1 \text{ where if} \\ 0 \leq j < n,\ i_j \text{ is a natural number and } \tau_j \in \mathit{Type}^6 \end{array}\right\}$$

(17) P $\rightarrow$ *for, in*

(18) COMP $\rightarrow \Delta$

(19) Tns $\rightarrow$ Pres, Past, *will*

Transformations (S-cyclic)

PRUNING CONVENTION

(i) Any node which exhaustively dominates a single element is itself deleted if that element is deleted.

(ii) Any node which immediately and exhaustively dominates a node with the same label is removed.

Transformations apply as many times as their structural description is met.

(20) Pronominalization (obligatory)

$$\begin{array}{llllll} \mathrm{NP} & u_i & \mathrm{X} & he_i & \mathrm{Y} & he_i \\ 1 & 2 & 3 & 4 & 5 & 6 \quad \Rightarrow \\ 1 & 2 & 3 & 4 & 5 & \left\{\begin{array}{ll} \text{a.} & he \\ \text{b.} & he_j \end{array}\right\} \end{array}$$

i,j are natural numbers

The choice of pronoun is b. iff the $\mathrm{N_{prop}}$ in 1 is he_j.

(21) NP-lowering (obligatory)

$$\begin{array}{llll} \mathrm{NP} & u_i & \mathrm{X} & [he_i]_{\mathrm{NP}} \\ 1 & 2 & 3 & 4 \quad \Rightarrow \\ \varnothing & \varnothing & 3 & 1 \end{array}$$

[6] It should be noted that the inelegance of the representation of pronouns here is due to the fact that we are representing them in deep structure by variables. If pronouns were to be interpreted directly (as in Cooper, 1975), variables would not be present in deep structure, but rather pronouns such as *he*, *she* and *it*.

(22) Affix Hopping (obligatory)

$$\begin{Bmatrix} \text{Pres} \\ \text{Past} \end{Bmatrix} \quad [X]_{\begin{Bmatrix} V_i \\ V_t \\ V_S \\ V_{t,PP} \end{Bmatrix}}$$

$$\begin{array}{lll} 1 & 2 & \Rightarrow \\ \varnothing & 2 + 1 & \end{array}$$

(23) Adv_{Aux} Movement (obligatory)

$$\begin{array}{ll} Adv_{Aux} & \text{will} \\ 1 & 2 \Rightarrow \\ \varnothing & 2 + 1 \end{array}$$

(24) Case Marking (obligatory)

$$\begin{Bmatrix} V_t \\ V_{t,PP} \\ P \end{Bmatrix} \quad he_{\alpha}$$

$$\begin{array}{lll} 1 & 2 & \Rightarrow \\ 1 & 2 & \\ & [+\text{accusative}] & \end{array}$$

α is a natural number or a subscript as specified in Rule 1b.

(25) Wh-Deletion (optional)

$$\begin{array}{lllllll} \text{Nom} & \text{Vbl} & [\Delta]_{COMP} & [X & wh & Y]_S & \\ 1 & 2 & 3 & 4 & 5 & 6 & \Rightarrow \\ 1 & \varnothing & that & 4 & \varnothing & 6 & \end{array}$$

wh is either *which* or *who*.

(26) Wh-Movement (obligatory)

$$\begin{array}{llllll} \text{Vbl} & [\Delta]_{COMP} & [X & wh & Y]_S & \\ 1 & 2 & 3 & 4 & 5 & \Rightarrow \\ \varnothing & 4 & 3 & \varnothing & 5 & \end{array}$$

wh as in (25).

We define the *set of sentences of the fragment* as the set of terminal strings of the surface structures generated by rules 1–26 (according to the usual conventions of transformational grammar) except:

(i) those surface structures which contain a node labeled *Vbl*

(ii) those surface structures which contain an occurrence of *who* or *which* not immediately dominated by a node labeled COMP

(iii) those surface structures which contain an occurrence of Δ.[7]

[7] In addition, we need to rule out those surface structures whose derivation involves a violation of the Complex NP Constraint. For an indication of how this might be formulated see Cooper 1975, (Ch. 6).

Translation

Notations of the Logic Which Differ from Those in PTQ

Let u, v, w, x, y, z represent $v_{1,e}$, $v_{3,e}$, $v_{5,e}$, $v_{7,e}$, $v_{9,e}$, $v_{11,e}$, respectively
Let u_i represent $v_{2i,e}$ where i is a natural number
Let K, J represent $v_{1,\langle e,t\rangle}$, $v_{3,\langle e,t\rangle}$, respectively
Let $\mathcal{K}$ represent $v_{1,\langle\langle e,t\rangle,t\rangle}$
Let P_i represent $v_{2i,\langle s,\langle e,t\rangle\rangle}$ where i is a natural number
Let S_i represent $v_{2i\langle e,\langle s,\langle e,t\rangle\rangle\rangle}$

If $\delta \in \mathrm{ME}_{\langle\langle s,\langle\langle e,t\rangle,t\rangle\rangle,\langle\langle e,t\rangle,\langle e,t\rangle\rangle\rangle}$ then δ_* is to be $\lambda u[\delta(^\wedge\lambda KK(u))]$

Adjustments to the Logic of PTQ

We add the following syntactic and semantic rules:

If $\alpha \in \mathrm{ME}_{\langle\langle s,t\rangle,t\rangle}$, then $\mathrm{OFT}(\alpha) \in \mathrm{ME}_{\langle\langle s,t\rangle,t\rangle}$

If $\alpha \in \mathrm{ME}_{\langle\langle s,t\rangle,t\rangle}$, then $[\mathrm{OFT}(\alpha)]^{\mathfrak{A},i,j,g}$ is that function k such that if $p \in \mathrm{D}_{\langle s,t\rangle,\mathrm{E},\mathrm{I}}$, then $k(p) = 1$ iff there are several j' such that $p(i,j') = 1$ *and* if $p(i,j') = 1$, then $\alpha^{\mathfrak{A},i,j,g}(p) = 1$.[8]

Translation of Terminal Elements of Deep Structures

who: $\lambda KK(z)$
everybody else$_i$*:* $\lambda K[\forall u[[person'(u) \wedge \sim[u = u_i]] \supset K(u)]]$
John: $\lambda KK(j)$
he$_i$*:* $\lambda KK(u_i)$
he$_{[i_0,\tau_0],\ldots,[i_{n-1},\tau_{n-1}]}$: $\lambda K\exists x[\forall y[{}^\vee f(v_{i_0,\tau_0}, \cdots, v_{i_{n-1},\tau_{n-1}})(y) \equiv y = x] \wedge K(x)]$

Where:

If $n > 1$, f is that n–place composition of the syntactic operation of functional application which obtains a meaningful expression of type

[8] This analysis of *often* is clearly inadequate in a number of ways, one of the most obvious being that *several* remains undefined. If *several* requires there to be more than one j', then it will be the case that a present tense sentence symbolized as

$$\mathrm{OFT}(\lambda P[{}^\vee P])(run'(j))$$

is logically false. This might appear unintuitive as a representation for *John often runs*. However, it should be borne in mind that the interpretation which is given to the simple present in this fragment is not the iterative interpretation that appears to be necessary in order to make sense of this sentence. If we compare it with *John is often running*, which more closely represents the interpretation we are providing, that is, a true present rather than an iterative or habitual reading, it seems that the addition of *often* should create at the least a necessarily false sentence.

As the analysis of tense is not the main concern of this paper, I will not try to present a more adequate treatment.

$\langle e,\langle s,\langle e,t\rangle\rangle\rangle$ consisting of $v_{i_0,\tau_0}, \cdots, v_{i_{n-1},\tau_{n-1}}$ (in that order) and parentheses (v_{i_k,τ_k} is the i_kth variable of type τ_k) *or* if there is no such composition, f is the n-place constant function whose value is P_0.

If $n = 1$, f is the identity function.[9]

the: $\lambda J \lambda K \exists u[\forall v[J(v) \equiv v = u] \wedge K(u)]$
a: $\lambda J \lambda K \exists u[J(u) \wedge K(u)]$
every: $\lambda J \lambda K \forall u[J(u) \supset K(u)]$
no: $\lambda J \lambda K \sim \exists u[J(u) \wedge K(u)]$
each: $\lambda J \lambda K \forall u[J(u) \supset K(u)]$
be: $\lambda \mathscr{K} \lambda u \mathscr{K}(\lambda v[u = v])$
u_i: u_i i is any natural number
often: OFT
will: $\lambda p[\mathrm{W}[{}^{\vee}p]]$
Pres: $\lambda p[{}^{\vee}p]$
Past: $\lambda p[\mathrm{H}[{}^{\vee}\mathrm{p}]]$
Δ: z

Otherwise any terminal element α translates into a constant represented by α', which is of type

$\langle e,t\rangle$ if α is generated from $\mathrm{N_{com}}$
$\langle\langle s,\langle\langle e,t\rangle,\langle e,t\rangle\rangle\rangle,\langle\langle s,\langle\langle e,t\rangle,t\rangle\rangle,\langle e,t\rangle\rangle\rangle$ if α is generated from $\mathrm{V_t}$
$\langle\langle s,t\rangle,\langle e,t\rangle\rangle$ if α is generated from $\mathrm{V_S}$
$\langle\langle s,\langle\langle e,t\rangle,t\rangle\rangle,\langle e,t\rangle\rangle$ if α is generated from $\mathrm{V_t}$
$\langle e,t\rangle$ if α is generated from $\mathrm{V_i}$

Translation of Branching Structures[10]

If α is a tree of the form

then α' (the translation of α) is the translation of β. (Rules 2b′, 3b′, 4a′, 7b′, 8–19.)

(7a′)

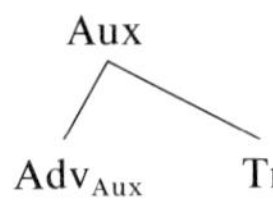

$\mathrm{Adv'_{Aux}}(\mathrm{Tns'})$

[9] It should be noted that the inelegant nature of this definition is an artifact of the requirement that we specify a translation for each member of the set of syntactic variables. If we associate pronouns directly with a set of translations (as in Cooper, 1975, Ch. 4), the characterization of the translation set would be that given in (53) in the text.

[10] Rule numbers correspond to those of the phrase structure rules.

(6′) 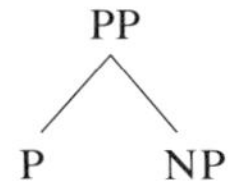

$P'(^\wedge NP')$

(5′) 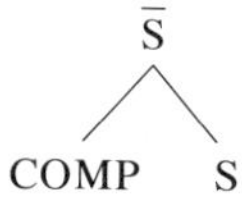

$\lambda COMP'[S']$

(4b′) 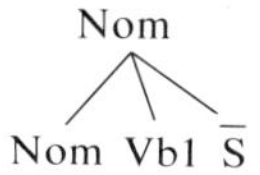

$\lambda Vbl'[Nom'(Vbl') \wedge S'(Vbl')]$

(3a′) 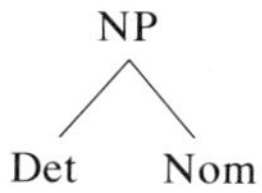

$Det'(Nom')$

(2e′) 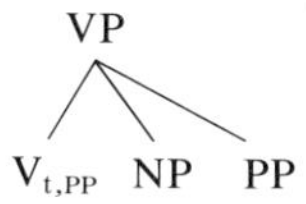

$V'_{t,PP}(^\wedge PP')(^\wedge NP')$

(2d′) 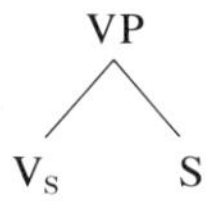

$V'_S(^\wedge S')$

(2c′) 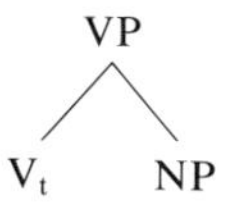

$V'_t(^\wedge NP')$

(2a′) 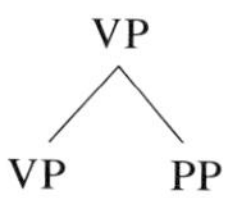

$PP'(VP')$

(1b′) 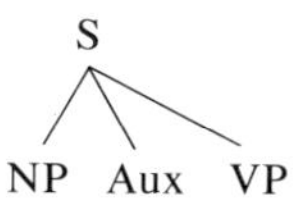

$Aux'([^\wedge NP'(VP)])$

(1a′) 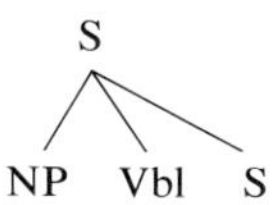

$NP'(\lambda Vbl'[S'])$

We restrict our attention at least to those interpretations for which the following meaning postulates hold, where R represents $v_{1,\langle e,\langle e,t\rangle\rangle}$, and Z, $v_{1,\langle e,\langle\langle e,t\rangle,\langle e,t\rangle\rangle\rangle}$ and $\mathfrak{R}$, $v_{1,\langle s,\langle\langle e,t\rangle,t\rangle\rangle}$ and $\mathfrak{I}$, $v_{3,\langle s,\langle\langle e,t\rangle,t\rangle\rangle}$

Meaning Postulate 1

$$\exists R \forall u \forall \mathfrak{R} \Box[\delta(\mathfrak{R})(u) \equiv [{}^\vee \mathfrak{R}](\lambda w[R(w)(u)])]$$

where δ translates *type, meet, own, beat*

Meaning Postulate 2

$$\exists R \exists Z \forall u \forall \mathfrak{R} \forall \mathfrak{I} \Box [\gamma(\mathfrak{I}(\delta(\mathfrak{R}))(u) \equiv [{}^{\vee}\mathfrak{I}](\lambda v[[{}^{\vee}\mathfrak{R}](\lambda w[Z(v)(R(w))(u)])])]$$

where γ translates *for*, δ, *type*

Meaning Postulate 3

$$\exists R \forall u \forall \mathfrak{R} \forall \mathfrak{I} \Box [\delta(\gamma(\mathfrak{I}))(\mathfrak{R}) \equiv [{}^{\vee}\mathfrak{R}](\lambda w[R(w)(u)])]$$

where δ translates *put*, γ, *in*

REFERENCES

Bennett, M. (1976). Accommodating demonstrative and indexical expressions in Montague grammar. Mimeograph, University of Pittsburgh, Pennsylvania.

Cooper, R. (1975). Montague's Semantic Theory and Transformational Syntax. Unpublished doctoral dissertation, University of Massachusetts, Amherst.

Cooper, R. (1977). Review of Montague (1974). *Language,* **53**(4), 895–910.

Cooper, R., and Parsons, T. (1976). Montague grammar, generative semantics, and interpretive semantics. In B. Partee (Ed.), *Montague Grammar.* New York: Academic Press.

Geach, P. (1962). *Reference and Generality: An Examination of Some Medieval and Modern Theories,* emended edition, 1968. Ithaca, New York: Cornell University Press.

Held, W. Jr. (1957). *The Hittite Relative Sentence.* Supplement to *Language,* **33**(4) (Part 2). Language dissertation no. 55.

Jackendoff, R. (1972). *Semantic Interpretation in Generative Grammar.* Cambridge, Mass: MIT Press.

Karttunen, L. (1969). Pronouns and variables. In R. Binnick *et al.* (Eds.), *Papers from the Fifth Regional Meeting of the Chicago Linguistic Society.* Chicago: Department of Linguistics, University of Chicago.

Karttunen, L. (1971). Definite descriptions with crossing coreference: A study of the Bach–Peters paradox. *Foundations of Language,* **7,** 157–182.

Montague, R. (1974), *Formal Philosophy: Selected Papers of Richard Montague,* edited and with an introduction by Richmond Thomason. New Haven: Yale University Press.

Parsons, T. (1972). A portion of a draft of a projected work entitled *A Semantics for English.* Mimeograph, University of Massachusetts, Amherst.

Partee, B. (1975). Deletion and variable binding. In Edward L. Keenan (Ed.), *Formal Semantics of Natural Language: Papers from a colloquium sponsored by the King's College Research Centre, Cambridge.* Cambridge, England: Cambridge University Press.

Quine, W. V. O. (1960). *Word and Object.* Cambridge, Mass: MIT Press.

Raman, C. (1973). The Old Hittite Relative Construction. Unpublished doctoral dissertation, University of Texas, Austin, Texas.

Thomason, R. (1974). On the semantic interpretation of the Thomason 1972 fragment. Unpublished mimeo, University of Pittsburgh, Pittsburgh, Pennsylvania.

HOW DO PRONOUNS DENOTE?[1]

ROLAND R. HAUSSER

1. INTRODUCTORY REMARKS

Departing from the transformational tradition of describing certain **anaphoric** pronouns on the basis of syntactically stipulated coreference, the following analysis attempts to describe pronouns in terms of a general approach to context dependency. In order to insure formal explicitness and compatibility, definitions implementing "contextual reconstruction" as well as detailed analyses of different pronouns will be presented in the form of an extension of Montague 1973, henceforth PTQ.

In Sections 2, 3, and 4, I will discuss the relevant data analyses making up the background of this paper. The next step is to give the definitions of contextual reconstruction (Section 5). Finally, Sections 6, 7, and 8 present syntactico–semantic analyses of reflexive and nonreflexive personal T pronouns, possessive T and CN pronouns, and the definite article. These

[1] I would like to thank Helmut Frosch, Frank Heny, Joachim Jacobs, Lauri Karttunen, and Theo Vennemann, who commented on an earlier version of this paper. I am especially indebted to Godehard Link and Dietmar Zaefferer for their help in developing the definitions of the context model. The responsibility for all mistakes and inadequacies is, of course, my own.

SYNTAX AND SEMANTICS
VOLUME 10

ISBN: 0-12-613510-X

analyses include a treatment of the so-called constraints on "backward pronominalization." It is shown that our new approach handles the pertinent data, including certain well-known counterexamples to the standard approach to "pronominalization."

2. OUTLINE OF THEORY

Pronouns in English are basic expressions belonging to certain syntactic categories and functioning according to their category. Their morphology exhibits both regular and (relatively) irregular paradigms. But the crucial question is: **How do pronouns denote?**

In transformational grammar, an attempt is made to answer this question at least for a subclass of pronouns, the so-called anaphoric pronouns: The denotation of anaphoric pronouns is inferred through **syntactically stipulated coreference** between the pronoun and its **antecedent.** An antecedent is a "full" expression occurring in the same sentence as the pronoun and somehow responsible for a certain possible interpretation of the pronoun.

However, it is quite common for pronouns like *he*, *she*, *it* to refer **without** an explicit antecedent. This is the **indexical** use of pronouns. Indexical pronouns depend on contextual information for their proper interpretation. Contextual information may be nonverbal; at any rate, contextual information does not always appear on the linguistic surface of the expression containing the pronoun in question. This means that any analysis which syntactically stipulates coreference in its account of anaphoric pronouns is bound to construe a principled syntactico–semantic contrast between the anaphoric and the indexical use of pronouns.

The transformational approach to anaphoric pronouns has resulted in a number of interesting observations, in particular, the so-called "constraints on backward pronominalization" (Lees and Klima, 1963; Chomsky, 1965, Postal, 1969; Langacker, 1969; Ross, 1969; Lakoff, 1968, etc.). These analyses, however, were never able to resolve certain technical problems arising with complex constructions such as Bach–Peters sentences, the donkey sentence, or the paycheck sentence (cf. Sections 3 and 4). At the center of these difficulties is the use of stipulated coreference.

In transformational grammar, coreference is usually stipulated on the basis of an identity principle: Two NPs are coreferential if they are identical in the "deep structure." In the writings of Richard Montague, however, coreference is stipulated in terms of underlying pronouns with an identical index. That Montague indeed adopted a syntactico–semantic concept of anaphoric pronouns is obvious from a comparison of the treatment of anaphoric pronouns in Montague 1970a (henceforth EFL), PTQ, and the quite different treatment of indexicals in Montague 1968.

In contrast to the "identical-NP approach" of transformational grammar,

Montague's proposal for stipulating coreference is at least based on well-defined semantics, namely, the independently motivated principle of **bound variables,** familiar from predicate calculus. The transformational approach, on the other hand, is linguistically more detailed: It handles several syntactic constellations of anaphoric pronouns, whereas Montague's rules permit only "forward pronominalization." The crucial point is that neither approach to the anaphoric use of a small subgroup of pronouns is naturally extendable to the indexical use of these same pronouns. Furthermore, neither approach is able to account for all pertinent data. Anaphoric pronouns which elude a given analysis are commonly called "dangling pronouns."

The alternative approach to pronouns developed in this paper finds no essential syntactico–semantic difference between indexical and anaphoric pronouns. Instead, we are guided by certain semantic similarities between definite descriptions and pronouns like *he, she, it, they*. Both kinds of terms may be used anaphorically as well as indexically, and both give rise to curiously restricted coreference phenomena. Concerning the semantics of definite descriptions, I will follow the analysis of Hausser (1974b, Ch. 5).

This analysis of the definite article shares certain intuitions with Vendler (1967, Ch. 2), where it is proposed that the proper denotation of definite descriptions should be insured syntactically by generating "underlying relative clauses" in the "deep structure" of the definite description in question; transformations using "recoverable deletions" are intended to reconcile the unexpanded surface structure with the widely expanded deep structure. A problem arises for this approach, however, because recoverable deletions are defined only with respect to the explicit linguistic surface. Vendler is aware of this problem and suggests that if there is no suitable previous discourse (that is, a discourse which contains the information necessary to recover the presumed deletions), then definite descriptions function like proper names. This assumption, however, is incompatible with the rigid-designator analysis of proper names presumed here (cf. Kripke, 1972, PTQ, and Hausser, 1974b). The point is that definite descriptions are context dependent in their denotation, whereas proper names are assumed to be context **in-**dependent.

Instead of a syntactic approach, let us implement Vendler's idea in the form of **contextual reconstruction.** Contextual reconstruction is a formal mechanism for filling in contextual information into the translations of context-dependent expressions. It is defined as an interpretational device: The surface syntax is not affected by contextual reconstruction.

This approach to context dependency is able to insure the proper denotation of definite descriptions by providing a formal basis for expanding the **descriptive content** of the definite description in question. The descriptive content of a definite description is the (basic or complex) property (corresponding to a CN phrase) which remains after subtracting the quantificational machinery that makes up most of the semantic representation of definite descriptions. For example, the descriptive content of *the man* is the property *man'*. Formally, the descriptive content of a given context-dependent term is extended by introducing additional properties provided by the context.

For this purpose, I will define a context model, which not only specifies the denotation of the constants, but in addition assigns a **stage description** to each point of reference. A stage description is a regimented, formal description of the presumed utterance situation. Furthermore, context-dependent expressions of English are translated in part by means of characteristic **context variables.** A reconstruction function applied to a context variable results in replacing the argument by a "suitable" contextual value (according to certain standards).

Contextual reconstruction is a general approach to phenomena of context dependency. It is not only able to handle the interpretation of definite descriptions and nonredundant answers (cf. Hausser, 1976c), but also various kinds of pronouns. Take, for example, such T pronouns as *I, you, he, she, it, we, they*. We determine the denotation of these pronouns simply by expanding their descriptive content through contextual reconstruction. The only difference between T pronouns and definite descriptions is that (i) one is syntactically derived whereas the other is basic, and (ii) one carries a great amount of descriptive content, whereas the other consists of nothing but a quantifier expression with a context variable (and sometimes a gender specification).

The constraints on backward pronominalization are naturally implemented as constraints on the "internal reconstruction" of third person T pronouns: The constraints exclude expressions in certain locations from being possible contextual arguments. The interpretation of Bach–Peters sentences, the donkey sentence, the paycheck sentence, etc., presents neither complication nor exception in this approach. Furthermore, no syntactico–semantic contrast is postulated between anaphoric and indexical use of certain expressions. According to my analysis, a pronoun (or definite description) is used **indexically** if it is reconstructed **externally,** and **anaphorically** if it is reconstructed **internally.** Some pronouns permit only exterior reconstruction (that is, reconstruction with respect to a stage description), for example, *I, you, we*. Other pronouns permit only interior reconstruction (that is, reconstruction with respect to the translation of the target sentence), for example, *himself.* And some pronouns permit both external and internal reconstruction, for example, *he, she, it, they.*

My approach is similar to that of Jackendoff (1969, 1972) in that I reject a syntactic analysis of coreference. Jackendoff's rules of "interpretation," however, do not determine the denotation of pronouns; they only specify whether a third person T pronoun may or may not receive a coreferential interpretation. Although the interpretive approach to pronouns avoids some of the problems of the standard transformational approach, it is plagued by problems of its own which have to do with Jackendoff's **Noncoreferentiality**

Convention (if a pair of NPs have not been marked coreferential by previous applications of the interpretive rules of coreference, they are to be marked as noncoreferential, [cf. Postal, 1972; see also Dougherty, 1969, and Wasow, 1973 for interpretive analyses of pronouns]). My approach differs from transformational accounts in general in that I develop a syntactico–semantic treatment of pronouns in a well-defined semantic framework (that is, through standard model-theoretic semantics).

3. TWO SYNTACTIC APPROACHES TO ANAPHORIC PRONOUNS

Pronouns occur in a number of different syntactic categories, and not only as noun phrases. Hence, the term "pronoun" is unfortunate. For example, the so-called nonreflexive personal pronouns *I, you, he, she, it, we, they,* the reflexive personal pronouns *myself, yourself, himself, herself, ourselves, themselves,* and the possessive pronouns *mine, yours, his, hers, its, ours, theirs,* are all of category T (noun phrase). The pronouns *one, ones,* however, are of category CN (common noun). Furthermore, there are possessive pronouns of category T/CN (quantifiers), namely, *my, your, his, her, its, our, your, their*. There are also the pronouns *here, there, therefore, then, thus,* which are of Category IAV (adverbials).

Syntactically, a pronoun is characterized by the category to which it belongs. For example, a T pronoun like *he* or *she* may serve as subject or object, and is assigned to different genders, numbers, and cases. That T pronouns function at least syntactically like full noun phrases is apparent from the following examples:

(1) a. *She kissed him.*
b. *Mary kissed the man who arrived yesterday.*

Similarily, CN pronouns may serve as head nouns of restrictive relative clauses, may be modified by adjectives (*green one*), and may combine with a quantifier to make a noun phrase (*the green one*). Pronouns of Category IAV are subject to the same subcategorization typical of all adverbs: *Here* and *there* are adverbs of **place** and **direction,** *therefore* is an adverb of **cause,** *thus* is an adverb of **manner,** and *then* is an adverb of **time.**

There are no special syntactic restrictions on the occurrence of pronouns. Compare, for example, (2) and (3):

(2) *John kissed Mary in front of the Eiffel Tower.*

(3) *I kissed her there.*

Syntactically, (2) and (3) are of the same structure, and in this sense the syntax of pronouns is trivial. The general syntactic rules apply to pronouns in exactly the same way as to other expressions of the corresponding category.

Semantically, on the other hand, pronouns differ from ordinary denoting expressions. Consider the difference in interpreting (2) as opposed to (3) with respect to an interpretation and a point of reference. In (2), the truth value depends on the denotation of *John, kiss, Mary,* and *in front of the Eiffel Tower* as specified by the model. In particular, *John, Mary,* and *in front of the Eiffel Tower* are to be defined as denoting particular individuals and a particular place (area). In (3), however, the situation is quite different insofar as it would be intuitively wrong to assign fixed denotations to the pronouns *I, her,* and *there.* The denotation of the pronouns in (3) depends on who utters (3), where, and with respect to whom. Ordinary denoting expressions like *John, Mary,* or *in front of the Eiffel Tower* differ from pronouns in that they are not context dependent in the indicated sense.

The assumption that pronouns are context-dependent expressions naturally suggests a linguistic treatment in terms of context dependency. Most of the recent literature, however, has concentrated on syntactic solutions in terms of stipulated coreference (including EFL and PTQ). Before turning to my alternative approach, I will discuss in the present and the following section the goals and inadequacies of three representative attempts to handle pronouns syntactically. Those readers who are familiar with this material may safely go on to Section 5. Let me mention, however, that the analysis to be developed in Sections 6, 7, and 8 will be evaluated with respect to the data discussed in Sections 3 and 4.

What is the difference between the standard transformational approach to pronominalization and the approach presented in EFL and PTQ? Consider example (4).

(4) ***A girl*** *got a cat which scratched* ***her.***

In PTQ, the pronoun *her* (on its anaphoric interpretation with *a girl* as antecedent) is derived from an **underlying pronoun.** The derivation of (4) in PTQ may be characterized by the following analysis tree.

(5) *a girl gets a cat such that it scratches her*

a girl — *he_0 gets a cat such that it scratches him_0*

he_0 — *get a cat such that it scratches him_0*

The pronouns he_0 (him_0) are translated into variables which are bound by the same quantifier:

(5′) $\bigvee x[girl'(x) \wedge \bigvee y[[cat'(y) \wedge scratch'(y, \hat{P}P\{x\})] \wedge get'(x, \hat{P}P\{y\})]]$

In other words, Montague realizes the suggestion (for example, in Quine, 1960) that "pronouns are variables" on the level of semantic representation, and utilizes the ready-made binding principle of predicate calculus in order to handle stipulated coreference semantically. In the categorial syntax of

PTQ, coreference is marked in terms of underlying coindexed pronouns; the antecedent is "quantified in."

On the other hand, within the standard theory of TG (for example, Chomsky, 1965), coreference is stipulated on the basis of underlying identical full noun phrases. As an example of a standard transformational derivation of an anaphoric pronoun, consider (6):

(6)

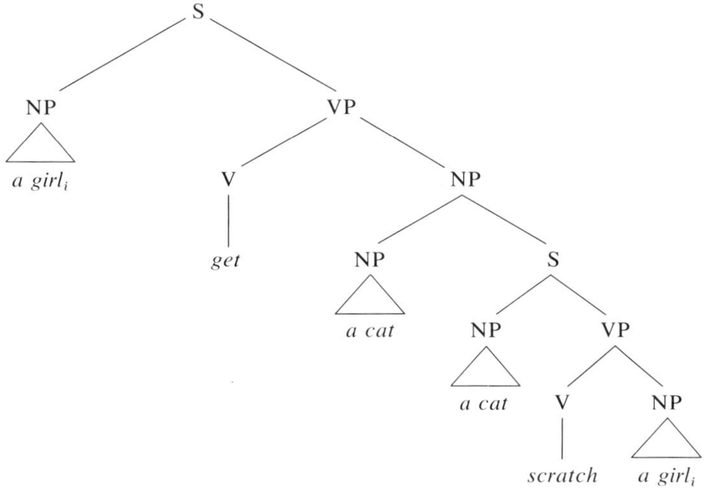

Obligatory pronominalization changes the second occurrence of *a girl* in (6) into the pronoun *her*.

The coindexed underlying pronoun approach of PTQ and the identical-NP-reduction approach of the standard theory are equally well-founded syntactic principles for stipulating coreference. But both approaches are subject to fundamental empirical limitations. Let us discuss a number of standard counterexamples to either or both of the two approaches. First, consider the following sentence from Partee 1975:

(7) *Every man who loves* **a woman** *loses* **her.**

On the dominant reading of (7), *every man* has wider scope than *a woman* (that is, on this reading *a woman* is interpreted nonspecifically, cf. Hausser, 1976b). This reading, however, cannot be derived in a standard derivation within the PTQ framework:

(7′) $\bigwedge x[man'(x) \wedge \bigvee y[woman'(y) \wedge love'(x, \hat{P}P\{y\})] \rightarrow lose'(x, \hat{P}P\{y\})]$

The point is that the variable y at the end of the formula is outside the scope of the existential quantifier. In other words, *her* on its coreferential reading (antecedent *a woman* on nonspecific interpretation) is not analyzable as an anaphoric pronoun within a framework that uses coindexed underlying pronouns in combination with the bound variable principle for handling stipulated coreference. Instead, *her* in (7) must—under the specified reading—be classified as a dangling pronoun (Partee uses the term ''pronoun of laziness'').

While sentence (7) raises a fundamental problem for the coindexed pronoun approach as implemented in PTQ, it is a straightforward case of ''forward pronominalization'' for the identical-NP-reduction approach of TG. Thus, whether a pronoun is classified as ''anaphoric'' or as ''dangling'' depends on the specific mechanism used for stipulating coreference.

Although the identical-NP-reduction approach is fine for examples like (7), it has its own special problems with examples like (8):

(8) *Every senator voted for himself.*

According to the original pronominalization approach of TG, the deep structure of (8) should look like (8a):

(8a)

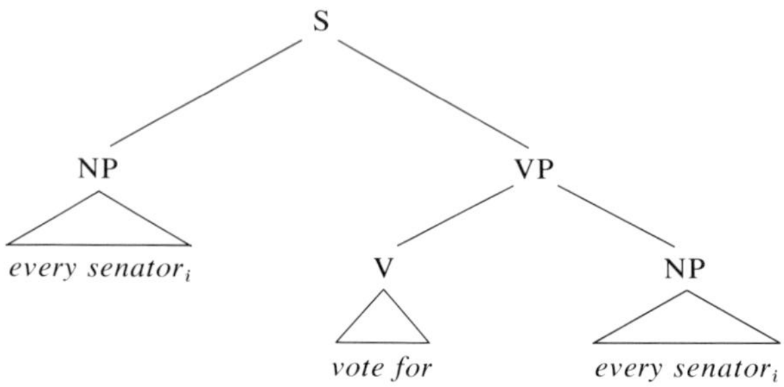

Obligatory pronominalization (reflexivization) transforms (8a) into the surface form (8). The problem is that (9)

(9) *Every senator voted for every senator.*

happens to be assigned the same deep structure as (8), yet (9) has a different meaning than (8). This is in conflict with the assumptions (i) that transformations are meaning preserving, and (ii) that the deep structure of a sentence represents its full meaning (deep structures are the sole input to the rules of semantic interpretation).

The problem that sentences like (8) and (9) raise for the identical-NP-reduction approach as defined in standard TG finds a natural solution in frameworks using the pronoun-as-variable approach. In PTQ, for example, (8) would translate into (8′), whereas (9) would translate into (9′):

(8′) $\bigwedge x[senator'(x) \rightarrow vote\ for'(x, \hat{P}P\{x\})]$

(9′) $\bigwedge x[senator'(x) \rightarrow \bigwedge y[senator'(y) \rightarrow vote\ for'(x, \hat{P}P\{y\})]]$

The truth conditions for (8′) and (9′), respectively, reflect the intuitive semantic difference between (8) and (9) adequately.

One kind of example, however,—involving the so-called Bach–Peters sentences—is unsolvable both for the identical-NP-reduction approach and the pronoun-as-variable approach. Consider (10):

(10) ***The man*** *who deserves* **it** *will get* **the prize** ***he*** *wants.*

A standard deep structure of (10) in TG is (11).

(11)

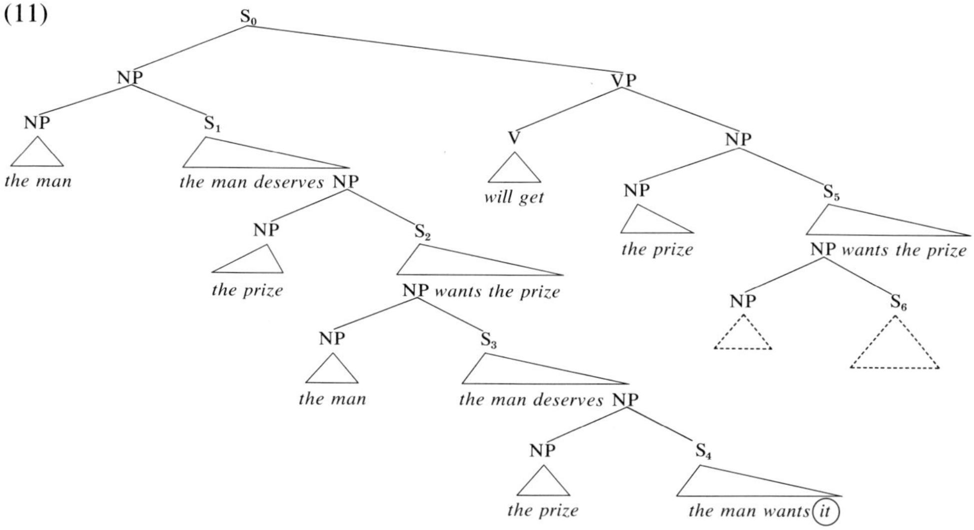

Let us assume that the left branch of embeddings in (11) is finite, and that pronominalization has applied once to yield the circled *it*. The problem is that *the prize* in cycle S_2 cannot pronominalize *the prize which the man wants* in S_3 (derived from *the man* [*the man wants it*]) because the two NPs in question are not **identical.** The problem arising in connection with example (10) is related to the assumption that the full-NP-reduction approach as defined in TG operates on identical NPs.

In PTQ, however, the Bach–Peters sentence in (10) raises a problem of a different kind. Consider analysis tree (12).

(12)

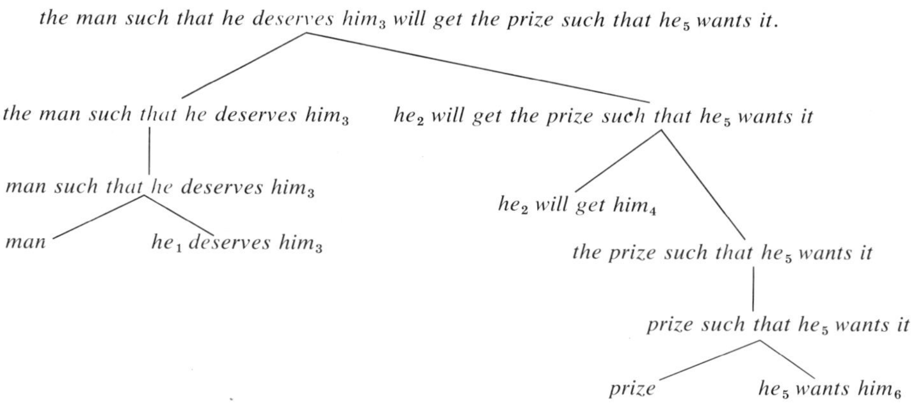

That the pronoun indices in the top line of (12) cannot be eliminated in the standard derivation of sentence (10) reveals the fact that Bach–Peters sentences present a variable-binding problem on the level of semantic representation. Remember that the problem raised by sentence (7) was similarly a variable-binding problem for PTQ.

It seems to me that the fundamental difficulty arising with the identical-NP approach, on the one hand, and that arising with the coindexed-pronoun approach, on the other, simply coincide in the case of Bach–Peters sentences. Furthermore, it seems possible that combining the full-NP-reduction approach and the pronoun-as-variable approach into one system may allow the generation of Bach–Peters sentences. But apart from the fact that such a solution [which would also take care of example (7)] is costly and results in a great number of alternative syntactic derivations which do not correspond to different readings, there are still further data which elude even such a combined approach. I refer to the so-called donkey sentence.[2]

(13) *Every man who marries a woman who owns a donkey*
that $\left\{\begin{matrix} he \\ she \end{matrix}\right\}$ *likes beats it.*

To show this in detail requires a long and tedious argument. (For a discussion see Cooper, this volume.)

A problem of a different nature is raised by example (14) (from Karttunen, 1969):

(14) *Every* man_1 *who gave* **his paycheck** *to* his_1 *wife is smarter*
than the man_2 *who gave* ***it*** *to* his_2 *mistress.*

The *it* in (14) is not coreferential with its antecedent! Is it the use of **stipulated** coreference which results in the indicated differences? To see that this is not the case, let us consider next an approach to pronouns which avoids stipulating coreference but retains a separate analysis of anaphoric pronouns. This "definiteness approach" has been pursued in various forms by a number of linguists.

4. A "DEFINITENESS" APPROACH TO ANAPHORIC PRONOUNS

It has been recognized for some time that there exists a close relation between T pronouns and definite terms (for example, Postal, 1969; Kuroda, 1968; Stockwell, Schachter, and Partee, 1973). Without going into the technical details of these earlier approaches, let us discuss a related approach to anaphoric pronouns which is based on the hypothesis that

(i) anaphoric pronouns are to be derived from an underlying definite article + CN pronoun (*one*),

[2] This example was pointed out to me by Barbara Partee. It is due to Ed Gethier (in response to a proposal by Terence Parsons).

(ii) coreference should be established in terms of definiteness.

This tentative definiteness approach to anaphoric pronouns still proceeds on the assumption that anaphoric and indexical pronouns are, in a semantically relevant sense, different phenomena, and that therefore a separate analysis of anaphoric pronouns is justified. The pronominalization mechanism employed, however, does **not** depend on stipulated coreference for determining the denotation of an anaphoric pronoun. Instead, anaphoric third person nonreflexive T pronouns are derived from underlying *the* + *one*.

With regard to the relation between T pronouns and definite terms, consider (16a) and (16b):

(16) a. *Every man who loves* ***a woman*** *loses* ***her.***
b. *Every man who loves* ***a woman*** *loses* ***the woman.***

The point is that (16a) and (16b) are synonymous on the respective coreferential readings. Thus, if we manage to relate the corresponding readings of (16a) and (16b) to one and the same semantic representation, we may solve the coreference problem in (16a) in terms of concrete semantic properties of definiteness (provided we have a semantic treatment of definiteness).

Definite terms are special in that they are equipped with a characteristic coreference mechanism. This is demonstrated in (17):

(17) *Every man who loves* ***a woman*** *loses* ***a woman.***

There is no reading of (17) in which the underlined terms can be interpreted as coreferential, which illustrates how artificial the identity principle of the "standard theory" is from an intuitive semantic point of view.

In order to realize our tentative definiteness approach to anaphoric pronouns we have to

(18) a. linguistically motivate the transition from *the woman* (in (16b)) to *her* [in (16a)].
b. formally capture the possible coreference between *a woman* and *the woman* in (16b) (that is, provide a **semantic** treatment of definiteness capable of explaining the phenomenon of possible coreference in (16b)).

As one possible solution to (18a), consider the surface syntactic approach exemplified by the following analysis tree (read from bottom up):

(19)

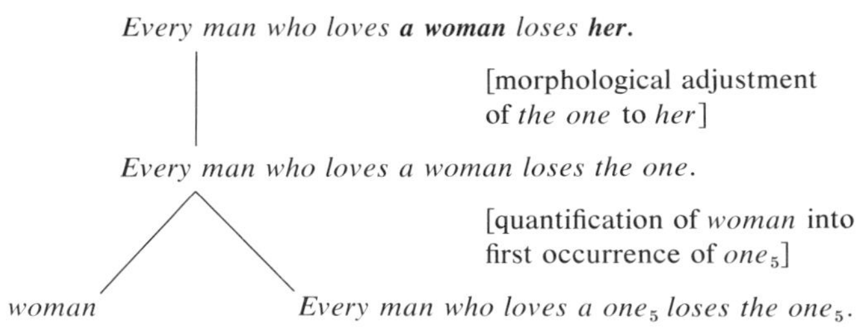

Let us assume that one_n is a CN pronoun which functions also in the derivation of such sentences as (20):

(20) *Bill bought a blue* ***book*** *and Frank bought a red* ***one.***

[CN quantification, $F_{10,4}$]

book *Bill bought a blue* $\boldsymbol{one_4}$ *and Frank bought a red* one_4.

The CN quantification rule in (20) puts a common noun in place of the first pronoun of a certain index and adjusts all other pronouns with the same index in the usual way. CN pronouns that have the same index may be called "codescriptive."

Semantically, let us treat one_n as a variable $P_n \in ME_{f(CN)}$. The semantic results of CN quantification replace all occurrences of P_n with the translation of the common noun in question. By deriving *her* from *the one* and by using CN pronominalization in (19), we translate (16a) and (16b) into the same semantic representation. Note that CN pronominalization is independently motivated by data like (20).

The morphological transition from underlying *the one* to the anaphoric pronoun *her* in analysis (19) is somewhat problematic. It may be argued, however, that the quantifier + **unmodified** CN pronoun is subject to an irregular morphology in general. Consider for example:

(21)
a one → *one*
my one → *mine*
your one → *yours*
his one → *his*
our one → *ours*
their one → *theirs*
etc.

The surface syntactic solution for deriving anaphoric pronouns indicated in analysis (19) does not stipulate coreference. Instead, we resort to **stipulating codescriptiveness** and appeal to certain (so far unspecified) semantic properties of definiteness.

What such a semantic analysis of definiteness will involve is captured in informal terms by the following quotation from O. Jespersen:

> The chief use of the definite article is to indicate the person or the thing that at the moment is uppermost in the mind of the speaker and presumably that of the hearer too. Thus it recalls what has just been mentioned—or else the whole situation is sufficient to show what is meant [1933, p. 162].

But if definiteness is a general context-dependency phenomenon—as Jespersen suggests—and anaphoric pronouns are explained through definiteness, then the semantic justification of the concept of anaphoric pronouns falls away: If anaphoric pronouns require a general context-dependency treatment anyway (definiteness has to be explained), then it is of no advantage to treat anaphoric and indexical pronouns separately. Also, since there are no morphological differences between indexically versus anaphorically used third person pronouns, a uniform treatment is highly desirable.

This tentative definiteness approach to anaphoric pronouns would capture certain important intuitions (given that the missing treatment of definiteness is supplemented). But it is already obvious that this approach is no more successful than the other two treatments of anaphoric pronouns discussed in Section 3. This is illustrated by the following difficulties:

(23)
a. The constraints on backward pronominalization apply only to third person T pronouns, but not to CN pronouns (see Stockwell, Schachter, and Partee, 1973). This raises serious problems if we want to incorporate the constraints into a system which derives anaphoric pronouns from codescriptive CN pronouns.
b. Assuming that proper names are **basic** T phrases, proper names can never serve as antecedents of anaphoric pronouns derived from complex definite terms (that is, definite article + CN pronoun) because codescriptiveness cannot be stipulated in these cases.
c. The problem in (23b) could be circumvented if we resort to a second pronominalization process (for example, the pronoun-as-variable approach as defined originally in PTQ), in addition to the definiteness approach in question. This inflationary move would at least permit the generation of Bach–Peters sentences, which otherwise could not be accounted for.
d. The donkey sentence [see (13)], however, would still elude the combined approach suggested in (23c).

The linguistic motivation for postulating a syntactico–semantic difference between indexical and anaphoric pronouns is that it permits us in this one case to answer the question "How do pronouns denote?" in terms of a syntactically stipulated coreference (or codescriptiveness) relation. This concept of anaphoric pronouns, however, leads into fundamental descriptive difficulties. This, in combination with the fact that anaphoric pronouns are not a morphologically distinct class and do not even relate to the whole class of nonreflexive personal T pronouns,[3] is reason enough to turn to a quite different approach to pronouns, based on a general treatment of context dependency. The first step is to define a **context-dependent extension** of the intensional logic of PTQ, which allows for a formal semantic description of indexicals in natural language.

[3] Only third person nonreflexive T pronouns are open to the anaphoric—indexical dichotomy, while first and second person are always indexical.

5. A NEW APPROACH TO CONTEXT DEPENDENCY

A number of logicians and philosophers have been interested in the semantic problem posed by indexicals. Montague (1968) notes that it was Bar Hillel (1954) who suggested that "pragmatics concern itself with what C. S. Peirce had in the last century called *indexical expressions*. An indexical word or sentence is one of which the reference cannot be determined without knowledge of the context of use; an example is the first person pronoun 'I' [p. 119]." Currently, the notion **pragmatics** may be understood as referring to either a subpart of semantics also called "indexical semantics," or to the theory of language *use* as opposed to semantics. Since Montague (1968) treats indexicals in a purely semantic way, it is clear that he used the notion pragmatics in the former sense.

A treatment of indexicals in a formal semantic framework requires

(24) a. specification of the formal nature of the **context**
b. specification of the formal nature of the **reference mechanism** by means of which the denotation of indexicals is determined via the context in question.

Before we begin to develop an analysis based on contextual reconstruction, it will be helpful to discuss briefly two earlier approaches to the treatment of context dependency in a model–theoretic framework.

One approach is the **coordinates approach** (Montague, 1968; Lewis, 1972). Besides the coordinates specifying a possible world and a moment of time, additional coordinates are defined for every single context-dependency phenomenon to be treated. Lewis (1972), for example, uses different coordinates for possible speakers (pronoun *I*), for possible hearers (pronoun *you*), possible places (pronoun *here*), possible indicated objects (pronoun *this*), and even for possible previous discourses. In short, the coordinates approach defines a context of use simply as an extended point of reference.

It seems to me, however, that the modal coordinate specifying a possible world and the tense coordinate specifying a moment of time represent phenomena which are intuitively different from the instances of context dependency to be characterized by the various other coordinates. Furthermore, to define a context of use as an arbitrary n-tuple does not capture the highly specific structural interaction between context-dependent expressions and a context. Yet the coordinates approach at least provides a clear, formal semantic answer to (24b).

An alternative treatment of context of use is the **proposition approach** (Stalnaker, 1968; Karttunen, 1974). The proposition approach defines a context of use simply as a **set of propositions.** This analysis was developed in connection with certain problems of counterfactual conditionals, and it is not clear how the denotation of indexicals is to be determined formally from an arbitrary set of propositions. Cresswell (1973) modified the proposition approach slightly by defining a context as a function from "tokens" into propositions (in the sense that a context is a property of tokens, where the property is captured in the form of a proposition). However, since neither Stalnaker nor Cresswell go beyond specifying domain and range of the crucial function postulated, neither system provides a satisfactory solution for (24b).

In order to interpret such context-dependent expressions as pronouns, definite terms, and nonredundant answers (cf. Hausser, 1976c) at a point of reference with respect to a context, a context has to be more than an n-tuple of arbitrary values or a set of uninterpreted propositions. It seems to me that the speaker interprets context-dependent expressions on the basis of very specific properties of the utterance situation (both verbal and nonverbal)—structures which none of the context concepts discussed are equipped to capture. And we cannot expect that a theory will render linguistically satisfactory formalizations of contextual phenomena if the grammar operates on a smaller basis of contextual information than the speaker uses in daily life.

What we need is a formal concept of context which is able to capture all the information necessary for the interpretation of context-dependent expressions. My proposal is to define a context of use as a **stage description.** A stage description is a regimented description of a certain time interval (representing the utterance situation) assigned to a point of reference. My program is to define a context-dependent intensional logic IL_2, which is interpreted with respect to a **context model** $\langle \mathfrak{A}, con \rangle$, where $\mathfrak{A}$ is an interpretation as defined in PTQ and *con* is a function which assigns exactly one stage description to each point of reference. I want to interpret the constants and variables of IL_2 as usual with respect to $\mathfrak{A}$ and a point of reference $\langle i,j \rangle$. The denotation of a context-dependent expression defined in IL_2 is to be determined via $con(\langle i,j \rangle)$, where $con(\langle i,j \rangle)$ is the stage description assigned to the point of reference $\langle i,j \rangle$.

This definition of a stage description is based on the notion of a **moment specification.** A moment specification is defined as an ordered pair $\langle j, \phi \rangle$, where j is a moment of time and ϕ is a declarative sentence of a certain context language IL_1 (which is a specification of the intensional logic defined in PTQ). I interpret $\langle j, \phi \rangle$ intuitively in the sense that the IL_1 formula ϕ *describes* the moment j. For example, (26)

(26) $$\langle j, enter'(\hat{\ }b) \wedge say'(\hat{\ }b, \hat{\ }m, \text{``}I\ love\ you\text{''}) \rangle$$

is a rough illustration of a moment specification which asserts that at moment j Bill comes in and says to Mary, "I love you."[4]

Formally, we shall introduce a moment specification as a member of the set M which is defined as follows:

[4] For obvious reasons, it would be more adequate to define interval specifications rather than moment specifications. However, a reasonably simple tense logic based on intervals is presently not known to me. The fact that we use moment specifications does not affect any of the basic ideas of our approach to context dependency.

(27) A set of moment specifications M is the Cartesian product $J \times ME_t^{\mathrm{IL}_1}$, where J is a nonempty set serving as the field of a simple ordering (J is regarded as a set of moments of time), and $ME_t^{\mathrm{IL}_1}$ is the set of formulas generated by a context language IL_1.

A stage description is defined as a sequence of moment specifications characterizing a certain interval of time:

(28) A stage description is an n-tuple $\langle \mu_1, \ldots, \mu_n \rangle$ of elements of $\mathrm{J} \times \mathrm{ME}_t^{\mathrm{IL}_1}$ for some $n \in \mathrm{N}$ such that for all its elements $\mu_\iota = \langle j', \phi \rangle$, $\mu_\kappa = \langle j'', \phi \rangle$, $\iota < \kappa$ if and only if $j' < j''$ $(1 \leq \iota, \kappa \leq n)$.

It follows that the first element of a stage description describes the earliest moment of the interval specified, while the rightmost element consists of the moment specification with the latest time index in the stage description. Furthermore, for each moment of time there is only one moment specification in a stage description.

Finally let us define a *context model* $\mathfrak{B}$. $\mathfrak{B}$ is an ordered pair, consisting of the PTQ interpretation $\mathfrak{A} = \langle A, I, J, \leq, F \rangle$, and the context-assignment function *con:*

(29) $$\mathfrak{B} =_{\mathrm{def}} \langle \mathfrak{A}, con \rangle$$

The context-assignment function *con* is defined as follows:

(30) Let I, J be nonempty sets, $\leq$ a simple order on J. Let $\mathfrak{M}$ be a set of stage descriptions defined with respect to J and $\mathrm{ME}_t^{\mathrm{IL}_1}$. *Con* is a function from $I \times J$ into $\mathfrak{M}$, such that for all $i \in I, j \in J$, if $con(\langle i,j \rangle) = \langle \langle j_0, \phi_0 \rangle, \langle j_1, \phi_1 \rangle, \ldots, \langle j_n, \phi_n \rangle \rangle$, then $j = j_n$.

A context model is defined with respect to a context language IL_1. *Con* assigns to each point of reference of the interpretation $\mathfrak{A}$ one and only one stage description defined with respect to J and $\mathrm{ME}_t^{\mathrm{IL}_1}$. The link-up condition in (30) stipulates that the final moment of the interval specified by a stage description **coincides** with the moment specified by the time index of the point of reference to which *con* has assigned the stage description in question.

An important kind of context model is what I call a **proper** context model:

(31) $\mathfrak{B}$ is a proper context model if and only if
 (i) $\mathfrak{B}$ is a context model $\langle \mathfrak{A}, con \rangle$, where $\mathfrak{A} = \langle A, I, J, \leq, F \rangle$, and
 (ii) it holds for all $i \in I, j \in J$ that if $con(\langle i,j \rangle) = \langle \langle j_0, \phi_0 \rangle, \langle j_1, \phi_1 \rangle, \ldots, \langle j_n, \phi_n \rangle \rangle$, then $\phi_0{}^{\mathfrak{A},i,j,g}$, $\phi_1{}^{\mathfrak{A},i,j_1,g}$, $\ldots$, and $\phi_n{}^{\mathfrak{A},i,j_n,g}$ are all 1 for arbitrary g

(where $j_n = j$ [according to (30)] and $j_0, j_1, \ldots,$ $j_n \in \mathrm{J}$, and $\phi_0, \phi_1, \ldots, \phi_n \in \mathrm{ME}_t^{\mathrm{IL}_1}$).

The interpretation of the IL_1 expressions $\phi_0, \ldots, \phi_n$ in a stage description with respect to $\mathfrak{A}$ and the specified points of reference [see (31ii)] is according to the truth conditions of IL as defined in PTQ. In a proper context model, possible values of $con(\langle i,j\rangle)$ are only those stage descriptions that happen to describe situations which are in accord with the "facts" as laid down by the interpretation $\mathfrak{A}$ with respect to the interval assigned to the point of reference $\langle i,j\rangle$.

The problem of interpreting a context-dependent[5] language is to assign proper denotations to the context-dependent expressions in question. I propose to translate various context-dependent expressions of English by means of different **context variables.** A context variable is a new kind of expression which is neither a constant nor a variable. It is not **interpreted** in the usual sense, but rather is **replaced** by a suitable denoting expression taken from a stage description. The replacement is performed by a **reconstruction function** G which maps context variables into denoting expressions of corresponding type derived from the stage description assigned to the point of reference under interpretation. Once all context variables in a given IL_2 expression are **contextually reconstructed,** the semantic interpretation of the reconstructed IL_2 expression runs standard. The auxiliary language IL_2 is a context-dependent intensional logic in the sense that it contains context variables. The syntax and semantics of IL_2 may be characterized as follows.

The vocabulary of IL_2 is the vocabulary of IL (that is, the intensional logic defined in PTQ) united with the set *C-var,* where C-var is defined as $\bigcup_{a \in \mathrm{Type}} \text{C-var}_a$. C-var_a is the set of context variables of type a. We assume that there are denumberably many context variables of each type, and if $n \in \mathrm{N}$ and $a \in \mathrm{Type}$, we shall understand by $\mathrm{cv}_{n,a}$ the nth context variable of type a.

By $\mathrm{ME}_a^{\mathrm{IL}_2}$ is understood the set of **meaningful expressions** of type a; this notion has the following recursive definition, which replaces (1)–(8) of PTQ, p. 256.

1. Every variable, constant, and context variable of type a is in $\mathrm{ME}_a^{\mathrm{IL}_2}$.

The other seven clauses correspond to (2)–(8) in PTQ, with "ME_a" replaced by "$\mathrm{ME}_a^{\mathrm{IL}_2}$."

Let us turn now to the semantics of IL_2. Suppose that $\mathfrak{B}$ is a context model $\langle \mathfrak{A}, con\rangle$ defined with respect to the context language IL_1, where $\mathfrak{A}$ is defined as $\langle A, I, J, \leq, F\rangle$ in the usual way. Suppose also that g is an $\mathfrak{A}$-assignment (of values to variables), that is, a function having as its domain the set of all variables (but not context variables), and such that $g(u) \in \mathrm{D}_{a,A,I,J}$

[5] A context-dependent language in the sense here defined is not to be confused with a **context-sensitive** language, that is, a language generated over context-sensitive rewriting rules as discussed in, for example, Lyons (1968, p. 235, fn.)

whenever u is a variable of type a. Suppose further that G is a reconstruction function having as its domain the set of all context variables and such that $G(\Gamma)$ is an expression of type a whenever Γ is a context variable of type a.

If α is a meaningful expression of IL_2, we shall understand by $\alpha^{\mathfrak{B},g,G}$ the **intension** of α with respect to $\mathfrak{B}$, g, and G; and if $\langle i,j \rangle \in I \times J$, then $\alpha^{\mathfrak{B},i,j,g,G}$ is to be the **extension** of α with respect to $\mathfrak{B}$, i, j, g, and G—that is, $\alpha^{\mathfrak{B},g,G}(\langle i,j \rangle)$, or the function value of the intension of α when applied to $\langle i,j \rangle$. These notions may be introduced by the following recursive definition:

1. If α is a constant, then $\alpha^{\mathfrak{B},i,j,g,G}$ is $F(\alpha)(\langle i,j \rangle)$
2. If α is a variable, then $\alpha^{\mathfrak{B},i,j,g,G}$ is $g(\alpha)$.
3. If α is a context variable, then $\alpha^{\mathfrak{B},i,j,g,G}$ is $G(\alpha)^{\mathfrak{B},i,j,g,G}$.

These clauses are followed by clauses corresponding to (2)–(10) in PTQ (1973, pp. 258–259), the only difference being that all indices $\mathfrak{A}$, i, j, g are replaced by $\mathfrak{B}$, i, j, g, G.

We call an IL_2 expression α **reconstructible** with respect to a context model, a point of reference, and a reconstruction function G if and only if for every context variable Γ occurring in α, $G(\Gamma)$ is defined (reconstruction functions are partial functions). If ϕ is an IL_2 formula (that is, a member of $ME_t^{IL_2}$) such that ϕ is reconstructible with respect to $\mathfrak{B}$, i, j, g and G, then ϕ is true with respect to $\mathfrak{B}$, i, j, and G, if $\phi^{\mathfrak{B},i,j,g,G}$ is **defined** and 1 for every $\mathfrak{A}$-assignment g, and false otherwise. If ϕ is not reconstructible with respect to $\mathfrak{B}$, i, j, g and G, however, then $\phi^{\mathfrak{B},i,j,g,G}$ is **undefined.** The build up of stage descriptions is treated on a par with the build up of a traditional interpretation or model (which fully continues its function as part of our extended context model).

6. ANALYZING PRONOUNS IN TERMS OF CONTEXTUAL RECONSTRUCTION

Having given the general definitions of the context-dependent intensional logic IL_2 and of a context model, let us turn now to the task of incorporating a linguistically satisfying description of pronouns into our extended system. To do this we have to

(32) a. motivate detailed translations for various pronouns, whereby the aspect of context dependency is to be captured in terms of characteristic context variables
b. specify the possible values for each context variable employed.

Clause (32b) will require further assumptions concerning the build up of "appropriate" stage descriptions. Based on these assumptions, we can specify the **position** of possible values for each context variable in any appropriate stage description (the **type** of possible values must agree with the type of the context variable in question).

Although the whole translation expression corresponding to a T pronoun is going to be of type $f(\mathrm{T})$, this does not determine the type of the context variable to be **contained** in the translation (representing the aspect of context dependency in T pronouns). As illustrations of two reasonable possibilities consider (33) and (34), which are alternative rough translations:

(33) $\hat{P}\bigvee x[male'(x) \wedge \Gamma\{x\} \wedge P\{x\}]$

(34) Γ^m

In (33), Γ is a context variable of type $\langle s, f(\mathrm{CN})\rangle$, ranging over **properties of individual concepts;** in (34), Γ^m is a context variable of type $\langle s, f(\mathrm{T})\rangle$, ranging over properties of properties of individual concepts (cf. PTQ, p. 259).

How are we going to decide between such alternatives? Note that there may be **indefinite** antecedents to T pronouns, as in (35):

(35) *Every man who loves* ***a woman*** *loses* ***her.***

Assuming we choose proposal (34) and translate *her* by means of a context variable of type $f(\mathrm{T})$, we would on one reading reconstruct (35) into an expression which is equivalent to the translation of (36):

(36) *Every man who loves* ***a woman*** *loses* ***a woman.***

The identical noun phrases in (36), however, may under no circumstances be interpreted as coreferential [also, proposal (36) would require the specification of context variables which are *sorted* with respect to gender and number].

For this reason, let us choose the proposal indicated in the tentative translation (33). This choice has the advantage of allowing us to reconstruct (35) automatically into an IL_2 expression which is *equivalent* to the translation of (37):

(37) *Every man who loves* ***a woman*** *loses* ***the woman.***

The sentences in (33) and (37) are synonymous on their respective anaphoric readings.

Our semantic representation of nonreflexive personal T pronouns should formally capture the characteristic numerically qualified, existential presupposition and (sometimes) gender. These semantic properties are jointly implemented in the following translations of the T pronoun *he*:[6]

[6] Note that $\hat{P}\bigwedge x \in [\ .\ .\ .\ ]^1 P\{x\}$ and $\hat{P}\bigvee x \in [\ .\ .\ .\ ]^1 P\{x\}$ are logically equivalent, assuming that $[\ .\ .\ .\ ]^1$ is defined as representing a **unit set.** Since *they* (like the plural of the definite article) must translate by a universal rather than an existential quantifier, the use of a restricted universal quantifier in the translation of *he* (as well as in the singular of the definite article) reduces the semantic difference between singular and plural of the terms (quantifier expressions) in question to a simple difference in the cardinality of the respective quantifier restrictions ($[\ .\ .\ .\ ]^1$ versus $[\ .\ .\ .\ ]^2$).

(38) *he* translates into $\hat{P}\bigwedge x \in [\bigwedge y[male'(y) \wedge \Gamma_3\{y\}] \leftrightarrow x = y]]P\{x\}$, which is abbreviated as $\hat{P}\bigwedge x \in [male'(x) \wedge \Gamma_3\{x\}]^1P\{x\}$

The $\bigwedge x \in [f(x)]g(x)$ notation in (38) represents **restricted quantification** as defined in Hausser 1976a (see also Belnap, 1970; Hausser, 1974b; Rescher, 1968; and Quine, 1963). The truth conditions for restricted quantification are such that $\bigwedge x \in [f(x)]h(x)^{\mathfrak{A},i,j,g}$ and $\bigvee x \in [f(x)]h(x)^{\mathfrak{A},i,j,g}$ are **bivalent** (that is, true or false, but not truth valueless) as long as there exists at least one variable assignment g' such that $f(x)^{\mathfrak{A},i,j,g'}$ is 1. On the other hand, if the existential presupposition concerning $f(x)$ is not fulfilled, the mentioned formulas are # (undefined) with respect to the point of reference in question ($\mathfrak{A}$ is an interpretation as defined in PTQ, (i,j) is a point of reference, and g is a variable assignment function). The truth conditions of restricted quantification and the usual nonrestricted quantification (for the latter, for example, $\bigwedge x[f(x) \rightarrow h(x)]$ and $\bigvee x[f(x) \wedge h(x)]$) **agree,** however, as long as $f(x)$ is true at the point of reference in question.[7]

Note that translation (38) induces the **presupposition** that *he* denotes one and only one individual concept (thus the uniqueness presupposition is fulfilled). What this amounts to in a more general context is exemplified in (39),

(39) *He walks.*

which would translate into (39a).

(39′) $\bigwedge x \in [male'(x) \wedge \Gamma_3\{x\}]^1 walk'(x)$

The expression (39′) is # (truth valueless, or rather, **undefined**) with respect to an interpretation $\mathfrak{A}$, a point of reference $\langle i,j\rangle$, a stage description $con(\langle i,j\rangle)$, and a reconstruction function G, **if** there is **no** or **more than one** individual concept which has the property *male′*, the property *walk′*, and the property $G(\Gamma_3)$—the contextual value for Γ_3.

Following these remarks, the definition of the (distributive) plural of the third person T pronouns *they* is fairly obvious:

(40) *they* translates into $\hat{P}\bigwedge x \in [\Gamma_3\{x\} \wedge \bigvee y[\Gamma_3\{y\}] \wedge \bigvee z[\Gamma_3\{z\}]]P\{x\}$, which is abbreviated as
$\hat{P}\bigwedge x \in [\Gamma_3\{x\}]^2P\{x\}$.

(I follow here the treatment of singular, distributive plural, and collective plural presented as an extension of PTQ in Hausser, 1974b, Ch. 2; for the sake of simplicity I will limit the following discussion to the singular and the

[7] Since our system combines the treatment of semantic presuppositions presented in Hausser 1976a with the system of contextual reconstruction, we combine a **strong** and a **weak** presuppositional logic in our grammar (cf. Rescher, 1969).

Although for the phenomena described in connection with *P*-inducers and cancelling contexts only a presuppositional logic of the strong type is adequate (as argued in Hausser, 1976a), the interpretation of context-dependent expressions clearly requires a weak presuppositional logic. Note that there are expressions which are context-dependent and P-inducing at the same time, as for example the definite article *the*.

distributive plural only.) Based on translation (40), example (41) will translate into (41′):

(41) *They walk.*

(41′) $\bigwedge x \in [\Gamma_3\{x\}]^2 walk'(x)$

(41′) will be # with respect to an interpretation $\mathfrak{A}$, a point of reference $\langle i,j \rangle$, and a stage description $con(\langle i,j \rangle)$, if Γ_3—which is to be reconstructed with respect to $con(\langle i,j \rangle)$—does not turn out to refer to **more than one** individual concept (**non**uniqueness presupposition). Note that a specification of a natural gender is not appropriate in translation (40) because *they* (in contrast to the **singular** third person) does not specify a gender.

Apart from number and gender, T pronouns differ regarding person. This semantic property will be represented by means of different context variables. Consider the following translations of all personal pronouns:

(42)
(i) *I* translates into $\hat{P}\bigwedge x \in [\Gamma_1\{x\}]^1 P\{x\}$.
(ii) *you* translates into $\hat{P}\bigwedge x \in [\Gamma_2\{x\}]^1 P\{x\}$.
(iiia) *he* translates into $\hat{P}\bigwedge x \in [male'(x) \wedge \Gamma_3\{x\}]^1 P\{x\}$.
(iiib) *she* translates into $\hat{P}\bigwedge x \in [female'(x) \wedge \Gamma_3\{x\}]^1 P\{x\}$.
(iiic) *it* translates into $\hat{P}\bigwedge x \in [neuter'(x) \wedge \Gamma_3\{x\}]^1 P\{x\}$.
(vi) *we* translates into $\hat{P}\bigwedge x \in [\Gamma_1\{x\}]^2 P\{x\}$.
(vii) *you* translates into $\hat{P}\bigwedge x \in [\Gamma_2\{x\}]^2 P\{x\}$.
(viii) *they* translates into $\hat{P}\bigwedge x \in [\Gamma_3\{x\}]^2 P\{x\}$.[8]

[8] In line with my analysis of scope ambiguities and scope restrictions (Hausser, 1976b), the nonreflexive T pronouns *I, you, he, she, it, we, you, they* should be marked [+w, −r] like the definite article *the*. As proof consider the following examples:

(i) *They read a book.*
(i′) $\bigwedge x \in [\Gamma_3\{x\}]^2 \bigvee y[book'(y) \wedge read'(x, \hat{P}P\{y\})]$
(i″) $\bigvee y[book'(y) \wedge \bigwedge x \in [\Gamma_3\{x\}]^2 read'(x, \hat{P}P\{y\})]$

(i) is ambiguous between a nonspecific (i′) and a specific (i″) reading regarding *a book*. With respect to negation, however, nonreflexive T pronouns are subject to a scope restriction:

(ii) *They didn't walk.*
(ii′) $\bigwedge x \in [\Gamma_3\{x\}]^2 \sim walk'(x)$
(ii″) $*\sim\bigwedge x \in [\Gamma_3\{x\}]^2 walk'(x)$

Expression (ii) is unambiguous and permits intuitively only the narrow scope negation reading (ii′). With respect to embedding, finally, nonreflexive T pronouns exhibit the scope restriction typical of presupposing terms:

(iii) *A girl who saw them panicked.*
(iii′) $\bigvee x[[girl'(x) \wedge \bigwedge y \in [\Gamma_3\{y\}]^2 see'(x, \hat{P}P\{y\})] \rightarrow panic(x)]$
(iii″) $*\bigwedge y \in [\Gamma_3\{y\}]^2 \bigvee x[[girl'(x) \wedge see'(x, \hat{P}P\{y\})] panic(x)]$

Thus, (iii) permits intuitively only the reading (iii′), where the relativized element has wider scope than the pronoun in the relative clause.

By marking nonreflexive T pronouns in the lexicon as [+w, −r], the undesired readings will be filtered, provided that our extension incorporates the syntactic refinements developed in Hausser 1976b.

In (42), we distinguish three different kinds of context variables (Γ_1, Γ_2, and Γ_3) all of type f(CN).[9] The reason for this distinction is that we want to characterize the difference between first, second, and third person T pronouns in terms of the **position of possible contextual arguments in a respective stage description.** As an example, consider (43), which, according to our definitions, translates into (43′):

(43) *I love you.*

(43′) $\bigwedge x \in [\Gamma_1\{x\}]^1 love'(x, \hat{P} \bigwedge y \in [\Gamma_2\{y\}]^1 P\{y\})$

The intended contextual reconstruction and interpretation of (43′) with respect to a context model $\mathfrak{B}$ (defined as $\langle \mathfrak{A}, con \rangle$, where $\mathfrak{A}$ is an interpretation $\langle A, I, J, \leq, F \rangle$ as specified in PTQ), a point of reference $(\langle i,j \rangle)$ $(i \in I, j \in J)$, a variable assignment g, and a reconstruction G is illustrated below:

(44)

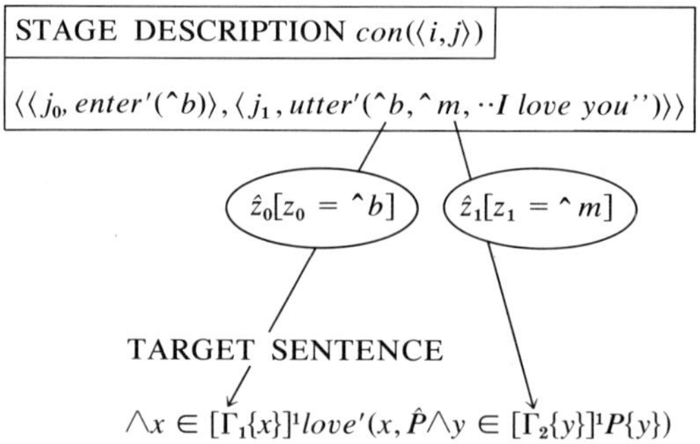

RECONSTRUCTION

$$\bigwedge x \in [\hat{z}_0[z_0 = \hat{}\boldsymbol{b}]\{x\}] love'(x, \hat{P} \bigwedge y \in [\hat{z}_1[z_1 = \hat{}\boldsymbol{m}]\{y\}] P\{y\})$$

(for the pertinent definitions, see Section 8) The reconstruction in (44) reduces via lambda-conversion to (45):

[9] In sentences that contain more than one pronoun of a certain "person," the context variables in the translations of the different pronouns need to be kept distinct in order to insure separate reconstruction of each pronoun. This may be done by adding an occurrence-index, thus creating an infinite number of context-variables of a given kind, for example, $\Gamma_{3,0}$, $\Gamma_{3,1}$, $\Gamma_{3,2}$, . . . , $\Gamma_{3,n}$ $(n \in N)$ for context variables representing the context-dependency aspect of third-person. A sentence like

He brings it and she kisses it.

would thus translate into, for example,

$$[\bigwedge x_0 \in [male'(x_0) \wedge \Gamma_{3,0}\{x_0\}]^1 bring(x_0, \hat{P} \bigwedge x_1 \in [neuter'(x_1) \wedge \Gamma_{3,1}\{x_1\}]^1 P\{x_1\})] \wedge [\bigwedge x_2 \in [female'(x_2) \wedge \Gamma_{3,2}\{x_2\}]^1 kiss(x_2, \hat{P} \bigwedge x_3 \in [neuter'(x_3) \wedge \Gamma_{3,3}\{x_3\}]^1 P\{x_3\})].$$

(45) $\bigwedge x \in [x = {}^\wedge b]^1 love'(x, \hat{P} \bigwedge y \in [y = {}^\wedge m]^1 P\{y\})$

The diagram (44) gives an example of a stage description which reports both nonlinguistic and linguistic events occurring in the given interval. Linguistic events are quoted in the stage description, that is, they occur as the third argument [of type $f(\mathrm{T})$] of certain **utterance predicates** of IL_1. For the sake of simplicity, let us assume that the first and second argument of utterance predicates in a stage description are to be **constants** of type $\langle s, e\rangle$.

A stage description $con(\langle i,j\rangle)$ is called **appropriate** for a given target sentence if an L surface of the target sentence is quoted in the last utterance predicate of the last moment specification of $con(\langle i,j\rangle)$. We can now specify the values of reconstruction functions for the context variables Γ_1 and Γ_2 as follows:

(46) Let ϕ be a meaningful IL_2 expression containing Γ_1 or Γ_2. If $con(\langle i,j\rangle)$ is an appropriate stage description with respect to ϕ, and α, β are the first and second argument of the last utterance predicate in $con(\langle i,j\rangle)$, then $\Gamma_1{}^{\mathfrak{B},i,j,g,G}$ and $\Gamma_2{}^{\mathfrak{B},i,j,g,G}$ are to be $\hat{x}[x = \alpha]$ and $\hat{y}[y = \beta]$, respectively. If $con(\langle i,j\rangle)$ is not an appropriate stage description for ϕ, then $\Gamma_1{}^{\mathfrak{B},i,j,g,G}$ and $\Gamma_2{}^{\mathfrak{B},i,j,g,G}$ are **undefined** for all $G \in \mathfrak{G}$ (where $\mathfrak{G}$ is a set of reconstruction functions).

Stage descriptions which are appropriate for a given target sentence ϕ describe by definition an **utterance** of the L-correlate of ϕ.

7. COREFERENCE PHENOMENA ARISING WITH DEFINITE TERMS

It is the defining property of context-dependent expressions that they depend on certain properties of the utterance situation for their interpretation. The use of context variables and the inductive development of progressively more detailed stage descriptions (and more specific restrictions on possible values of reconstruction function) is well-suited to describe these linguistic phenomena in a differentiated formal fashion. Before discussing further details of the contextual reconstruction of different context variables, however, let me explain the handling of coreference phenomena (arising with T pronouns and definite descriptions) in this system.

T pronouns are characterized by the fact that they are

(47) a. **basic** terms
b. **context-dependent** terms.

On the one hand, proper names and T pronouns have in common that they are basic terms. (They differ, however, in that proper names are context independent.) Definite descriptions and T pronouns, on the other hand, have in common that they are context dependent. They differ, however, in that definite descriptions are derived (that is, nonbasic) terms. It follows that [as far as (47a) and (47b) are concerned] T pronouns do not differ from so-called full NPs. On the contrary, it is my hypothesis that T pronouns are equipped with sufficient syntactico–semantic properties to denote *without* any additional mechanism designed only for pronouns (for example, stipulated coreference, stipulated codescriptiveness). T pronouns function no differently from ordinary full NPs.

To further support this hypothesis, let me compare the semantics of *definite descriptions* and T pronouns. One goal is to find a basic semantic principle which jointly explains the possible coreference of the boldface italic terms in (16a) and (16b), respectively:

(16) a. *Every man who loves **a woman** loses **her**.*
b. *Every man who loves **a woman** loses **the woman**.*

I will assume the semantics of the definite article developed in Hausser (1974b, Ch. 5), where definiteness is treated as a combination of two independently motivated semantic principles, namely,

(48) a. presupposition of existence
b. context dependency

Regarding the phenomenon of context dependency arising with definite descriptions, compare examples (49) and (50), which contain a definite and an indefinite term, respectively:

(49) ***The dog** is outside.*

(50) ***A dog** is outside.*

In what way do the truth conditions of (49) and (50) differ? A necessary and sufficient condition for the truth of (50) is that there is something outside which fits the description of a dog. The existence of a dog which is outside is also a necessary condition for the truth of (49), but it is not a sufficient condition. It is the existence of a special "intended" (a "unique") dog, which is required for the truth of (49). In other words, (49) refers not to any old dog, but, for example, to the dog we have been talking about, the dog our neighbors own, etc. Thus, the denotation of definite descriptions is specified not only in terms of the explicit descriptive content of the term (for example, the property *dog'*), but in addition by certain implicit properties, which are to be supplied by the context.

The semantic properties of being an "existential" P-inducer and of being context dependent are formally treated in the following translations of the singular and the distributive plural of *the:*

(51) a. *the* (sg.) translates into $\hat{Q}\hat{P}\bigwedge x \in [Q\{x\} \wedge \Gamma_0\{x\}]^1 P\{x\}$.

b. *the* (d.pl.) translates into $\hat{Q}\hat{P}\bigwedge x \in [Q\{x\} \wedge \Gamma_0\{x\}]^2 P\{x\}$.

The numerically qualified existential presupposition induced by singular *the* (uniqueness presupposition) and plural *the* (nonuniqueness presupposition) is captured in the form of restricted quantification, whereby the **cardinality** of the set restricting the range of the quantifier expression is indicated in form of the superscripts "^{1}" and "^{2}". These superscripts are abbreviations for longer expressions of standard predicate calculus. For further details on the semantics of the singular, as well as the distributive and collective plurals presumed here, see Hausser (1974a and 1974b, Ch. 2).

The aspect of context dependency, furthermore, is treated in form of the context variable Γ_0 contained in the above translations of the definite article. The combination of a numerically qualified presupposition and a context variable leads to a characteristic semantic effect, which is illustrated with our example (50), translating into (50′):

(50) *The dog is outside.*

(50′) $\bigwedge x \in [dog'(x) \wedge \Gamma_0\{x\}]^1 outside'\{x\}$

Only those reconstructions of Γ_0 (with respect to a stage description) may be called successful which are **compatible** with the numerically qualified existential presupposition.

What are the syntactico–semantic restrictions on possible reconstructions of Γ_0 in the environment of (50′)? Possible values of any given reconstruction $G(\Gamma_0)$ will have to be of type f(CN). Furthermore, possible values will have to be semantically compatible. For the sake of illustration, let us assume that the f(CN) expressions $\hat{}dog'$, $\hat{}cat'$, and $\hat{z}(own'(\hat{}m, \hat{P}P\{z\}))$ are the values of the possible reconstructions $G'(\Gamma_0)$, $G''(\Gamma_0)$, and $G'''(\Gamma_0)$, respectively, presuming that we interpret (50a) with respect to a context model $\mathfrak{B}$, a point of reference ($\langle i,j\rangle$), a variable assignment g, and the reconstruction function G', G'', or G'''.

Without going into the definitions of the interpretation in question, we can say already that **neither** $G'(\Gamma_0)$ nor $G''(\Gamma_0)$ are likely to turn out to be **successful** reconstructions (that is, reconstructions which do not directly cause lack of truth value). The reason is that $[dog'(x) \wedge \Gamma_0\{x\}]^{1\,\mathfrak{B},i,j,g,G'}$ will correspond to the empty set except with respect to "one-dog" worlds. $[dog'(x) \wedge \Gamma_0\{x\}]^{1\,\mathfrak{B},i,j,g,G''}$, on the other hand, will correspond to the empty set except with respect to worlds where there is a unique animal which is a dog and a cat at the same time. $[dog'(x) \wedge \Gamma_0\{x\}]^{1\mathfrak{B},i,j,g,G'''}$, however, will correspond to a unit set provided that $\bigvee x[dog(x) \wedge \bigwedge y[[dog'(x) \wedge own'(\hat{}m, \hat{P}P\{y\})] \leftrightarrow x = y]]^{1\,\mathfrak{B},i,j,g,G}$ (for arbitrary G) is true.

The requirement of compatibility drastically limits the number of possible contextual reconstructions which may be called **successful** (at any given point of reference). The narrowing of the range of possible denotations (resulting from the interaction of a numerically qualified existential presupposition and a context variable) presents the formal semantic essence of "definiteness."

Regarding the semantic similarity of third person nonreflexive T pronouns and definite descriptions, compare the following translations:

(52) a. *the dog* translates into $\hat{P}\bigwedge x \in [dog'(x) \wedge \Gamma_0\{x\}]^1 P\{x\}$.
b. *he* translates into $\hat{P}\bigwedge x \in [male'(x) \wedge \Gamma_3\{x\}]^1 P\{x\}$.

(53) a. *the dogs* translates into $\hat{P}\bigwedge x \in [dog'(x) \wedge \Gamma_0\{x\}]^2 P\{x\}$.
b. *they* translates into $\hat{P}\bigwedge x \in [\Gamma_3\{x\}]^2 P\{x\}$.

According to our analysis, the only semantic difference between definite descriptions and T pronouns is that the descriptive content of T pronouns is comparatively limited. This explains why the number of possible successful reconstructions is usually higher in the case of T pronouns than in the case of definite descriptions (compatibility is satisfied more easily in the case of T pronouns).

Let us turn now to the phenomenon of possible coreference arising with definite descriptions of third person T pronouns and definite descriptions alike. Consider (54):

(54) *John told Bill that he was sick.*

The pronoun *he* in (54) may either refer to John, Bill, or some other male individual specified by the context (for example, Bill's son). Consider now, what it would mean to call (54) **syntactically ambiguous.** To say that (54) is two-ways ambiguous amounts to maintaining the contrast between anaphoric and indexical use of pronouns, which I reject. To call (54) indefinitely many ways ambiguous, however (because of indefinitely many possible indexical uses) reduces the notion of syntactic ambiguity to absurdity.

According to the approach to context dependency proposed here, (54) is not syntactically ambiguous. Instead, the pronoun *he* is treated as being semantically unspecified in a certain well-defined respect. The specific semantic incompleteness of *he* is equated with the characteristic aspect of context dependency inherent in this pronoun; formally, it is represented in terms of the context variable Γ_3. Consider the translation of (54):

(54′) $tell'(\hat{}j, \hat{P}P\{\hat{}b\}, \hat{}\bigwedge x \in [male'(x) \wedge \Gamma_3\{x\}]^1 sick'(x))$

(I presume here a treatment of TV/t-verbs such as *tell-that*.)

Translation (54′) represents the one and only reading of (54). The various reference possibilities of *he* in (54) are treated where they belong, namely in the realm of interpretation. In this particular example, we have to distinguish between two different types of possible reconstructions:

(i) exterior reconstructions
(ii) interior reconstructions

The possible interior reconstructions of (54′) are formally indicated in (55):

(55)

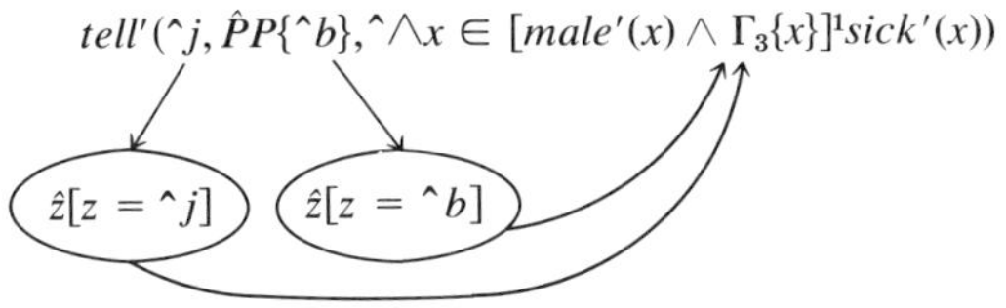

RESULTS:

$$tell'(^\wedge j, \hat{P}P\{^\wedge b\}, ^\wedge\bigwedge x \in [male'(x) \wedge x = \left\{\begin{matrix} ^\wedge j \\ ^\wedge b \end{matrix}\right\}]\, sick'(x))$$

The two interior reconstructions indicated represent the reading in which *he* is coreferential with John, and the reading in which *he* is coreferential with Bill.

As an example of an exterior reconstruction of (54′) with respect to $\mathfrak{B}$, i, j, g, and G, consider (56):

(56)

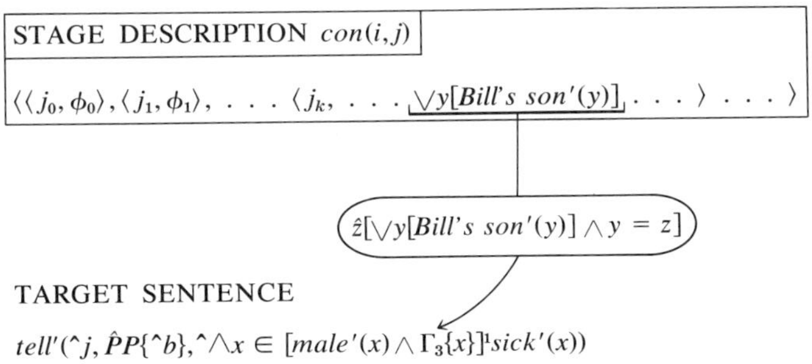

Result of Indicated Exterior Reconstruction

$$tell'(^\wedge j, \hat{P}P\{^\wedge b\}, ^\wedge\bigwedge x \in [male'(x) \wedge \bigvee y[Bill's\ son'(y)]\ ^\wedge y = x]]^1 sick'(x)$$

The distinction between interior and exterior reconstruction relates to the concepts of anaphoric and indexical use of certain pronouns. I propose to call those interpretations of context-dependent expressions *anaphoric* which are based on interior reconstruction, whereas interpretations of context-dependent expressions based on exterior reconstruction may be called **indexical.** The crucial point is that our approach avoids the use of different reference mechanisms for handling anaphoric and indexical use. Thus, we assign uniform semantic representations to both uses of T pronouns, which corresponds to the fact that they exhibit a uniform surface morphology.

It is an important feature of our analysis of nonreflexive personal T pronouns that the context variables are of type f(CN) rather than f(T) or $\langle s, e\rangle$. It is for this reason that we are able to simultaneously

(57) a. implement numerically qualified existential presuppositions directly into the semantic representation of nonreflexive personal T pronouns,
b. employ the requirement of compatibility for the characterization of successful reconstructions,
c. account for the general phenomenon of coreference arising with reconstructed T pronouns and definite descriptions alike,
d. account for those special cases where a reconstructed T pronoun is **not** coreferential with its "antecedent."

Since we have already discussed the points specified in (57a), (57b), and (57c), it remains to deal with (57d). The standard example of a noncoreferential anaphoric T pronoun is the so-called paycheck sentence:

(58) *The young man who gave* ***his paycheck*** *to his wife was smarter than the old man who gave* ***it*** *to his mistress.*

On the relevant reading of (58), *it* refers to a different paycheck than its assumed antecedent, *his paycheck.*

In order to formally explain the indicated interpretation of (58) in terms of contextual reconstruction, it is necessary to provide detailed translations of possessive T/CN pronouns:

(59) *my* translates into $\hat{Q}\hat{P}\bigwedge x \in [Q\{x\} \wedge \bigwedge y \in [\Gamma_1\{y\}]^1 relate'(x, \hat{P}P\{y\}]^1 P\{x\}$.
your translates into $\hat{Q}\hat{P}\bigwedge x \in [Q\{x\} \wedge \bigwedge y \in [\Gamma_2\{y\}]^1 relate'(x, \hat{P}P\{y\}]^1 P\{x\}$.
his translates into $\hat{Q}\hat{P}\bigwedge x \in [Q\{x\} \wedge \bigwedge y \in [male'(y) \wedge \Gamma_3\{y\}]^1 relate'(x, \hat{P}P\{y\}]^1 P\{x\}$.
her translates into $\hat{Q}\hat{P}\bigwedge x \in [Q\{x\} \wedge \bigwedge y \in [female'(y) \wedge \Gamma_3\{y\}]^1 relate'(x, \hat{P}P\{y\}]^1 P\{x\}$.
its translates into $\hat{Q}\hat{P}\bigwedge x \in [Q\{x\} \wedge \bigwedge y \in [neuter'(y) \wedge \Gamma_3\{y\}]^1 relate'(x, \hat{P}P\{y\})]^1 P\{x\}$.
our translates into $\hat{Q}\hat{P}\bigwedge x \in [Q\{x\} \wedge \bigwedge y \in [\Gamma_1\{y\}]^2 relate'(x, \hat{P}P\{y\})]^1 P\{x\}$.
your translates into $\hat{Q}\hat{P}\bigwedge x \in [Q\{x\} \wedge \bigwedge y \in [\Gamma_2\{y\}]^2 relate'(x, \hat{P}P\{y\})]^1 P\{x\}$.
their translates into $\hat{Q}\hat{P}\bigwedge x \in [Q\{x\} \wedge \bigwedge y \in [\Gamma_3\{y\}]^2 relate'(x, \hat{P}P\{y\})]^1 P\{x\}$.

According to these translations, a term like *my car* is analyzed as *the car which stands in a relation to me.* Note that (59) specifies *singular* quantifiers. While *my car* translates into

$$\hat{P}\bigwedge x \in [car'(x) \wedge \bigwedge y \in [\Gamma_1\{y\}]^1 relate'(x, \hat{P}P\{y\}]^1 P\{x\},$$

my cars would translate into

$$\hat{P}\bigwedge x \in [car'(x) \wedge \bigwedge y \in [\Gamma_1\{y\}]^1 relate'(x, \hat{P}P\{y\}]^2 P\{x\}.$$

As an illustration of the interpretation of T/CN possessives consider (60).

(60) *Mary sold my car.*

(60′) $sell(\hat{}m, \hat{P}\bigwedge x \in [car'(x) \wedge \bigwedge y \in [\Gamma_1\{y\}]^1 relate'(x, \hat{Q}Q\{y\})]^1 P\{x\})$

Example of a reconstruction of (60′):

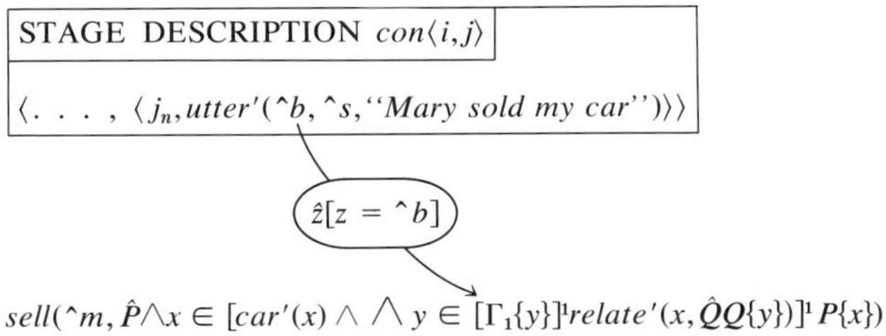

Note that the translations of T/CN possessives contain the same context variables as nonreflexive personal T pronouns (namely, Γ_1, Γ_2, and Γ_3; compare with (42) Section 6). It follows that *my*, *your*(sg.), *our*, and *your*(pl.) permit only exterior reconstruction, while *his*, *her*, *its*, and *their* permit both interior and exterior reconstruction. As an example, consider (61):

(61) *John sold* ***his*** *car.*

His in (61) may either relate to *John* (interior reconstruction of Γ_3) or to some other male person specified in the stage description assigned to the point of reference under interpretation.

Besides the T/CN possessives there are also T possessives, namely, *mine*, *yours*, *his*, *hers*, *its*, *ours*, *yours*, and *theirs*. In light of our discussion, the semantic representation of T possessives should be as follows:

(62) *mine* translates into $\hat{P}\bigwedge x \in [\Gamma_0\{x\} \wedge \bigwedge y \in [\Gamma_1\{y\}]^1 relate'(x, \hat{Q}Q\{y\})]^1 P\{x\}$.
yours translates into $\hat{P}\bigwedge x \in [\Gamma_0\{x\} \wedge \bigwedge y \in [\Gamma_2\{y\}]^1 relate'(x, \hat{Q}Q\{y\})]^1 P\{x\}$.
his translates into $\hat{P}\bigwedge x \in [\Gamma_0\{x\} \wedge \bigwedge y \in [male'(y) \wedge \Gamma_3\{y\}]^1 relate'(x, \hat{Q}Q\{y\})]^1 P\{x\}$.
hers translates into $\hat{P}\bigwedge x \in [\Gamma_0\{x\} \wedge \bigwedge y \in [female'(y) \wedge \Gamma_3\{y\}]^1 relate'(x, \hat{Q}Q\{y\}]^1 P\{x\}$.
its translates into $\hat{P}\bigwedge x \in [\Gamma_0\{x\} \wedge \bigwedge y \in [neuter'(y) \wedge \Gamma_3\{y\}]^1 relate'(x, \hat{Q}Q\{y\}]^1 P\{x\}$.
ours translates into $\hat{P}\bigwedge x \in [\Gamma_0\{x\} \wedge \bigwedge y \in [\Gamma_1\{y\}]^2 relate'(x, \hat{Q}Q\{y\})]^1 P\{x\}$.

yours translates into $\hat{P}\bigwedge x \in [\Gamma_0\{x\} \wedge \bigwedge y \in [\Gamma_2\{y\}]^2 relate'(x, \hat{Q}Q\{y\})]^1 P\{x\}$.
theirs translates into $\hat{P}\bigwedge x \in [\Gamma_0\{x\} \wedge \bigwedge y \in [\Gamma_3\{y\}]^2 relate'(x, \hat{Q}Q\{y\})]^1 P\{x\}$.

(The above definitions specify only the singular of T possessives.) Our analysis captures the fact that T possessives are two-way context dependent. One context variable in the translation of T possessives represents the aspect of person (Γ_1, Γ_2, or Γ_3); the other is Γ_0, which also appears in the translation of the definite article. As an example of the interpretation of a T possessive consider (63):

(63)

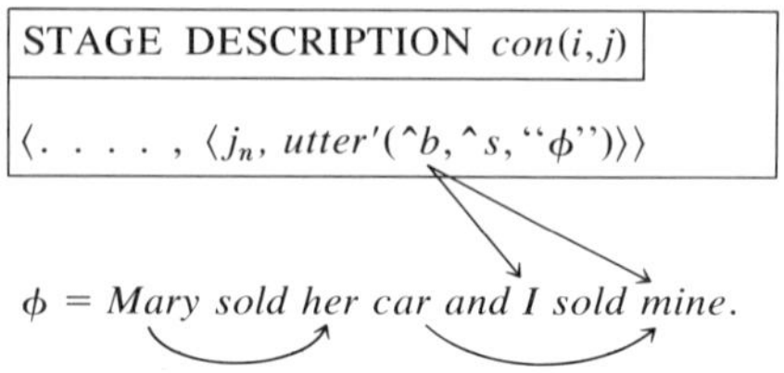

The interpretation of *mine* in (63) is based on an exterior as well as an interior reconstruction.

Let us return now to the paycheck sentence (58):

(58) *The* ***young man*** *who gave* ***his paycheck*** *to his wife was smarter than the* ***old man*** *who gave* ***it*** *to his mistress.*

Reconstructing *it* into *his paycheck'* results in an expression **which is itself context-dependent.** While the context variable in the original occurrence of *his paycheck'* can be internally reconstructed with respect to *young man,* the context variable in the reconstruction of *it* (that is, the Γ_3 in the reconstruction *his paycheck'*) relates to *old man. His paycheck* in (58) is what we will call the **primary antecedent** of *it*. The **secondary antecedent** of the primary reconstruction of *it* (that is, *his paycheck'*) is *old man.* The indicated interpretation of *it* in (58) is an example of an interior two-step reconstruction which does not result in coreference with the primary antecedent. The reason is obvious: The descriptive content of the reconstructed primary antecedent does not intersect with the descriptive content of the two-step reconstruction of the pronoun.

Two-step reconstructions also play an important role in the interpretation of Bach–Peters sentences. Consider (12):

(12) *The* ***man who deserves it*** *will get the* ***prize he wants.***

The primary antecedent of *he* is *man who deserves it* [remember that the context variable in nonreflexive personal T pronouns is of type *f*(CN)]. The secondary antecedent of *he* is *prize,* which relates to the *it* occurring in the primary reconstruction of *he*. The primary antecedent of *it* occurring explicitly in (12) is *prize he wants.* The secondary antecedent (relating to *he* occurring in the primary reconstruction) is *man*. In other words, the indicated reconstruction of (12) is equivalent to:

(12a) *The man who deserves* ***the prize the man wants*** *will get the prize which* ***the man who deserves the prize*** *wants.*

This reading of (12) (based on two internal two-step reconstructions) includes a coreferential interpretation of the two pronouns with their respective primary antecedents. The reason is that the descriptive content of the two-step reconstruction of the pronouns overlaps with the descriptive content of the respective reconstructed antecedents.

The principle employed here, of establishing the denotation of definite terms through the reconstruction of the descriptive content of the terms in question, accounts not only for the coreferential readings in (12) as well as the lack of coreference in (58), but also for "pronominal epithets" (Jackendoff, 1972). Consider for example:

(64) *I wanted* ***Charley*** *to help me, but* ***the bastard*** *wouldn't do it.*

In (64), coreference between the marked terms may be introduced by reconstructing *the bastard* into (roughly) *the bastard who is Charley.*

8. CONSTRAINTS ON RECONSTRUCTIONS

The descriptive power of our analysis resides to a large extent in the fact that we can specify the set of possible contextual values at any point of reference and for any given context variable (that is, any aspect of context dependency) in any degree of syntactico–semantic detail. For example, in Section 6, (46), we specified the respective reconstructions of I' and *you'*: For all $G \in \mathfrak{G}$ (where $\mathfrak{G}$ is a set of reconstruction functions), $G(\Gamma_1)$ and $G(\Gamma_2)$ are defined as $\hat{x}[x = {}^\wedge\alpha]$ and $\hat{y}[y = {}^\wedge\beta]$, where α, β are of type e and α, β are the first and second argument, respectively, of the last utterance predicate in the stage description under interpretation (provided that it is an "appropriate" stage description).

In specification (46), various different functions of a contextual reconstruction are intertwined: the constraints on types and positions of possible sources, the derivation of contextual values from possible source expressions, and the actual reconstruction. For the sake of a more perspicuous formulation, however, it seems advisable to separate the derivation of sets of suitable values from the contextual reconstruction proper. More specifically, we define a reconstruction function G ($G \in \mathfrak{G}$) as a function from the set of all context variables of IL_2 into the set of contextual values *Val*. Consequently we specify that for all $G \in \mathfrak{G}$ and all context variables Γ_n, $\Gamma_n{}^{\mathfrak{B},i,j,g,G} = val(\alpha, \Gamma_n, (i,j))$, where α is an IL_1 or an IL_2 source and *val* is a function generating contextual values for any context variable from any possible source available at any given point of reference. In other words, *val* is a function from $(\text{SOURCE}^{(i,j)} \cup TS) \times C \times (I \times J)$ into *val* [where $\text{SOURCE}^{(i,j)}$ is the set of all meaningful expressions occurring in the stage description $con(\langle i,j \rangle)$, TS is the set of meaningful expressions contained in the target sentence, C is the set of all context variables, and $I \times J$ is the set of all reference points]. For example, $val(\alpha, \Gamma_n, (i,j))$ specifies the contextual value α', where α' is derived from the source α, α' is a possible value for Γ_n, and α is an element of $\text{SOURCE}^{(i,j)}$, or of TS. Consider (65) as an illustration.

(65)

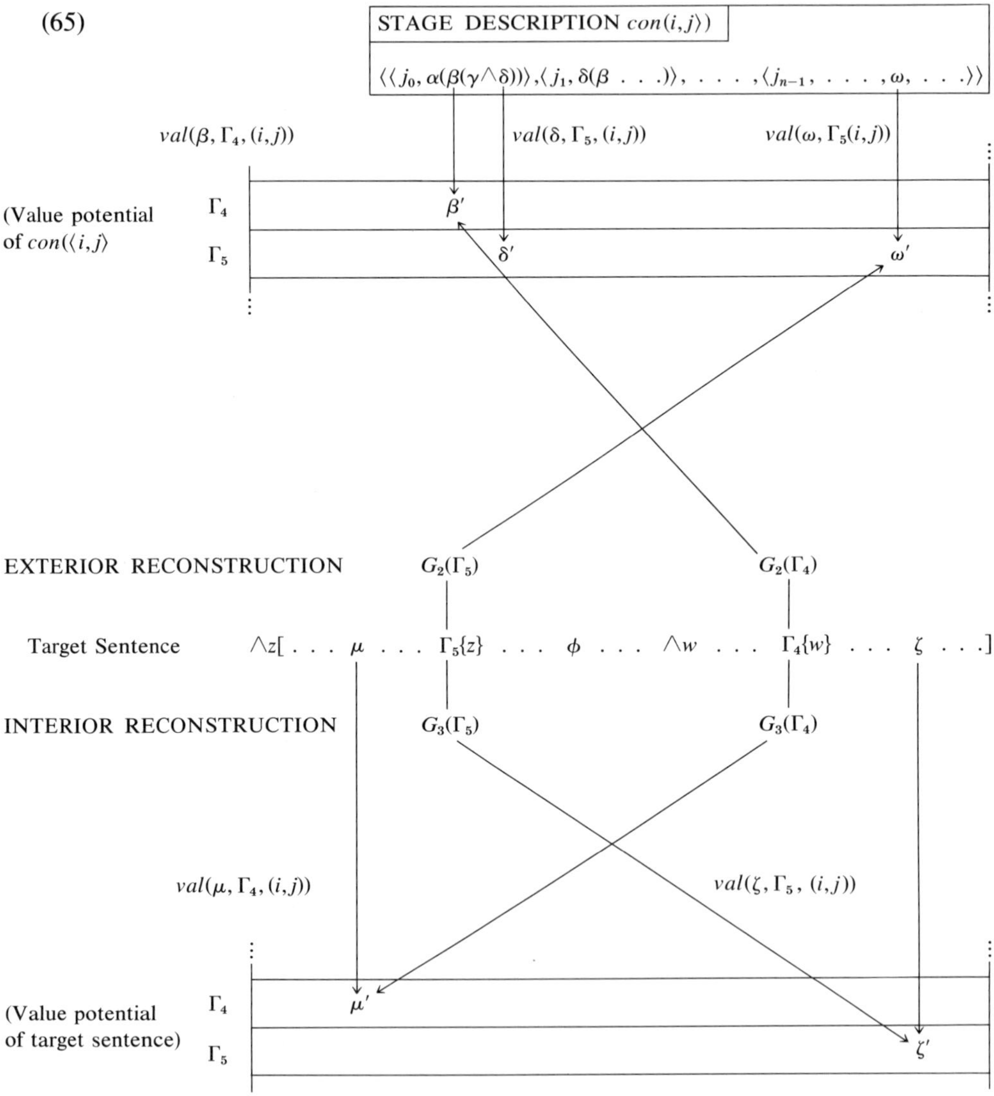

The elements β', δ', ω', μ', ζ' are members of the set $Val^{(i,j)}$ of all contextual values generated at (i,j), sorted with respect to different context variables.

A reconstruction $G(\Gamma_n)$ at a point of reference (i,j) is *successful* if and only if

(66) a. $G(\Gamma_n) = \alpha'$ is defined for some $\alpha' \in Val^{(i,j)}$,
b. $val(\alpha, \Gamma_n, (i,j)) = \alpha'$ is defined for a source expression α,
c. α' is semantically compatible with its reconstruction environment $[\ldots \Gamma_n \ldots]$.

If at a given point of reference (i,j) there is *more than one* possible contextual argument for a given context variable Γ_n (occurring in the target sentence ϕ under interpretation) then ϕ is **pragmatically ambiguous at** (i,j) [compare, for example, the target sentence in (65)]. Pragmatically different interpretations of a context-dependent expression ϕ correspond to different reconstruction functions G. Thus, if there are only three contextual values α', β', and γ' at (i,j) for the context variable Γ_n occurring in the target sentence ϕ, and $G_1(\Gamma_n) = \alpha'$, $G_2(\Gamma_n) = \beta'$, and $G_3(\Gamma_n) = \gamma'$, $(G_1, G_2, G_3 \in \mathfrak{G})$ are defined at (i,j), then for all other reconstruction functions G_m ($G_m \in \mathfrak{G}, m \neq 1,2,3$), $G_m(\Gamma_n)$ is undefined at (i,j). Thus, the denotation of a context-dependent expression ϕ depends on the particular reconstruction function employed in the interpretation in question.

The existence of a reconstruction function G, $G \in \mathfrak{G}$, such that $G(\Gamma_n)^{\mathfrak{B},i,j,g,G}$ is *defined* for a given Γ_n, depends on whether the set Val contains at least one element $val(\alpha, \Gamma_n(i,j))$ which is defined for α, Γ_n, and (i,j). The definedness of $val(\alpha, \Gamma_n, (i,j))$, in turn, depends on whether α is a *source expression* at (i,j) which corresponds to the *value specification* Γ_n. As examples consider the specifications on values for Γ_1 and Γ_2:

(67) If α is of type e, and α is the first argument of the last utterance predicate of $con(\langle i,j\rangle)$, then $val(\alpha, \Gamma_1, (i,j))$ is $\hat{x}[x = {}^\wedge\alpha]$; otherwise $val(\alpha, \Gamma_1, (i,j))$ is undefined at (i,j).

(68) If β is of type e, and β is the second argument of the last utterance predicate of $con(\langle i,j\rangle)$, then $val(\alpha, \Gamma_2, (i,j))$ is $\hat{y}[y = {}^\wedge\beta]$; otherwise $val(\beta, \Gamma_2, (i,j))$ is undefined at (i,j).

These are specifications on the value-generating function val. They reformulate specification (46), which was a specification on the reconstruction function G.

What is the specification concerning Γ_3, the context variable representing "third person"? We know that externally the reconstruction of Γ_3 will be basically unrestricted, whereas internally, the reconstruction of Γ_3 is subject to the constraints on backward pronominalization. These limit the relations that may hold between a noun phrase serving as antecedent and an anaphoric pronoun. We reformulate this question as: What are the positional restrictions on the internal reconstruction of Γ_3?

In order to formulate the restrictions on $val(\alpha, \Gamma_3, (i,j))$, we need the auxiliary notion of a **coargument** of Γ_3. A coargument of a context variable Γ_3 is an individual concept which stands in a relation (in intension or extension) with the referent of the *argument* of Γ_3. The pertinent kind of environment characterizing a coargument of a given Γ_3 in our extension is specified in (69):

(69) If $\delta \in ME_{f(TV)}$, and ϕ is $\delta(y, \hat{P}\bigwedge x \in [\ldots \Gamma_3\{x\}] \ldots)$, then y is the coargument of $\Gamma_3\{x\}$ in ϕ.

We may now formulate the restriction on the reconstruction of Γ_3 [or rather $\Gamma_{3,n}$ ($n \in$ N), cf. Footnote 9], which covers the constraints on backward pronominalization as formulated by Langacker 1969 or Ross 1969 without impeding the interpretation of such examples as Bach–Peters sentences, the paycheck sentence, or the donkey sentence.[10]

(70) If $\alpha \in ME_e$ and $\phi \in ME_t$, then

$val(\alpha, \Gamma_3, (i,j))$ is $\hat{x}[x = {^\wedge}\alpha]$,
$val(\phi, \Gamma_3, (i,j))$ is $\hat{x}[\phi \wedge \mu = x]$, for all $\mu \in ME_{\langle s,e \rangle}$ occurring in ϕ,

provided that α, ϕ do not arise in the scope of $\bigwedge x \ldots \Gamma_3\{x\} \ldots$ on the dominant reading of the target sentence ψ, and α, μ are not coarguments of $\Gamma_3\{x\}$ in ψ'.[11]

What is the linguistic motivation for the provision in (70)? The function of a Γ_3 reconstruction is to sharpen the descriptive content of the quantifier expression binding the argument of the Γ_3 in question. The provision in (70) excludes precisely those possible sources of internal reconstruction which are potentially unsuited to perform this function because their own denotation may depend on the reconstruction of the Γ_3 under consideration. The source expressions excluded from the internal reconstruction of a given Γ_3 are all expressions in the scope of $\bigwedge x \ldots \Gamma_3\{x\} \ldots$ as well as basic or derived coarguments of $\Gamma_3\{x\}$.

In order to illustrate this requirement of relative structural independence of a potential internal source from the Γ_3 it is supposed to reconstruct, let us consider a number of examples.

[10] Specification (70) does not yet handle the fact that "following antecedents" have to be definite, and that noun phrases embedded under certain prepositions may not serve as an antecedent (Lakoff, 1968). I leave these details to a later analysis.

[11] It is understood in general that possible source expressions of a given context-variable in a given target sentence at a given point of reference (i,j) arise either in the stage description $(\langle i,j \rangle)$ or in the target sentence. Furthermore, in order to maintain certain binding relations (consider (7) and (80) in this respect), simultaneous substitution of variables in an internal source is *suspended* until after the reconstruction is completed.

(71) *The girl sees her.*

(71′) $\bigwedge x \in [girl'(x)]^1 see'(x, \hat{P}\bigwedge y \in [female'(y) \wedge \Gamma_3\{y\}]^1 P\{y\})$[12]

The expression $girl'(x)$ in (71′) may not reconstruct the Γ_3 in question because x happens to be a coargument of $\Gamma_3\{y\}$.

(72) She sees the girl.

(72′) $\bigwedge y \in [female'(y) \wedge \Gamma_3\{y\}]^1 see'(y, \hat{P}\bigwedge x \in [girl'(x)]^1 P\{x\})$

In (72′), $girl'(x)$ may not reconstruct Γ_3 because $girl'(x)$ happens to be in the scope of $\bigwedge y \ldots \Gamma_3\{y\} \ldots$. Note that Γ_3 specification (70) excludes identical expressions which are on the *dominant* reading in the scope of $\bigwedge y \ldots \Gamma_3\{y\} \ldots$. The dominant reading of a surface sentence is understood as that reading which mirrors the surface order of the quantifiers and other operators involved in scope ambiguities on the level of semantic representation (cf. Hausser, 1976b).

(73) *The girl who kissed her loves Mary.*

(73′) $\bigwedge x \in [girl'(x) \wedge kiss'(x, \hat{P}\bigwedge y \in [female\cdot'(y) \wedge \Gamma_3\{y\}]^1 P\{y\})]^1 love'(x, \hat{P}P\{{}^\wedge m\})$

In (73′), x is a coargument of $\Gamma_3\{y\}$, which excludes $girl'(x)$ from being a possible source for the internal reconstruction in question. The constant m, however, is neither a coargument nor in the scope of $\bigwedge y \ldots \Gamma_3\{y\} \ldots$, and thus qualifies as a source.

(74) *She kissed the girl who loves Mary.*

(74′) $\bigwedge y \in [female' \cdot (y) \wedge \Gamma_3\{y\}]^1 kiss'(y, \hat{P}\bigwedge x \in [girl'(x) \wedge love'(x, \hat{P}P\{{}^\wedge m\})]^1 P\{x\})$

In (74′), both $girl'(x)$ and m are on the dominant reading in the scope of $\bigwedge y \ldots \Gamma_3\{y\} \ldots$ and therefore do not qualify as a possible source for the interior reconstruction of the Γ_3 in question.

(75) *The girl who kissed Mary loves her.*

(75′) $\bigwedge x \in [girl'(x) \wedge kiss'(x, \hat{P}P\{{}^\wedge m\}]^1 love'(x, \hat{P}\bigwedge y \in [female'(y) \wedge \Gamma_3\{y\}]^1 P\{y\})$

In (75′), x is a coargument of $\Gamma_3\{y\}$, which excludes $girl'(x)$ from being a possible source. However, m is neither in the scope of $\bigwedge y \ldots \Gamma_3\{y\} \ldots$ nor a coargument, and thus qualifies as a possible source.

(76) *Mary kissed the girl who loves her.*

[12] For the sake of simplicity, I will translate the definite article in the following examples without its characteristic context variable.

(76′) $kiss'(\hat{}m, \hat{P}\bigwedge x \in [girl'(x) \wedge love'(x, \hat{Q}\bigwedge y \in [female'(y) \wedge \Gamma_3\{y\}]^1 Q\{y\})]^1 P\{x\})$

In (76′), x is a coargument of $\Gamma_3\{y\}$, which excludes $girl'(x)$ from being a possible source. (It holds in general that the head noun of a relative clause is a coargument of a pronoun in that relative clause, except for cases where the pronoun is further embedded). However, m is neither in the scope nor a coargument of $\bigwedge y \ldots \Gamma_3\{y\} \ldots$ and qualifies as a possible source.

(77) *The girl who kissed her knows the lady who treated Mary.*

(77′) $\bigwedge x \in [girl'(x) \wedge kiss'(x, \hat{P}\bigwedge y \in [female'(y) \wedge \Gamma_3\{y\}]^1 P\{y\})]^1 know'(x, \hat{P}\bigwedge z \in [lady'(z) \wedge treat'(z, \hat{P}P\{\hat{}m\})]^1 P\{z\})$

In (77′), x is a coargument of $\Gamma_3\{y\}$, which excludes $girl'(x)$ from being a possible source. Neither z nor m are coarguments or in the scope of $\bigwedge y \ldots \Gamma_3\{y\}$; therefore, both $lady'(z)$ and m qualify as possible sources.

(78) *The girl who kissed Mary knows the lady who treated her.*

(78′) $\bigwedge x \in [girl'(x) \wedge kiss'(x, \hat{P}P\{\hat{}m\})]^1 know'(x, \hat{P}\bigwedge z \in [lady'(z) \wedge treat'(z, \hat{P}\bigwedge y \in [female'(y) \wedge \Gamma_3\{y\}]^1 P\{y\})]^1 P\{z\})$

In (78′), z is a coargument of $\Gamma_3\{y\}$, which excludes $lady'(z)$ from being a possible source (Compare with (77)). However, x and m are neither coarguments nor in the scope of $\bigwedge y \ldots \Gamma_3\{y\}$. Therefore, both $girl'(x)$ and m qualify as possible sources.

The reference to the **dominant reading** in specification (70) accounts also for the fact that *Mary* may be coreferential with *her* in (79), but not with *she* in (74) (where (79) is the passive of (74)):

(79) *The girl who loves Mary was kissed by her.*

(79′) $\bigwedge x \in [girl'(x) \wedge love'(x, \hat{P}P\{\hat{}m\})]^1 \bigwedge y \in [female'(y) \wedge \Gamma_3\{y\}]^1 kiss'(y, \hat{P}P\{x\})$

In (79′), x is a coargument of $\Gamma_3\{y\}$, which excludes $girl'(x)$ from being a possible source. The constant m, however, is outside the scope of $\bigwedge y \ldots \Gamma_3\{y\} \ldots$ and is not a coargument of Γ_3. Thus m is a possible source for Γ_3 in (79′)—though not in (74′). The reason is that the order of the quantifiers on the dominant reading of (79) is reversed in comparison to (74).

The indicated treatment of the constraints on backward pronominalization in terms of constraints on the internal reconstruction of Γ_3 differs from transformational accounts in that it is neither based on rules of possible coreference nor on rules of obligatory noncoreference. Rather, coreference, or lack of it, is treated entirely as a consequence of the (internally) reconstructed **descriptive content** of the pronoun under interpretation. Note that the value specification for Γ_3 stated in (70) does not impede the "anaphoric" interpretations of such standard puzzles as examples (10), (13), and (14). The proof on the basis of formal IL_2 translations is left as an exercise to the reader. But even without formal translations, it is obvious that the respective source expressions are neither coarguments nor in the scope of the respective pronouns on the dominant readings.

The restrictions on the **position** of possible source expressions of Γ_3 stated in the $val(\alpha, \Gamma_3, (i,j))$ under specification (70) are crucially complemented by the requirement of **semantic compatibility,** briefly discussed in section 7, and based on the treatment of numerically qualified existential presuppositions developed in Hausser 1976a. A contextual value α is said to lack semantic compatibility in a given reconstruction of a context variable if the reconstruction results directly in a presupposition failure. Thus, lack of semantic compatibility of an otherwise possible contextual argument explains in many instances why the source expression in question can not be an antecedent of the pronoun or definite description under interpretation. Consider, for example, the fact that on the *nonspecific* reading of (80) there is no possible interpretation of *it* as being coreferential with *a car.*[13]

(80) *Bill doesn't have* **a car** *and* **it** *is in the garage.*

Why is there no successful interior reconstruction of *it'* with respect to *no car Bill has'*? The nonspecific reading of (80) translates into (80′):

$$(80') \quad \sim\bigvee x[car'(x) \wedge have'(\hat{}b, \hat{P}P\{x\})] \wedge \bigwedge y \in [neuter'(y) \wedge \Gamma_3\{y\}]^1 in\ garage'(y)$$

If ϕ is the translation of the first conjunct of (80), and we interpret (80′) with respect to a given $\mathfrak{B}$, i, j, g, G, then $val(\phi, \Gamma_3, (i,j))$ will be the contextual value given under (80″):

$$(80'') \quad \hat{z}[\sim\bigvee x[car'(x) \wedge have'(\hat{}b, \hat{P}P\{x\}) \wedge x = z]]$$

Replacing Γ_3 in (80′) by (80″) results in the following reconstruction of the pronoun (after lambda conversion):

$$(81) \quad \hat{P}\bigwedge y \in [neuter'(y) \wedge \sim\bigvee x[car'(x) \wedge have'(\hat{}b, \hat{P}P\{x\}) \wedge x = y]]^1 P\{y\}$$

The reconstruction (81) does not result in a proper quantifier restriction, because under the assumption that the first conjunct of (80) is **true** under the nonspecific reading, then **all** individual concepts will have the property of "not being Bill's car." Since the indicated contextual reconstruction does not narrow the quantifier restriction in (81) as compared to the unreconstructed pronoun translation in (80′), the quantifier restriction will not corre-

[13] This example was pointed out to me by Lauri Karttunen.

spond to a unit set. This results in a presupposition failure (violation of the uniqueness condition indicated by the superscript "1").

On the **specific** reading of (80) (with respect to *a car*), however, *it* may be interpreted as being coreferential with *a car*. This corresponds to the fact that the respective reconstruction may result in a proper restriction of the reconstructed pronoun denotation: Under the indicated interpretation *it* will refer to a certain car which Bill does not have.

Another notable feature of the analysis of pronouns in terms of contextual reconstruction proposed here, is that a sentence like (82) is not treated as being **syntactically ambiguous** between a reading where *he* refers to *Bill* and a reading where *he* refers to *John*.

(82) *John told Bill that he was in danger.*

Presuming a treatment of TV/t verbs like *tell-that,* (82) translates unambiguously into (82′):

(82′) $tell\text{-}that'(\hat{}j, \hat{P}P\{\hat{}b\}, \hat{}\bigwedge x \in [male'(x) \wedge \Gamma_3\{x\}]^1 in\ danger'(x)$

The use of context variables permits us to treat the various reference possibilities of the pronoun in (82) where they belong, namely in the realm of interpretation. Treating (82) as syntactically ambiguous would amount to analyzing pronouns as **indefinitely many times ambiguous**—unless we return to the standard anaphoric/indexical dichotomy. Our analysis of pronouns as being semantically unspecified in certain well-defined respects (presented by a characteristic context variable) avoids either of these unattractive alternatives, and allows for a restricted and meaningful notion of the term "syntactic ambiguity."

The misguided attempts to treat anaphoric pronouns as being syntactically ambiguous has resulted in "puzzles" which find a completely natural solution in this framework. Karttunen (1977), for example, construes a contrast between (82) and (83) in terms of syntactic ambiguity:

(82) *John told Bill that he was in danger.* [*he* = *John* or *he* = *Bill*]

(83) *If Mary has* ***a car*** *or John has* ***a bicycle, it*** *is in the garage.*

According to Karttunen, in (82) "the pronoun requires a unique antecedent; the ambiguity . . . is due to the fact that there are two possible antecedents for the pronoun." In (83), however, "the availability of two antecedents does not result in any ambiguity." "The interesting fact about (5) [my example (83)] is that the pronoun *it* is perceived as unambiguous. There is no reading of (5) under which *it* is interpreted as definitely referring to a car or to a bicycle. Instead *it* seems to mean 'a car or a bicycle, whichever of these it is that exists.'"

According to our analysis, **both** (82) and (83) **translate unambiguously** into formulas of intensional logic. The translation of (83) is (83′):

(83′) $$[\bigvee x[car'(x) \wedge have'(\hat{}m, \hat{P}P\{x\})] \vee \bigvee y[bicycle'(y) \wedge have'(\hat{}j, \hat{P}P\{y\}) \rightarrow [\bigwedge z \in [neuter'(z) \wedge \Gamma_3\{z\}]^1 in\ the\ garage'(z)$$

On the level of contextual reconstruction, we may now explain Karttunen's observation that *it* in (83) refers to **whichever** of the possible antecedents exist (assuming **internal** reconstruction) on the basis of the familiar assumptions of our approach. Consider the following truth table (84), which characterizes the semantic interaction of the components of (83′) within the here presumed presuppositional logic, where *A* stands for *Mary has a car, B* stands for *John has a bicycle,* and *C* stands for *it is in the garage.*

(84)

A	B	C	$A \vee B$	$(A \vee B) \rightarrow C$
1	1	1	1	1
1	0	0(#)	1	1(#)
0	1	1(#)	1	1(#)
0	0	#	0	1
1	1	0	1	0
1	0	0(#)	1	0(#)
0	1	0(#)	1	0(#)
0	0	#	0	1

The undefined values (#) in the rightmost row of (84) represent interior reconstructions of *C* which are unacceptable because of presupposition failure. For example, if *B* (=*John has a bicycle*) is 1, while *A* (=*Mary has a car*) is 0, then the antecedent of (83), $A \vee B$, will be 1. If we now reconstruct *it* in *C* (=*it is in the garage*) with respect to Mary's car, then the consequent *C* will be # (undefined), and thus the whole sentence $(A \vee B) \rightarrow C$ will be # according to Kleene's definition of "→" (cf. Rescher, 1968, and Hausser, 1976a). However, if we reconstruct *it* with respect to John's bicycle, then the consequent will be bivalent, which in turn insures that the whole sentence will come out as bivalent. Thus Karttunen's intuitions regarding the interpretation of (83) find a straightforward explanation in truth functional terms: Only the "whichever" reading results in a defined truth value of (83).

9. ON REFLEXIVE PRONOUNS

Our treatment of pronouns is characterized by the assumption that expressions like *I, you, he, she, it, we, you, they, mine, yours, his, hers, its,* are full noun phrases which happen to be both basic and context-dependent (and similarily for the T/CN possessives *my, your, his, her, its, our, your, their,* which are treated as context-dependent quantifiers). The aspects of context dependency connected with the various pronouns are formally represented by characteristic context variables which serve as place holders for the semantic material to be provided by the context. The replacement of a given context variable by denoting expressions from the target sentence (interior reconstruction) or the stage description assigned to the point of reference under interpretation (exterior reconstruction) is defined as a highly restricted procedure. In order to be a suitable contextual value, an IL expression has to correspond to specifications regarding semantic type and position (in the target sentence or the stage description), and fulfill the requirement of semantic compatibility.

Rather than stipulating coreference syntactically in the case of anaphoric use, coreference is treated as a consequence of the similarity of the descriptive content of a reconstructed pronoun and its internal source (antecedent). I have shown that this approach to coreference offers solutions to long-standing puzzles such as the paycheck sentence, Bach–Peters sentence, etc.

The basic motivation for treating the pronouns mentioned above in terms of contextual reconstruction is to provide a uniform treatment for both anaphoric and indexical use. When we turn to the so-called **reflexive** T pronouns, however, this motivation does not apply because reflexives may be used **only** anaphorically. Thus, a treatment in terms of syntactically stipulated coreference and syntactic occurrence restrictions would be feasible in the case of reflexive pronouns. On the other hand, a treatment in terms of contextual reconstruction would be appealing because it would subsume reflexive pronouns under a uniform, general approach to all pronouns.

Since a syntactic treatment of reflexive pronouns in a Montague framework has already been presented (Bennett, 1974), let me discuss the ramifications of a treatment of reflexives in terms of contextual reconstruction. Consider (85) and (86):

(85) *John shaved him.*

(86) *John shaved himself.*

In (85), *him* cannot be interpreted as being coreferential with *John,* as accounted for by the restrictions on Γ_3 reconstruction stated under (70). In other words, *him* in (85) may be reconstructed only **externally.** In (86), however, the reflexive pronoun *himself* **must** refer to *John*. And it holds in general that reflexive pronouns permit **only** interior reconstruction; they are thus anaphoric pronouns *par excellence*.[14]

Although the contextual values of nonreflexive T pronouns are for various reasons defined as being of type f(CN), this would not work for the contextual values of reflexive T pronouns. Consider for example (87):

(87) *Every senator voted for himself.*

[14] Indeed, the preoccupation with "anaphoric pronouns" in transformational grammar started with an analysis of reflexive pronouns (Lees and Klima, 1963).

If we treated reflexive T pronouns as nonreflexive T pronouns, then (87) would reconstruct into something corresponding to (88),

(88) $$\textit{Every senator voted for} \left\{ \begin{array}{c} \textit{a} \\ \textit{the} \end{array} \right\} \textit{senator(s)}.$$

(depending on whether we translate the reflexive pronoun on the basis of restricted or nonrestricted quantification). However, (88) is intuitively not equivalent to (87). What is required instead is a reconstruction of (87) into something equivalent to (89).

(89) $$\bigwedge x \in [senator'(x)]^2 vote'(x, \hat{P}P\{x\})$$

Regarding the reconstruction of reflexive pronouns, furthermore, we have to account for the so-called sentence condition. In (90), for example,

(90) *John thinks that Bill shaves himself.*

the reflexive pronoun can be coreferential only with *Bill,* but not with *John.* The reason, according to Lees and Klima (1963), is that *John* and *himself* do not arise in the same "simplex."

The indicated characteristics of reflexive pronouns are accounted for by the following IL_2 translations (cf. (91)) in combination with the specification on the reconstruction of reflexive pronouns (cf. (92)). In the latter, the sentence-condition will be reformulated as a restriction on the interior reconstruction in terms of the already familiar notion of a coargument.[15]

Let $\Gamma_{\mathrm{rfl}.i,n}$ ($i = 1, 2$, or 3, $n \in \mathrm{N}$) be a context variable of type $\langle s, e \rangle$. Reflexive T pronouns are to be translated as follows:

(91) *myself* translates into $\hat{P}[P\{\Gamma_{\mathrm{refl}.1,n}\}]$
yourself translates into $\hat{P}[P\{\Gamma_{\mathrm{refl}.2,n}\}]$
himself translates into $\hat{P}[male'(\Gamma_{\mathrm{refl}.3,n}) \wedge P\{\Gamma_{\mathrm{refl}.3,n}\}]$
herself translates into $\hat{P}[female'(\Gamma_{\mathrm{refl}.3,n}) \wedge P\{\Gamma_{\mathrm{refl}.3,n}\}]$
itself translates into $\hat{P}[neuter'(\Gamma_{\mathrm{refl}.3,n}) \wedge P\{\Gamma_{\mathrm{refl}.3,n}\}]$
ourselves translates into $\hat{P}[P\{\Gamma_{\mathrm{refl}.1,n}\}]$
yourselves translates into $\hat{P}[P\{\Gamma_{\mathrm{refl}.2,n}\}]$
themselves translates into $\hat{P}[P\{\Gamma_{\mathrm{refl}.3,n}\}]$

[15] Although the intuitive notion of a coargument is straightforward, its formal characterization is complicated by the fact that the semantic type of arguments in a PTQ-like grammar **varies** depending on the particular semantic build up of the argument as well as on the extensional reductions. In (69) we specified the coargument of Γ_3 (where $\Gamma_3 \in ME_{f(\mathrm{CN})}$ is the characteristic context variable of a nonreflexive third person T pronoun) as an individual concept standing in a relation with the **argument** of Γ_3. However, since the context variable $\Gamma_{\mathrm{refl}.k}$ translating reflexive pronouns is of type $\langle s, e \rangle$, let us define a coargument of $\Gamma_{\mathrm{refl}.k}$ as an individual concept standing in a relation with $\Gamma_{\mathrm{refl}.k}$ itself.

The specification of possible values for $\Gamma_{\mathrm{refl}.k,n}$ ($k = 1, 2$, or 3, and $n \in \mathrm{N}$) is stated in (92):

(92) If α is of type $\langle s, e \rangle$, then

$$val(\alpha, \Gamma_{\mathrm{refl}.k,n}, (i,j)) \text{ is } \alpha$$

provided that (i) α is a coargument of $\Gamma_{\mathrm{refl}.k,n}$, and (ii) it holds that $\Gamma_k\{\alpha\}$ (that is, α has the property of being first, second, or third person in the nonreconstructed translation of the target sentence, depending on whether k in $\Gamma_{\mathrm{refl}.k,n}$ is 1, 2, or 3).

The reconstruction of reflexive T pronouns is illustrated below:

(93) *I shave myself.*

(93′) $\wedge x \in [\Gamma_1\{x\}]^1 shave'(x, \hat{P}[P\{\Gamma_{\mathrm{refl}.1,n}\}])$

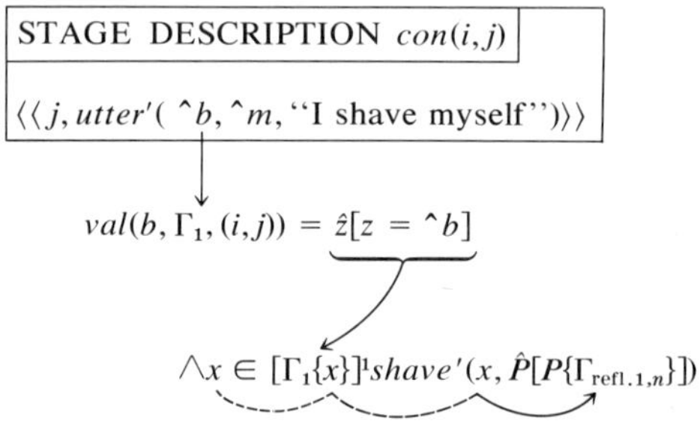

Reconstructed Formula

$\wedge x \in [x = {}^\wedge b]^1 shave'(x, \hat{P}[P\{x\}])$

(94) *Every senator voted for himself.*

(94′) $\wedge x \in [senator'(x)]^2 vote\ for'(x, \hat{P}[male'(\Gamma_{\mathrm{refl}.3,n}) \wedge P\{\Gamma_{\mathrm{refl}.3,n}\}$

Reconstructed Formula

$\wedge x \in [senator'(x)]^2 vote\ for'(x, \hat{P}[male'(x) \wedge P\{x\}])$

(95) *John tries to shave himself.*

(95′) $try\text{-}to({}^\wedge j, \hat{y}(shave'(y, \hat{P}[male'(\Gamma_{\mathrm{refl}.3,n}) \wedge P\{\Gamma_{\mathrm{refl}.3,n}\}])))$

Reconstructed Formula

$try\text{-}to({}^\wedge j, \hat{y}(shave'(y, \hat{P}[male'(y) \wedge P\{y\}])))$

By treating reflexive T pronouns on the basis of context variables of type $\langle s, e \rangle$, we are able to exploit the principle of bound variables without stipulating coreference between the reflexive pronoun and its antecedent syntactically. Furthermore, by reconstructing reflexive T pronouns with respect to their coargument, we can handle examples like (95) without reference to underlying simplex sentences.

We also have to account for the unacceptability of examples like (96) and (97):

(96) **John was shaved by himself.*

(97) **Himself was shaved by John.*

The apparent reason why (97) is unacceptable is that reflexives do not occur in the nominative case (as reflected morphologically by the absence of, for example, **heself* in English). Thus, generation of (97) may be prevented in terms of surface syntactic restrictions along the lines of the treatment of scope restrictions proposed in Hausser 1976b.

Sentence (96), on the other hand, could be ruled out on the basis of restrictions on the reconstruction of $\Gamma_{\text{refl}.k,n}$. Compare (96′), which is the presumable translation of (96), with the translation of (98):

(96′) $shave'(\Gamma_{\text{refl}.3,n}, \hat{P}P\{{}^{\wedge}j\})$

(98) *John shaved himself.*

(98′) $shave'({}^{\wedge}j, \hat{P}P\{\Gamma_{\text{refl}.3,n}\})$

By restricting the reconstruction of $\Gamma_{\text{refl}.k,n}$ to "left" coarguments, the interpretation of (96) may be prevented without impairing the interpretation of (98′). The general principle underlying the restrictions on the occurrence of reflexive pronouns seems to be that reflexives are excluded from surface as well as underlying **subject** positions.

This treatment of reflexives is subject to one difficulty, however, concerning the treatment of number agreement between a reflexive pronoun and its antecedent. Consider the following examples:

(94) *Every senator voted for himself (*themselves).*

(99) *John and Bill voted for themselves (*himself).*

(100) *After they (*he) arrived, every senator voted.*

Although the nonreflexive pronoun in (100) reflects the **cardinality** of its antecedent (presuming that the pronoun in (100) is to be coreferential with *every senator*), the reflexive pronouns in (94) and (99), respectively, have to agree with the **syntactic number** of their antecedents.

Since the number agreement between reflexives and their antecedents is purely surface syntactic, there is no straightforward way to account for the resulting occurrence restrictions in terms of restrictions on the reconstruction (interpretation) of reflexive pronouns. Thus, the treatment of reflexives outlined above is in its present form only suited to account for languages which do **not** mark syntactic number in reflexives (for example, German).

A similar problem arises in connection with **gender agreement** between a pronoun and its antecedent in languages which operate with syntactic rather than natural gender. Since English uses natural gender, we are justified in treating the gender of pronouns as a semantic property of the referents. Consider, for example, (101):

(101) *The man who saw Robin recognized him.*

(101′) $$\bigwedge x \in [[man'(x) \wedge see'(x, \hat{P}P\{{}^\wedge r\})]^1 recognize'(x, \hat{P}\bigwedge y \in [male'(y) \wedge \Gamma_3\{y\}]^1 P\{y\})$$

The internal reconstruction of Γ_3 with respect to r will be successful only if the referent of r has the property of being male. However, apart from the fact that the definition of this property is not without difficulty, this treatment of gender could not be applied in languages like Latin or German, which use syntactic gender.

One way to handle this general type of problem (which is language dependent in its various realizations, as we have indicated) would be to copy the relevant surface syntactic features into the translation formulas. Assume, for example, that the translations of English noun phrases which are syntactically singular are marked by a simple underlining, while the translations of noun phrases in the syntactic plural receive a double underlining. Thus (102) would translate into (102′):

(102) **Every senator voted for themselves.*

(102′) $$\underline{\bigwedge x \in [senator'(x)]^2} vote(x, \underline{\underline{\hat{P}P\{\Gamma_{\mathrm{refl}.3,n}\}}})$$

We may now rule out the reconstruction of $\Gamma_{\mathrm{refl}.3,n}$ with respect to x in (102′) by adding the requirement to the specification of $val(\alpha, \Gamma_{\mathrm{refl}.k,n}, (i,j))$ (cf. (92)) that the context variable and the source expression must have the same kind of underlining, and similarly for a treatment of the gender–agreement between a pronoun and its source (antecedent) in languages which use syntactic gender.

The difficulty we have encountered with number agreement in connection with reconstructing reflexive pronouns points to a general question: What is the nature of the relation between the level of surface syntax and the level of semantic representation? More specifically, is a universal semantic representation for all natural languages feasible?

The idea of a universal semantic representation in terms of a highly developed intensional logic is certainly attractive. The particular properties of a given natural language could be treated in terms of an individual surface syntax to be combined with a set of individual translation rules, which relate the surface structures under consideration to the universal semantic representation. We may even characterize the syntactico–semantic properties of different natural languages by defining a typology on the different kinds of sets of translation rules.

But what about contextual reconstruction? Since these procedures are part of the interpretation (their outcome depends on the point of reference), they are part of the semantics of the language in question. The need to refer to surface syntactic features like gender, number, etc. (depending on the natural language under consideration) in order to properly specify the restrictions on the internal reconstruction of pronouns indicates, however, that at least the specifications on the internal reconstruction of context variables cannot be treated in universal terms. However, the use of context variables (and of the apparatus required to interpret them) may provide the necessary measure of flexibility to allow for a semantic representation which is universal **except** for the specifications on the reconstruction of context variables through the **linguistic** context.

REFERENCES

Bach, E. (1970). Problominalization. *Linguistic Inquiry,* **1,** 121–122.

Bar-Hillel, Y. (1954). Indexical expressions. *Mind,* **63,** 359–379.

Belnap, N. (1970). Conditional assertion and restricted quantification. *Nous,* **4,** 1–12.

Bresnan, J. (1970). An argument against pronominalization. *Linguistic Inquiry,* **1,** 122–124.

Chomsky, N. (1957). *Syntactic Structures.* The Hague: Mouton.

Chomsky, N. (1965). *Aspects of a Theory of Syntax.* Cambridge, Mass.: MIT Press.

Chomsky, N. (1970). Deep structure, surface structure, and semantic interpretation. In D. Steinberg and L. Jakobovits (Eds.), *Semantics: An Interdisciplinary Reader in Philosophy, Linguistics, and Psychology.* Cambridge, England: Cambridge University Press.

Cresswell, M. J. (1973). *Logics and Languages.* London: Methuen.

Daily, M. (1975). *Anaphora and Coreference in English.* Unpublished doctoral dissertation, University of Texas, Austin, Texas.

Davidson, D., and Harman, G. (Eds.) (1972). *Semantics of Natural Language.* Dordrecht: Reidel.

Dougherty, R. (1969). An interpretive theory of pronominal reference. *Foundations of Language,* **5,** 488–519.

Hausser, R. (1974a). Syntax and semantics of plural. In Michael W. Lagaly *et al.* (Eds.), *Papers from the Tenth Regional Meeting of the Chicago Linguistic Society.* Chicago: Department of Linguistics, University of Chicago. Pp. 234–247.

Hausser, R. (1974b). Quantification in an Extended Montague Grammar. Unpublished doctoral dissertation, University of Texas, Austin, Texas.

Hausser, R. (1976a). Presuppositions in Montague grammar. *Theoretical Linguistics,* **3,** 245–280.

Hausser, R. (1976b). Scope ambiguities and scope restrictions. *Amsterdam Papers in Formal Grammar,* Vol. 1. Centrale Interfaculteit, Universiteit van Amsterdam.

Hausser, R. (1976c). The logic of questions and answers. To appear.

Hausser, R., and Zaefferer, D. (1977). Questions and answers in a context-dependent Montague grammar. To appear in the *Proceedings of the Bad Homburg Workshop on Formal Semantics.* Dordrecht: Reidel.

Jackendoff, R. (1969). *Some Rules of Semantic Interpretation for English.* Unpublished doctoral dissertation, MIT, Cambridge, Massachusetts.

Jackendoff, R. (1972). *Semantic Interpretation in Generative Grammar.* Cambridge, Mass.: MIT Press.

Jespersen, O. (1933). *A Modern English Grammar on Historical Principles,* London: George, Allen and Unwin.

Karttunen, L. (1969). Pronouns and variables. In R. Binnick *et al.* (Eds.), *Papers from the Fifth Regional Meeting of the Chicago Linguistic Society.* Chicago: Department of Linguistics, University of Chicago. Pp. 108–116.

Karttunen, L. (1974). Presuppositions and linguistic context. *Theoretical Linguistics* **1,** 182.

Karttunen, L. (1977). Whichever antecedent. Mimeo submitted to the Chicago Linguistic Society Squib Anthology.

Kripke, S. (1968). Naming and necessity. In D. Davidson and G. Harman (Eds.), *Semantics of Natural Language.* Dordrecht: Reidel.

Kuroda, S-Y. (1968). English relativization and certain related problems. *Language* **44,** 244–266.

Lakoff, G. (1968). Pronouns and reference. University of Indiana Linguistic Club, Bloomington, Indiana.

Langacker, R. (1969). Pronominalization and the chain of command. In D. Reibel and S. Schane (Eds.), *Modern Studies in English.* Englewood Cliffs, New Jersey: Prentice-Hall.

Lee, K. (1974). Some Extensions of Montague Grammar. Unpublished doctoral dissertation, University of Texas, Austin, Texas.

Lees, R. B., and Klima, E. S. (1963). Rules for English pronominalization, *Language* **39,** 17–28. [Reprinted in Reibel and Schane (1969).]

Lewis, D. (1972). General semantics. In D. Davidson and G. Harman (Eds.), *Semantics of Natural Language.* Dordrecht: Reidel.

Lyons, J. (1968). *Introduction to Theoretical Linguistics.* Cambridge, England: Cambridge University Press.

McCawley, J. (1970). Where do noun phrases come from? In R. Jacobs and P. Rosenbaum (Eds.), *Readings in English Transformational Grammar.* Waltham, Mass: Ginn and Company.

Montague, R. (1968). Pragmatics. In R. Thomason (Ed.), *Formal Philosophy: Selected Papers of Richard Montague.* New Haven: Yale University Press.

Montague, R. (1970a). English as a formal language. In R. Thomason (Ed.), *Formal Philosophy: Selected Papers of Richard Montague.* New Haven: Yale University Press.

Montague, R. (1970b). Universal grammar. In R. Thomason (Ed.), *Formal Philosophy: Selected Papers of Richard Montague.* New Haven: Yale University Press.

Montague, R. (1973). Proper treatment of quantification in ordinary English. In R. Thomason (Ed.), *Formal Philosophy: Selected Papers of Richard Montague.* New Haven: Yale University Press.

Montague, R. (1974). *Formal Philosophy: Selected Papers of Richard Montague,* edited and with an introduction by Richmond Thomason. New Haven: Yale University Press.

Partee, B. (1975). Montague grammar and transformational grammar. *Linguistic Inquiry* **6,** 203–300.

Postal, P. (1966). On so-called "pronouns" in English. *Georgetown University Monograph Series on Language and Linguistics,* No. 19, 177–206. [Reprinted in Reibel and Schane (1969).]

Postal, P. (1972). Some further limitations on interpretive theories of anaphora. *Linguistic Inquiry* **3,** 349–370.

Quine, W. V. O. (1960). *Word and Object,* Cambridge, Mass.: MIT Press.

Quine, W. V. O. (1963). *Set Theory and its Logic.* Cambridge, Mass.: Belknap Press of Harvard University Press.

Reibel, D., and Schane, S. (Eds.) (1969). *Modern Studies in English.* Englewood Cliffs, New Jersey: Prentice-Hall.
Rescher, N. (1969). *Many Valued Logic.* New York: McGraw-Hill.
Robbins, B. (1962). The transformational status of the definite article in English, TDAP No. 38.
Ross, J. R. (1967). Constraints on Variables in Syntax. Unpublished doctoral dissertation, MIT, Cambridge, Massachusetts.
Ross, J. R. (1969). On the cyclic nature of English pronominalization. In D. Reibel and S. Schane (Eds.) *Modern Studies in English.* Englewood Cliffs, New Jersey: Prentice-Hall.
Russell, B. (1903). On denoting. *Logic and Knowledge,* 1971.
Stalnaker, R. C. (1968). A theory of conditionals. In N. Rescher (Ed.), *Studies in Logical Theory,* Oxford: Basil Blackwell.
Steinberg, D., and Jakobovits, L. (Eds.). *Semantics: An Interdisciplinary Reader in Philosophy, Linguistics, and Psychology.* Cambridge: Cambridge University Press.
Stockwell, R., Schachter, P. and Partee, B. (1973). *The Major Syntactic Structures of English.* New York: Holt, Rinehart and Winston.
Vendler, Z. (1967). *Linguistics in Philosophy.* Ithaca, New York: Cornell University Press.
Wasow, T. (1973). *Anaphora in Generative Grammar.* Unpublished doctoral dissertation, MIT, Cambridge, Massachusetts.

A GENERAL THEORY OF THE MEANING OF ANAPHORIC PRONOUNS

TIMOTHY C. POTTS

1. THE HYPOTHESIS

Pronouns appear to have two distinct uses in general-purpose languages. They are sometimes used deictically, as when one person nudges another and says, looking or pointing at someone else, *He's the King of Ruritania*. But they are more commonly used in a linguistic context in which their meanings are related to those of proper names or quantifying expressions which occur in that context. Linguists call this the **anaphoric** use of pronouns. A similar distinction may be drawn between the uses of demonstratives (*this, that,* and sometimes *the*), which may be followed by a count or mass noun. Demonstratives are most commonly used deictically, as in *That (man) is the King of Ruritania,* but can also be used anaphorically, as in *Socrates kicked a dog and the dog bit him,* and *There's a snack in the refrigerator and that will keep the wolf from the door until dinner time*. This paper is restricted to the anaphoric use of pronouns, but proposes a general theory of their meaning in that it brings reflexive, personal, relative, and possessive pronouns under a single account, together with certain analogous expressions which I call "pro-verbs." It should, in my view, be possible to extend an account of the anaphoric use of pronouns to cover the anaphoric use of demonstratives, but

SYNTAX AND SEMANTICS
VOLUME 10

ISBN: 0-12-613510-X

an assumption which is made here for heuristic reasons prevents me from attempting this.

The paper applies, to sentences containing anaphoric pronouns, the linguistic theory defended in Potts (1977); it also makes use of a categorial grammar based upon Fregean principles (compare Potts, 1973). The guiding principles of the theory are:

1. that the phenomena of ambiguity and paraphrase (or translation) make it necessary to posit at least two types of structure which may be assigned to meaningful sentences: those in terms of which their meanings are understood (semantic structures), and those in terms of which they qualify as sentences of a particular language (phonetic structures; ''phonetic'' being taken broadly to include graphic realizations).
2. that the constituents of semantic and phonetic structures are not the same, some of the **interrelationships** between the constituents of semantic structures being correlated with **constituents** of phonetic structures.
3. that semantic structures are multidimensional, whereas phonetic structures are linear, the latter being projections of the former, which consequently have to employ special conventions, differing from one language to another, in order to compensate for the loss of dimensions (cp. classical perspective).
4. that representations of semantic structures, being limited to two or at most three dimensions, will necessarily be partial, and will often allow for equivalent alternatives. I therefore speak of representing *a* semantic structure for a sentence, not of representing *the* semantic structure. In what follows, it will also be assumed that such representations are always partial.

In what does the meaning of a pronoun consist? The most common view is that a pronoun is akin to a proper name in having a bearer, but that the same pronoun can have different bearers in different contexts. Thus, pronouns are similar to deictic expressions, but differ from them in that their bearers are fixed by the bearers of other names occurring in the sentence, rather than by ostension, for example. By analogy with the two-name theory of predication (cf. Geach, 1972, pp. 49–53, 289–301), I call this a two-name theory of pronouns. There are several variants upon it, for example, that a pronoun ''refers back'' to another expression with which it is paired. The use of the terms ''reference'' and ''coreference'' is widespread in the literature, ''reference'' being a rather indeterminate notion; ''bearer'' is more precise and may help us to see the inadequacy of a two-name theory. In this paper, however, I shall not be able to give a critique of current theories and must concentrate on expounding an alternative account.

My hypothesis is:

1. that pronouns are not semantic constituents of sentences, though they are phonetic constituents. We need them only because phonetic structures are linear.

2. that their function is to reduce by one the polyadicity of the expressions upon whose meaning they work.
3. that, in addition, they do this by identifying (making identical) what exercises two or more case roles associated with verbs (cp. Fillmore, 1968, for the notion of case role).

Points 2 and 3 call for expansion.

In order to spell out Point 2 in more detail, it is convenient to make use of the Fregean system of categories. For the purposes of this paper, I shall accept Frege's assumption that the basic semantic categories are those of proper name and proposition, the former including most proper nouns and the latter consisting of such sentences as are accounted either true or false (It is this assumption which prevents me from extending the account to include anaphoric uses of demonstratives). The simplest propositions will then consist of an arrangement or pattern of proper names, the verb and any prepositions or inflections of the proper names forming part of that pattern. Following Frege, I call those propositional patterns, schemas, or structures whose parts or constituents are proper names, schemas of **first level.** We can obviously classify schemas of first level according to the number of proper names that each interrelates (allowing a single proper name as a limiting case of interrelationship). The following notation, adapted from Ajdukiewicz 1935, will express this classification: 1PN, 2PNN, 3PNNN, . . . etc. An example will show how it may be understood: Read 2PNN as "the interrelationship of two constituents into a proposition, namely, a proper name and a proper name." The initial number, in this notation, tells us the polyadicity of the schema. At present, the list of constituents at the end appears to be redundant; this will, however, come into its own later.

The hypothesis, then, is that a pronoun works upon a schema of first level in order to reduce its polyadicity by one. So, if we apply a pronoun to a schema of category 2PNN, the result should be a schema of category 1PN. For example, take the schema *1 killed 2;* if we suppose the result of applying a reflexive pronoun to this schema to be the new schema *1 killed itself,* then we have indeed reduced the polyadicity of the original from two to one: *1 killed 2* contains two schematic symbols, for each of which a proper name may be substituted; *1 killed itself* contains only one such symbol. If we generalize the category notation introduced above, it will also be possible to assign a category to pronouns. The category 2PNN is a schema that will turn two proper names into a proposition (when they are substituted for its schematic symbols); so, in general, $n\alpha\beta_1 \ldots \beta_n$ will be the category of a schema which will turn n constituents of categories $\beta_1, \ldots, \beta_n$, respectively, into an expression or schema of category α. Thus, in the case above, a single schema of category 2PNN (β_1) is turned into a schema of category 1PN (α); substituting, then, in $1\alpha\beta_1$, the category of the reflexive pronoun comes out as 11PN2PNN.

It is permitted, in this notation, to add together **contiguous** numbers; thus, instead of 11PN2PNN, we may write 2PN2PNN. This gives us an alternative way of describing the function of a pronoun or, more precisely, the propositional pattern of which a pronoun, phonetically regarded, is a part. It is a pattern which interrelates a proper name and a first level schema of category 2PNN into a proposition. But this description is clearly too narrow. After all, we can also combine reflexive pronouns with first level schemas of category 3PNNN, for example, *1 bought 2 for himself.* Is the pronoun then of category 3PNN3PNNN? If so, then its meaning must be different from its meaning in the schema *1 dressed himself,* where its category will be 2PN2PNN. But that is totally implausible.

Fortunately, this difficulty is not as formidable as it first seems. It is a question of what combinations of schemas of different categories are possible. If the only type of combination which we allow is that licensed by the rule

(R1) $1\alpha\beta\ \beta \rightarrow \alpha$

where any category name may be substituted for α and any category name for β, then we shall indeed have to assign *himself* in *1 bought 2 for himself* to a different category from *himself* in *1 dressed himself.* But, as I have argued elsewhere (Potts, 1973), this rule alone will not justify many other combinations which we regard as perfectly legitimate. It must be supplemented by further, recursive rules. Of these, the one which concerns us here can be stated, in a simplified form, as:

(R2) if $\gamma\ \delta \rightarrow \alpha$, then $\gamma\ 1\delta\beta \rightarrow 1\alpha\beta$

The effect of this rule is that, for certain series of categories, a schema belonging to one term of the series will suffice (with respect to possible combinations) for any schema belong to a subsequent term of the series.

Take, for example, the series 1PP, 2PN1PN, 3PNN2PNN, ... Negation belongs to the first of these categories, 1PP; yet, very often, a sign of negation occurs in a sentence in a position where it cannot be construed as working upon the meaning of a subordinate sentence, for example, *At least one man in the crowd did not cheer.* In this case, negation has been applied to the schema *1 cheered,* which is of category 1PN, as is the result of the application, *1 did not cheer.* Now, without **(R2)**, we should have to assign negation to category 2PN1PN in this example, because it is used to turn a 1PN into another 1PN. But, by **(R1)**, we have:

1PP P $\rightarrow$ P

and so, taking this as the antecedent of **(R2)** and substituting N for β:

1PP 1PN $\rightarrow$ 1PN

If we now take this, in turn, as the antecedent of **(R2)**, and again substitute N for β, we shall get:

1PP 2PNN $\rightarrow$ 2PNN,

showing that, like a schema of category 3PNN2PNN, a schema of category 1PP will turn a 2PNN into another 2PNN. Of course, this will not work the other way around. A schema of category 2PN1PN, for instance, will turn a 2PNN into another 2PNN but will not turn a P into another P.

The series which concerns us here is 1P1PN, 2PN2PNN, 3PNN3PNNN, ... It is to be observed that I am assigning pronouns to the *second* term of this series, not to the first. To the first belong the schemas of which quantifying expressions, for example *every man, a dog, ten cows,* are the linguistic parts. These schemas can turn a 1PN into a P, as in *A dog barked,* and also a 2PNN into a 1PN, as in *I loves a Greek boy*. By contrast, we cannot allow a pronoun schema to turn a 1PN into a P; *It barked* may be a sentence, but it is not a proposition because, in order to be able to determine its truth value, we should need an answer to the question: ***What*** *barked?* (This could, of course, be supplied by the context.) Worse still, **Itself barked* is not even a sentence. These inconveniences are avoided by assigning pronoun schemas to category 2PN2PNN, which prevents a pronoun from being applied to a 1PN, but allows it to be applied to a 2PNN, a 3PNNN, and so on; the result, in each case, being reduction of the latter's polyadicity by one, just as with a quantifying schema. A pronoun, on this account, is thus a pattern which interrelates one proper name and a 2PNN into a proposition, or two proper names and a 3PNNN into a proposition, etc. The rules for combination of schemas of different categories are set out, with some examples, in Appendix 1.

This second point of the hypothesis has involved some rather technical considerations, and its full intuitive force will probably only be brought out in conjunction with the third point. Polyadicity alone is not enough to characterize the function of pronouns. We need, in addition, the idea introduced by Fillmore (1968) that it is part of the meaning of a verb to assign different roles to the bearers of the proper names which (in the simplest case) it interrelates; roles such as agent and patient, source and goal, instrument and location. Now the **same** thing can sometimes play more than one of these roles simultaneously, either with respect to a single action or state, or with respect to different actions or states. Pronouns are used to signify this; reflexive in the first case, personal or relative in the second. In other words, the way in which they reduce the polyadicity of a schema of first level is by identifying, in the sense of making identical, two of the places in the latter occupied by schematic symbols for which proper names may be substituted. This allows one proper name to fill more than one of these places at once; that is, places in the **semantic** structure of the proposition. Of course, this is not possible for the **phonetic** structure, because that is constrained by the linear ordering requirement.

The connection of pronouns with identity is easily shown. Instead of *Socrates kicked a dog and it bit him,* we can say *Socrates kicked a dog and* ***the same dog*** *bit him*. The other day I saw, in a post office, a wallet for keeping an old-age pension book in. On it was printed: *Finder of this book should* *return same to the nearest post office*. I should not myself regard this substitution of *same* for *it* as in accordance with the Queen's English, but it is fairly common in a number of dialects and, the decisive point, is immediately comprehensible even to one who speaks only the Queen's English. A full account of the use of pronouns should explain this use of *the same*, and how it is related to its use in, for example, *Lewis Carroll was the same man as Charles Dodgson*. I shall not be able to undertake this here; identity is too difficult a topic to sew up in a few paragraphs.

2. NOTATION

By contrast with a two-name theory, I am suggesting that the identification effected by a pronoun occurs *within* the semantic structure of the proposition, not in the connection between the proposition and what it describes. This will explain the possibility of intentional identity being signified by pronouns (cp. Geach, 1972, pp. 146–153), which creates an enormous difficulty for a two-name theory. As usual, mistaken accounts of the meaning of pronouns have arisen from assuming that semantic structures are far more like phonetic structures than is really the case. However, the greater the difference between the two types of structures, the more necessary it is to show how they are related. If semantic structures are multidimensional and phonetic structures linear, with some of the interrelationships between elements of the former being "hypostatized" as elements of the latter, then we may expect phonetic structures to be linear projections of semantic structures, in which various conventions are employed in order to mark distinctions in the latter which would otherwise be obscured. But this poses an awkward problem of exposition, for, if semantic structures are multidimensional, then *ex hypothesi* it will be impossible to represent them, except by resort to conventions, in one, two, or even three dimensions—the most that a perceptible representation can afford.

This difficulty cannot be overcome, but we can go some way toward meeting it through the use of two-dimensional notations. If a semantic structure contains more than two dimensions, we can at least isolate one from the rest by this device, though only one at a time, so that it may be necessary to vary the dimension chosen according to the subject matter. For this reason, as well as the space which it takes up, a two-dimensional notation is unlikely to be popular for setting out proofs or for working in generally. It is not a working tool. All the same, it can serve a very important theoretical function by showing us what is essential and what inessential in our representations more clearly than a linear notation can do. It can help us to distinguish the multiplicity which is essential to language for it to convey meaning from the multiplicity which arises from the linearity of speech.

Frege used a two-dimensional notation, but only for propositional logic, writing

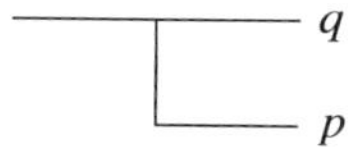

for implication, that is, instead of the linear notations $p \supset q$, $p \rightarrow q$, and Cpq, which are used today. By combining with this sign a small vertical stroke to represent negation, he could also represent (for classical logic) conjunction and disjunction:

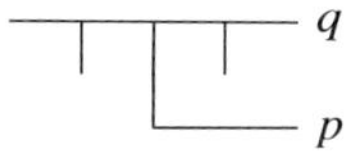

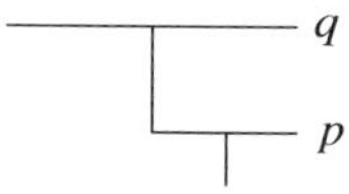

pq, $p \cdot q$, $p \,\&\, q$, $p \wedge q$, Kpq $\qquad$ $p \vee q$, Apq

Where the negation stroke occurs to the left of the junction between the line connecting p and q, it negates the whole implication; where it occurs to the right of the junction, it negates only the formula at the end of the line to which it is attached. So the first diagram above represents $NCpNq$, which is equivalent to Kpq in classical logic, and the second, $CNpq$, which is equivalent to Apq.

Frege's choice of implication was motivated by the system of proof that he adopted, in which he started with axioms and used substitution and detachment as rules. The representation of detachments is then extremely clear. For my purposes this is not important, and it will be more convenient to take a sign for conjunction as basic; I shall also rotate the diagram through 90 degrees and modify it slightly, but retain the small strokes for negation, thus obtaining:

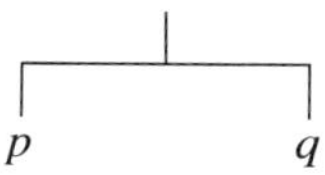

Conjunction
(Kpq)

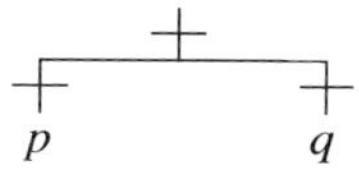

Disjunction
($NKNpNq$)

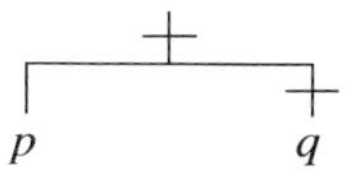

Implication
($NKpNq$)

For quantification, Frege retained a linear notation, representing universal quantification ($\bigwedge_x Fx, \forall_x Fx, (x)Fx, \Pi_x Fx$) by:

while existential quantification ($\bigvee_x Fx, \exists_x Fx, \Sigma_x Fx$) was represented by combining this with negation:

 that is, $\quad N\Pi_x NFx$

I now propose to extend Frege's two-dimensional notation to quantification. The first step is to introduce a two-dimensional representation of first level interrelationships, that is, of those which, in the simplest case, interrelate proper names into a proposition. We can regard the proper names (with a single proper name as the limiting case) as inputs to these signs, each of which will then have just one propositional output, for each is capable of turning a given number of proper names into a proposition. For present purposes, it will be necessary to represent the name inputs and the propositional output in the same (vertical) dimension, distinguishing them by their direction, which can be indicated by arrows, thus:

for 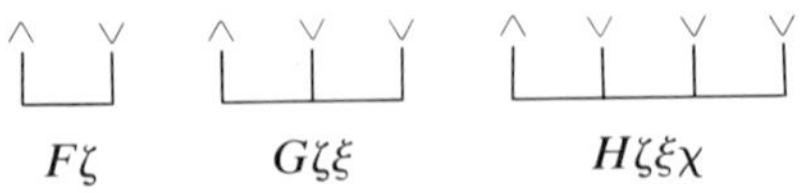

, etc.

In presenting these signs, I have followed the ordering convention of the linear notation. However, the linear position of the propositional output in the horizontal dimension, relative to the name inputs, is immaterial, because the arrow is enough to show which it is.

This is not so for the name inputs. Suppose, for example, that we were to adopt the convention that the order of name inputs in the horizontal dimension is to be the same as that of the proper names which are supplied to them in an English sentence. Then we should have to represent *John gave Fido to Mary* and *Mary was given Fido by John* by

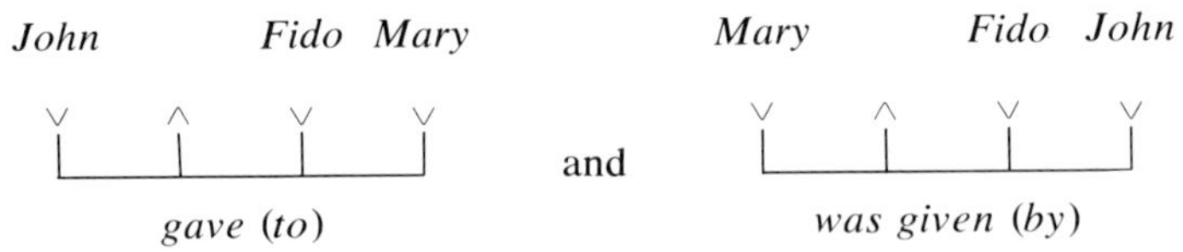

respectively. This is unsatisfactory on two counts: First, because the two sentences have the same meaning, at least to the extent that they have the same truth conditions; second, because we are incorporating into our **semantic** representation a feature of English **phonetic** structures, their word order convention. Thus, we are forced to represent the active and passive voices of *give* by distinct signs which are, moreover, only distinguished by the label written below them, so that we should have to resort to "meaning rules," *à la* Carnap, in order to relate the two signs semantically.

Given an extra dimension, it is easy to devise a representation of what is common to the meanings of these sentences, for example,

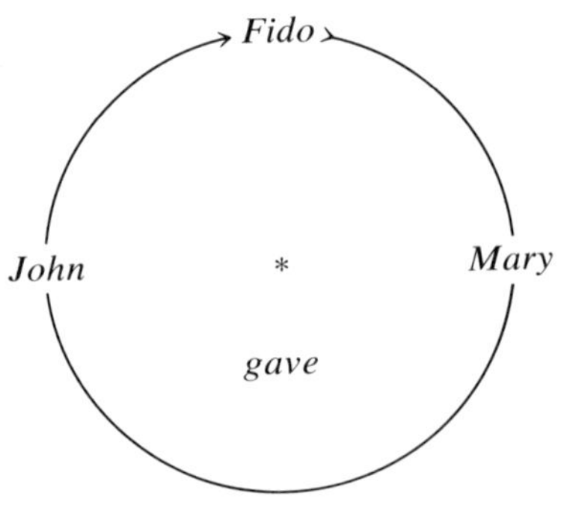

We are to imagine, here, that the propositional output comes from the center of the circle, at right angles to the page; similarly, that the three proper names are input to the circumference of the circle from above. In addition, the arrows on the circumference on either side of *Fido* indicate a direction, with respect to ownership, going from *John* to *Mary*. Now, seen from above, this diagram will convey the same meaning to us whether we hold the paper sideways, upside down, or the right way up (supposing that we can read the words written on it whatever the position of the paper). If, however, instead of looking at the diagram from above, our viewpoint were restricted to some position on the surface of the page, the circle would appear as a straight line and the proper names would be strung out horizontally above it. Moreover, as we moved our position, the order of the proper names would change. By placing ourselves below the diagram, we should see something like the first of the two-dimensional representations given above; but, by placing ourselves above it, we should see something like the second (reading, in both cases, from left to right). Thus, the effect of losing one dimension is that there are alternative ways of representing the same meaning which did not exist with the extra dimension.

I hope that this will elucidate the sense of "dimension" in my hypothesis that semantic structures are multidimensional. If two sentences have the same meaning, then by definition they have the same semantic structure; and even if there is only something in common to their meanings, part of the semantic structure of each will be the same. In an ideal notation, two sentences having the same meaning should receive the same semantic representation, to which (in that notation) there are no alternatives. But, as representations are perceptible, they can deploy at most three dimensions. Hence, the claim that semantic structures are multidimensional is equivalent to saying that no ideal notation for representing them is possible; while the existence of a given dimension is shown if, by introducing an extra dimension into our notation, we can eliminate some of the alternative representations which are unavoidable without it. What Frege called "the manner of presentation" of a thought can be seen as a choice from among the alternatives which are necessary in a linear notation.

The dimension that must be isolated in order to obtain a clear view of the use of anaphoric pronouns, however, is that of propositional inputs and outputs and (proper) name inputs and outputs, so we shall have to sacrifice the possibility of showing how proper names can be interrelated into propositions. Even so, we can introduce a convention which will make the *order* of name inputs in a first level sign inconsequential, and this, while not eliminating alternatives, will considerably simplify the representation of semantic structures for sentences containing pronouns. The arrows on the circumference of the circle in the preceding diagram can be seen as indicating the case roles associated with each name input. Thus, the direction ***from John to Mary***, marks out *John* as the Source and *Mary* as the Goal of the transaction, to use two of Fillmore's names for the case roles, while writing *Fido* between the two arrows shows that *Fido* is the Object. This suggests that we might modify the original notation for first level signs by labeling each name input with a name for the case role associated with it. It is not necessary, for present purposes, to be very exact about case roles, so long as we have enough names for them to prevent any label from occurring more than once in a first level sign; so I shall just borrow from Fillmore, using the following abbreviations:

a:	agent	s:	source
o:	object	g:	goal
i:	instrument	p:	place

The original examples now become:

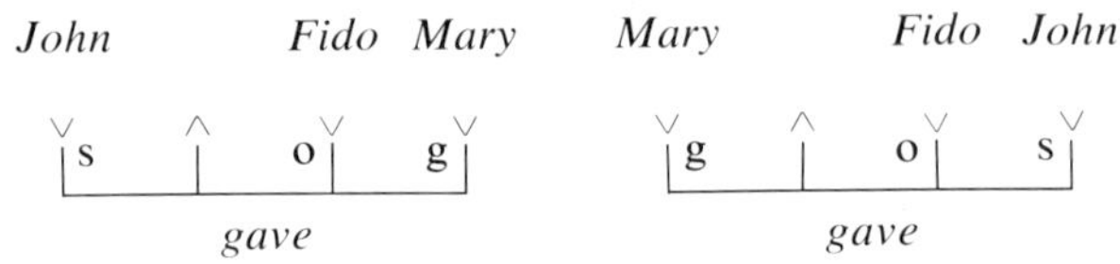

respectively. It is no longer necessary to distinguish the voice of the verb, and the two representations, though different, are to be accounted equivalent just as *Cpq* and *ANpq* are different representations of the same truth function of *p* and *q* in classical logic, and are equivalent. These first level signs are already two-dimensional projections from three (or perhaps more) dimensions. In order to obtain a phonetic structure, we have to eliminate the second dimension. Now, in eliminating the third dimension, we introduced a linear ordering of the proper names. Let us assume the ordering is that which they are to have in the phonetic structure, and that (for English) we put the propositional output in the second linear position. We then make a linear projection as follows:

1. Under each name input, write the proper name above it, prefixed by its case role label.
2. Under the propositional output, write the verb (or other expression written below the two-dimensional sign) prefixed by all of the case role labels in the order in which they occur in the two-dimensional sign.

Projection of the two signs above by this method then yields, respectively:

s-*John* sog-*gave* o-*Fido* g-*Mary*
g-*Mary* gos-*gave* o-*Fido* s-*John*

The **case role** prefixes to the proper names are next converted into **cases,** according to the language with which we are concerned; this depends partly upon the position of each name relative to the verb. Similarly, the prefix to the verb determines its voice, for example active for sog- and passive for gos- in the examples above. We then obtain:

NOM-*John* ACT-*gave* ACC-*Fido* DAT-*Mary*
DAT-*Mary* PASS-*gave* ACC-*Fido* ABL-*John*

(It should not be necessary to spell out these case labels.) In English, the cases are marked by prepositions, but nominative and accusative are unmarked, and the preposition belonging to the name in first position is deleted. So DAT- in the first example becomes *to* and ABL- becomes *by*, while all the rest disappear. We then have the correct English sentences, replacing the voice indicators of the verbs by the appropriate English forms.

This is only a very rough sketch of the method of projection for signs of first level, which does not, as such, belong to the topic of this paper, but must be presupposed in order that the reader may understand the method of projection for signs of second level, with which we shall be directly concerned. I have, however, worked out the method for signs of first level in some detail and hope to publish it eventually. Meanwhile, here I shall only give the first stage, in which the case role labels still appear. Moreover, I shall assume the equivalence of first level signs, which differ only in the horizontal order of their name inputs and propositional outputs and, therefore allow us the freedom to reorder them in making linear projections.

Frege's two-dimensional notation for propositional logic may now be extended to first order logic, if we observe what he says about the sign for the universal quantifier, namely, that it is not Π or Π_x (to use a modern equivalent), but $\Pi_x \varphi x$: That is to say, the index letters, together are integral parts of the single sign (1893, sec. 8). This quantifier may be combined with any sign of first level, whose polyadicity it will reduce by one, by taking the place of one of the proper names which, in the simplest case, the first level sign interrelates into a proposition. The index letters are necessary to show **which** of these proper names it replaces. Where the polyadicity of the first level sign is itself 1, therefore, combination with a quantifier will yield a propositional sign; so the category of the latter is 1P1PN.

We can think of a first level sign as having a certain number of name inputs, corresponding to its polyadicity, and a propositional output, in that it converts signs for names into a propositional sign. A quantifier, in order to combine with it as indicated, will then be seen as having not only a propositional input and output, but also a name **output.** We can represent this combination two-dimensionally as follows:

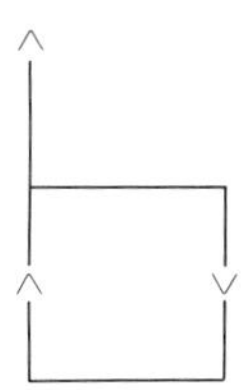

F

The first level sign is now a name input joined to a propositional output, and the quantifier joins a propositional input and a name output to a propositional output. Propositional outputs and inputs are connected by arrows pointing upward, name inputs and outputs by arrows pointing downward. We can think of the lines and arrows as together constituting pathways through the structure; and I shall call any propositional sign in this notation a **network.** In Frege's original notation, there are only propositional pathways; here, there are also name pathways, and so the arrows are necessary to distinguish them from propositional pathways.

It is to be observed that, in this notation, the index letters of the quantifiers have disappeared. This shows that they are neither constituents nor essential features of semantic structures (as, indeed, we might have guessed from their absence from general-purpose languages) but are, rather, necessitated by the linear notation and scope convention of modern logic. That notation employs the convention that a sign for an interrelationship (function) is written to the left of the signs of what it interrelates (its arguments); I use the "Polish" linear notation here because it is the only one that observes this convention consistently. It is in terms of this convention that the notion of *scope*, which is so important in the theory of meaning, is defined. But scope is equally clear in the two-dimensional notation; if we call each of the signs consisting of solid lines a **connection,** then any connection lies within the scope of each connection which is above it on the same propositional path way.

The connections effected by name pathways are analogous to scope, and are equally important in the theory of meaning. The two-dimensional notation brings this out, whereas the linear notation conceals it, because, once having used the linear ordering to define the propositional pathways through the structure, it is not available to represent the name pathways, which must then be indicated by a quite different convention, that of index letters. What the two-dimensional notation shows, by using the same type of representation for both types of pathway, is that the role of the index letters in the notation is structural, telling us how the constituents (or interrelationships) are combined, one with another. I have been tempted to call this "back scope" because, in the linear notation, it runs in the opposite direction from (propositional) scope.

Linguists are familiar, from the "trees" of transformational grammar, with a two-dimensional representation of something approximating propositional scope. This alone, however, is not enough to give a correct account of the meanings of sentences containing more than one quantifying expression —or, we might add, containing pronouns. The classic counterexample is the passive transformation, of

At most a hundred people speak every Slavonic language.

into

Every Slavonic language is spoken by at most a hundred people.

a palmary example which I owe to Professor P. T. Geach. Generative semanticists, accordingly, have resorted to index letters in addition (cf., for example, McCawley, 1971). All I am doing here is to extend the "tree" principle to name pathways. Naturally, though, it will not be possible without loss of information to "translate" these networks into the trees of transformational grammar, in which the connections between the name inputs of first level connections and those of quantifying connections cannot be shown by pathways.

Frege always used unrestricted quantification; his quantifier sign representing *everything* in English. He dealt with quantifying expressions like *every man* by treating the count noun as a disguised first level connection, that is, by paraphrasing *every man* as *of everything, if it is a man, then* . . . . As I am not concerned with quantification as such in this chapter, I shall simplify this aspect of semantic representation by writing the count noun over the horizontal bar of the quantifying connection; only if it is omitted will the sign mean *everything*. Thus, I am, in effect, using restricted quantification. The existential quantifier will be represented by Frege's technique, using negation signs; other types of quantifier will be indicated by ad hoc signs written at the side of the vertical stroke. A few examples of commonly encountered first order formulas may help the reader to absorb the notation:

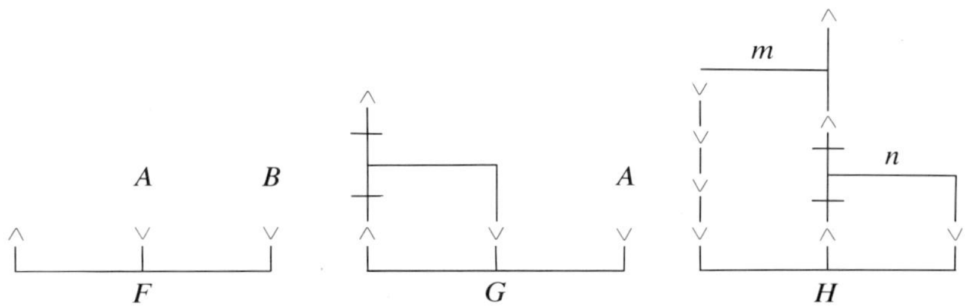

Fab $\Sigma_x Gxa$ $\Pi_x \Sigma_y Hxy$ with *x*s restricted to *m*s, and *y*s to *n*s

The last of these three networks could be used to represent a structure for *Every boy loves some girl* by putting *m* = *boy*, *n* = *girl*, and *H* = *loves*. It will also be seen from this network that the (horizontal) linear position of the propositional output from a first level connection is not significant.

If we use first order logic to represent a semantic structure for a proposition containing a pronoun or pronouns, then there will be no part of the logical formula which corresponds to the pronoun. The occurrence of a pronoun in the proposition, however, can be correlated with a certain *fact* about the formula, namely, that either the same schematic letter, of the type for which a proper name may be substituted, occurs more than once in it, or it contains a triple, quadruple, etc. of index letters of the same quantifier. Thus, we should represent a structure of *A politician killed himself* by the formula

$\Sigma_x Fxx$, and a structure of *Socrates kicked a dog and it bit him* by the formula $\Sigma_x KFaxGxa$. In both cases, the index letter of the quantifier occurs three times. Now, if the category of the existential quantifier $\Sigma_x \varphi x$ is 1P1PN, that of $F\zeta\zeta$ and $KFa\zeta G\zeta a$ must be 1PN, not 2PNN. This is why Frege says that the meaning of *killed himself* is a function of *one* argument (1879, §9). Hence, if we write $F\zeta\zeta$ instead of $F\zeta\xi$, or $KFa\zeta G\zeta a$ instead of $KFa\zeta G\xi a$, thereby signifying that the *same* proper name or quantifier must be substituted for each occurrence of ζ, we have reduced the polyadicity of each of these first level signs by one, since they would otherwise be of category 2PNN. But our hypothesis is that the function of a pronoun is, precisely, to reduce the polyadicity of a first level interrelationship by one, by making two of its name inputs identical.

Although first order logic contains no sign which corresponds to a pronoun in a general-purpose language, it would be possible to introduce one without either extending or restricting the expressive power of the system. We should require that the index letters of quantifiers occur in pairs and add to the standard vocabulary a sign of category 2PN2PNN, for example, $I_{xy}\varphi xy\zeta$, which conforms to the requirements of Frege's linear notation. Then $\Sigma_x Fxx$ would become $\Sigma_x I_{yz} Fyzx$ and $\Sigma_x KFaxGxa$ would become $\Sigma_x I_{yz} KFayGzax$ and, in general, any part $\alpha(x, x)$ of a formula which contains two occurrences of x would be replaced by $I_{yz}\alpha(y, z)x$. It is to be observed that nothing simpler than this would answer; for instance, we could not use $I_x\varphi x\zeta$, because then we should not know, given a first level sign of three or more places, **which** two were being made identical: In $I_x Fxab$, is the first place identified with the second or with the third?

The addition of $I_{xy}\varphi xy\zeta$ to the vocabulary of first order logic and the prohibition, in formulas, of triples, quadruples, etc. of index letters naturally makes formulas of this system much more complicated than those of standard first order logic. It would be extremely cumbersome as a notation for proofs, and since Frege's aim was to provide a clear proof notation (he might more appropriately have called it a *Beweisschrift* than a *Begriffsschrift*), he did well to use a different convention. Moreover, he would soon have been in difficulties with a two-dimensional notation for quantification because it was necessary, for his purposes, to go above the second level. As we shall see later, although the two-dimensional notation works smoothly for first order formulas (= second level), it has to be modified in order to provide equivalent representations of second order ones (= third level).

Here, however, proof is a secondary consideration, meaning the primary one. A notation with the minimum of different conventions then has the advantage, and the two-dimensional notation allows us to represent the structural relationships which I have called name pathways by the same method as for the more familiar propositional or scope pathways. Moreover, although we can understand the sign $I_{xy}\varphi xy\zeta$ as **saying** that two name pathways are joined into a single one, we can actually *show* it in a very natural way, given two dimensions. This sign has one propositional input and one propositional output, but one name input and two name outputs. It can therefore be represented two-dimensionally as:

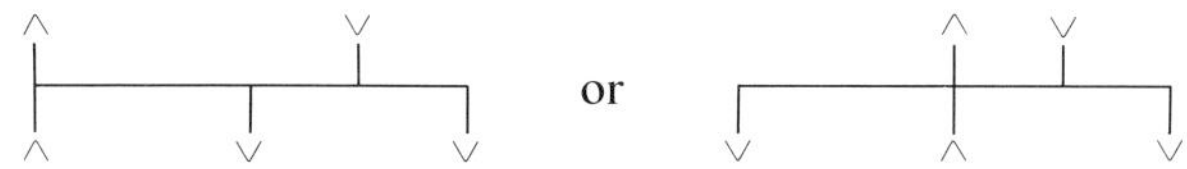

I shall call this a *name connection.*

3. REFLEXIVE PRONOUNS

The term "pronoun of laziness" has been introduced by Geach for "any pronoun used in lieu of a repetitious expression [1972, p. 98]." There are, I think, problems with the notion when it is applied to expressions other than proper names, such as definite descriptions, but the distinction is clear when it is used to contrast, for example, the occurrence of the reflexive pronoun *himself* in the following pair of sentences.

(1) *Canning killed himself.*

(2) *A politician killed himself.*

For (1) has the same truth conditions as *Canning killed Canning,* whereas (2) does not have the same truth conditions as *A politician killed a politician;* in the latter, the killer need not be the same politician as the one killed. Some people object that no one would use the sentence *Canning killed Canning* instead of *Canning killed himself* unless Canning, the killer, was a different Canning from Canning, the killed. I do not find this objection convincing; a rather pedantic person, asked, *Who killed Canning?* might well reply, *Canning killed Canning* rather than *Canning killed himself* in order to preserve the form of the question as nearly as possible in his answer. What the objection does bring out, though, is that the repetition of a proper name in a sentence does not, of itself, guarantee that each occurrence has the same bearer. But this does not affect Geach's distinction, since he proposes a test which takes the sentence containing a pronoun as its starting point, and in applying which it may therefore be assumed that the repeated proper name has the same bearer as the original occurrence.

In first order logic, we take *Canning killed Canning,* rather than *Canning killed himself,* as the canonical form, using the formula *Faa* to represent the structure of both indifferently. In the two-dimensional notation, as well as in the linear notation containing the *I*-operator, however, we have a choice between

(3) and (4)

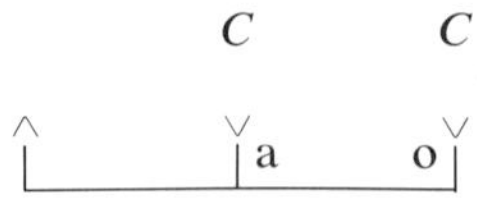

killed

Faa

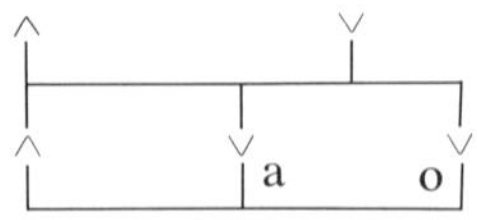

killed

$I_{xy}Fxya$

If we were to join the name connection to the first level connection in network (4), eliminating the arrowheads between them, the result, regarding only its inputs and outputs, would be equivalent to a first level connection with only the name input:

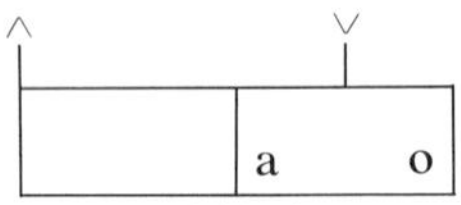

killed

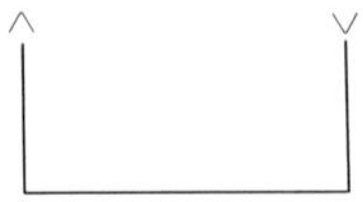

killed itself

Eliminating the internal complexity of this sign, the schema *1 killed itself* is then represented by the standard sign for a first level connection having one name input, as in the right-hand diagram above. So the effect of applying the name connection to a first level connection having two name inputs has been to reduce the latter's polyadicity by one. The advantage of the two-dimensional notation over a linear one, whether $I_{xy}Fxya$ or one which uses category names, that is,

$$2PN_1 2P^2N_3N_4\ 2P^2N_3N_4\ N_1$$

is that the two-dimensional notation shows, where a linear notation does not, that the name inputs have been **connected.** In the linear notation, the two name inputs of the first level sign might just have been **replaced** by a single, new name input. The super- and subscripts of the categorial notation, when joined in pairs, correspond to the arrows of the two-dimensional notation, but there is nothing there or in $I_{xy}Fxya$ to show us the course followed by the propositional and name pathways **within** each sign. Moreover, it is more difficult not to see each category name (or capital letter of the Fregean notation) as representing a **constituent** of the structure; the connections effected by the super- and subscripts in the categorial notation, and the relative linear positions of the capital letters together with the connections effected by the index letters in the Fregean notation, constituting the interrelationships between the constituents.

The question now arises whether the distinction between semantic structures implied by networks (3) and (4) is justified. Or are the two networks simply equivalent alternatives? It seems not, for how could the introduction of extra dimensions absorb these differences? The use of *Faa* in first order logic to represent a structure both for *Canning killed Canning* and for *Canning killed himself* rests upon the convention that each proper name in our vocabulary has just one bearer. Hence, if the same proper name occurs more than once in a proposition, it is guaranteed that it has the same bearer for each occurrence. But this cannot be guaranteed by the sign itself, independently of any convention. The mere equiformity of two names cannot, as such, determine that they have the same bearer, and we know quite well that, from time to time, confusion arises in a general-purpose language through the use of a proper name that has more than one bearer. We can, indeed, resolve the confusion ad hoc by attaching different indexes to the name. Thus, there are two Peters in my department, so we often have to use their surnames as indexes: Peter Geach and Peter Long. But, in another context, we might discover that there were two Peter Geaches or two Peter Longs, and then we should need further indexes. Nor is there any point in the process of adding indexes at which it would be impossible for the name to have more than one bearer.

The same applies to network (3), considered as a propositional sign. We can always assign different bearers to each of the two occurrences of *C* and, if we do so, the meaning of the sign will not be the same as that of network (4), in which there is only one occurrence of *C*. The two networks, indeed, correspond to two different histories of the formula *Faa* in standard first order logic. We can obtain the formula either from $F\zeta\xi$, in which we may, but need not, replace both of the Greek letters by *a* (this corresponds to network (3)), or from $F\zeta\zeta$, in which we **must** replace both of the Greek letters by the same schematic symbol (this corresponds to network (4)). Now, in constructing a theory of inference, we shall need the logician's convention that each proper name has just one bearer, but in giving a theory of meaning, the two propositional signs above must be distinguished. Thus, there will be a valid inference in first order logic from (3) to (4) and conversely, but, still, they are not equivalent signs: They are deductively equivalent within a context which presupposes the logician's convention, but not semantically equivalent because (3) could always be given a meaning different from (4).

In first order logic, it is assumed that only extensional expressions are to be substituted for the first level signs. If we go beyond this to consider the meanings of propositions containing intentional expressions, there is a further argument for distinguishing between semantic structures for those in which proper names are repeated and for those which contain pronouns instead. An example with a reflexive pronoun occurs in a story related by William James (1890, I.):

> In half-stunned states self-consciousness may lapse. A friend writes me: "We were driving back from ______ in a wagonette. The door flew open and X, alias 'Baldy', fell out on the road. We pulled up at once, and then he said 'Did anybody fall out?' or 'Who fell out?'—I don't exactly remember the words. When told that Baldy fell out, he said, 'Did Baldy fall out? Poor Baldy!' " [pp. 273–274, fn.]

Having received an answer to his question, then, Baldy knew that Baldy had fallen out of the wagonette but, as his response shows, did not know that he himself had fallen out. Now, *1 knew that 2 had fallen out of the wagonette* is a first level schema of category 2PNN, because it will interrelate two proper names into a proposition, so network (3) should represent a semantic structure for *Baldy knew that Baldy fell out of the wagonette* and network (4) a semantic structure for *Baldy knew that he himself fell out of the wagonette*. In this case, although both occurrences of *Baldy* in the first proposition have the same bearer, there is no valid inference from the first proposition to the second, so *a fortiori* the two propositions cannot have the same meaning. Nor does it take much imagination to envisage the converse situation in which a person has forgotten his own name but knows something about himself, for example, that he has fallen out of a wagonette. There would then be no valid inference from *Baldy knows that he himself has fallen out of the wagonette* to *Baldy knows that Baldy has fallen out of the wagonette*. It is thus essential that we should be able to distinguish between the semantic structures of pairs of propositions of this type.

The inferences from (3) to (4) and from (4) to (3) are, however, all that we need to give a theory of inference for extensional propositions containing pronouns. Given a natural deduction system for first order logic, such as Gentzen's NK (1935), we only have to add an introduction rule and an elimination rule for the name connection or the *I*-operator. As most people will prefer to carry out proofs in a linear notation, I state the rules in terms of the latter:

$$\text{I+} \quad \frac{\alpha(a, a)}{I_{xy}\alpha(x, y)a} \qquad \text{I}- \quad \frac{I_{xy}\alpha(x, y)a}{\alpha(a, a)}$$

where $\alpha(a, a)$ is any formula in which there are at least two occurrences of a, and $\alpha(x, y)$ is obtained from $\alpha(a, a)$ by replacing one occurrence of a by x and another by y. In this system, the formation rule is to be observed that index letters of quantifiers may only occur in formulas in pairs, and not in triples, quadruples, etc.; moreover there is the usual assumption that each proper name which may be substituted for the schematic letters a, b, c, etc. has just one bearer. As the introduction and elimination rules for the *I*-operator are exact converses, it has no logical force; for this reason, it is quite properly omitted from the vocabulary of standard first order logic.

Representation of a structure for sentence (2) shows how the quantifier rules for NK block invalid inferences:

(5)

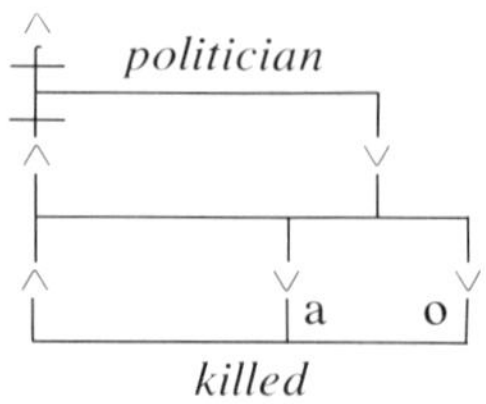

$\Sigma_x I_{yz} Fyzx$

The rule for eliminating the existential quantifier is:

$$\Sigma- \qquad \frac{\Sigma_x \alpha(x) \qquad \begin{matrix}[\alpha(a)] \\ \vdots \\ \beta\end{matrix}}{\beta}$$

where $\alpha(a)$ is any formula containing at least one occurrence of a and $\alpha(x)$ is obtained from $\alpha(a)$ by replacing just one occurrence of a by x. Proofs in NK are set out in the form of trees; at the top of each branch must be either a premise or an assumption of the proof. Some of the rules allow us to drop assumptions in the course of the proof. The square brackets around $\alpha(a)$ and the dots beneath it indicate that $\alpha(a)$ must occur at the top of at least one of the branches that lead to β. Although set out here as a schema, this rule is therefore really a disguised **thema,** that is, a rule which allows us to transform one **deduction** into another deduction, not a formula or formulas into another formula. We can bring this out by expressing it in the form:

$$\frac{\Gamma, \alpha(a) \quad \vdash \beta}{\Gamma, \Sigma_x \alpha(x) \vdash \beta}$$

That is, if β follows from $\alpha(a)$ and any further set of premises Γ, then β also follows from $\Sigma_x \alpha(x)$ and the same set of premises. In any system of proof, either all of the rules must be set out as schemas, or all of them as themas; but if we adopt the idea of having one introduction rule and one elimination rule for each operator, a mixture of schemas and themas is necessary, so either way some of the rules have to be dressed up to look like rules of the other type.

There are three restrictions on the use of $\Sigma-$: a must not occur

1. in $\Sigma_x \alpha(x)$,
2. in β,
3. in any premise or assumption upon which β depends, except $\alpha(a)$ (though it may occur in assumptions from which β is derived, provided that they have been dropped en route to β).

The rule for introducing the existential quantifier is simply

$$\Sigma+ \quad \frac{\alpha(a)}{\Sigma_x \alpha(x)}$$

with no restrictions. Hence, from *A politician killed himself,* we can validly infer *A politician killed a politician,* but not conversely. The proof of the former is:

$$\dfrac{\overset{1}{\Sigma_x I_{yz} Fyzx} \quad \dfrac{\dfrac{\dfrac{\overset{2}{I_{yz}Fyza}}{Faa}\ \text{I}-}{\Sigma_y Fay}\ \Sigma+}{\Sigma_x\Sigma_y Fxy}\ \Sigma+}{\Sigma_x\Sigma_y Fxy}\ \Sigma - 2$$

The numeral 2 written after the label of $\Sigma-$ on the last line indicates that assumption 2 is dropped at that point as permitted by $\Sigma-$; so the final conclusion depends upon premise 1 only. The three restrictions upon the use of $\Sigma-$ are all observed.

If we try to construct a proof of the converse inference, we find that it violates restriction 1:

$$\dfrac{\overset{1}{\Sigma_x\Sigma_y Fxy} \quad \dfrac{\overset{2}{\Sigma_y Fay} \quad \dfrac{\dfrac{\overset{3}{Faa}}{I_{yz}Fyza}\ \text{I}+}{\Sigma_x I_{yz} Fyzx}\ \Sigma+}{\Sigma_x I_{yz} Fyzx}\ \Sigma - 3^*}{\Sigma_x I_{yz} Fyzx}\ \Sigma - 2$$

The formula substituted for $\Sigma_x\alpha(x)$ in the first use of $\Sigma-$ contains a; in order to conform to the restriction, we should have to choose a new letter instead of a for the a of the rule, say b. But then assumption 3 would have been Fab, and we could not have used I+ below it.

The introduction and elimination rules for the universal quantifier are

$$\Pi+ \quad \frac{\alpha(a)}{\Pi_x\alpha(x)} \qquad \text{and} \qquad \Pi- \quad \frac{\Pi_x\alpha(x)}{\alpha(a)}$$

with the restriction on $\Pi+$ that a must not occur in $\Pi_x\alpha(x)$ nor in any premise or assumption on which $\Pi_x\alpha(x)$ depends. I leave it for the reader to check for himself that this allows us to infer validly from *Every politician loves every politician* to *Every politician loves himself,* but not conversely. With the rules for introducing and eliminating the *I*-operator, we have all that is needed for a theory of inference for pronouns which are of category 2PN2PNN, so I shall have nothing new to say on this matter subsequently. From time to time, however, it may help the reader if I mention the applications of the rules to more complex examples. It seems fair to conclude, then, that in principle, the theory of inference for these pronouns is trivial, in extensional contexts.

In intentional logic, of course, neither I+ nor I− will hold. But this applies equally to the rules for identity which can be added to first order (extensional) logic. Thus, for example, a rule

$$\frac{\alpha(a) \quad a = b}{\alpha(b)}$$

will not be valid if *John knows that Lewis Carroll wrote* ***Alice in Wonderland*** is substituted for $\alpha(a)$ and *Lewis Carroll* = *Charles Dodgson* for $a = b$. Intentional logic has not yet reached the stage where we can profitably consider what rules would be appropriate for identity and the *I*-operator. But, by distinguishing formulas of the type $\alpha(a, a)$ from those of the type $I_{xy}\alpha(x, y)a$, we are at least providing for the possibility of a theory of inference for intentional propositions containing pronouns, whereas their assimilation in standard first order logic leaves no room for subsequent maneuvering.

When a (multidimensional) semantic structure is projected onto a (linear) phonetic structure, two things happen. First, some of the **interrelationships** between the semantic constituents are hypostatized so that they become **constituents** of the phonetic structure, and, second, these, together with phonetic constituents corresponding to the semantic constituents, are arranged in a linear order. The projection is effected by seeing the sentence as organized in a different way, by looking at its structure from a different point of view. Projection is not computation. One can compare it with seeing a building from different positions, or with inspecting a building and then looking at the architect's drawings: first the plan, then the elevations, then the sections. Of course, the comparison is not exact: In architectural drawings, only one dimension is lost and some of the information from that dimension is omitted in each drawing taken separately; whereas, in linguistic projection, more than one dimension is lost, but all of the information is (normally) preserved by means of special conventions.

In specifying a method of projection from networks to English sentences, it must be remembered that the networks are already only two-dimensional and so must incorporate many of the conventions belonging to phonetic structures. They can only represent certain aspects of semantic structure, and can never be wholly "purified" from phonetic "contamination." So the method of projection, because it relates **representations** in the first instance, will only correspond to part of what is involved in the linear projection of a semantic structure. For expository purposes, it will be clearer to set out the method in stages, beginning with hypostatizations, just as we might teach someone classical perspective by giving him a series of steps to perform in making a perspective drawing. This makes projection appear to be a computation, although it is not; any more than seeing, for example, a landscape as a two-dimensional picture is a computation.

In devising the two-dimensional notation, I have done my best to find a way of representing first level schemas and name connections so that they can be seen either as patterns of pathways and, thus, as interrelationships, or as constituents. Network (4), viewed as a representation of a semantic structure, is to be regarded as displaying interrelationships. But we could also see the pattern of solid lines as constituting junction boxes connected to each other by the inputs and outputs linked by the arrows. This is the first step in projecting a network onto a linear representation of a phonetic structure. Thus, network (4) may be seen as having three constituents: the proper name, the name connection, and the first level schema.

Once we have determined the phonetic constituents, it remains to well-order them. I shall call this part of the method of projection **linear insertion,** because the clearest way I can find to describe it consists of inserting the hypostatizations of second level connections into schemas of first level. For the simplest case, in which we have a network consisting only of one first level connection and proper names, the method of linear insertion has already been described in the last section. The next case is where we have a network consisting of one first level connection and quantifying connections. Here, the hypostatizations of the latter, namely, quantifying expressions, are treated exactly like proper names for the purposes of linear insertion. Thus, phonetic structure does not distinguish between quantifying expressions and proper names and, indeed, it is a regular feature of phonetic structure that it assimilates whole groups of semantic categories into a single phonetic category.

At this stage, we must make a tentative foray into phonetic syntax. Traditional grammar and its more recent offshoots will doubtless have much to teach us here, but we cannot take any of it on trust, because our method is exactly the opposite of that of most grammarians. By and large, they have worked from phonetics to syntax and thence to semantics; we distinguish sharply between semantic syntax, on the one hand, and phonetic syntax, on the other, and work from semantics to phonetics. This, too, is why our foray into phonetic syntax must remain tentative; it is largely a waste of time to develop a phonetic syntax until one has worked out a satisfactory theory of meaning (at least for a certain area of language) because phonetic syntax, although it does not guarantee meaning, is nevertheless the servant of meaning.

I think it is fairly evident that among the most powerful determinants of word order in English are its verbs or, more generally, its first level schemas. If we call any first level schema a *Verb* (with a capital "V," to distinguish it from the narrower, more traditional sense of "verb"), and any expression which may, in a phonetic structure, replace one of its schematic numbers a *Term,* then the simplest English phonetic structures will be:

T_1 V
T_1 V T_2
T_1 V T_2 T_3

For simplicity, I am omitting from these representations the schematic symbols for which inflections (case and voice) must be substituted; strictly speaking, the second formula in the series above should be:

i-T_1 k-V j-T_2

It will not be possible, initially, to define *Term* and, for the purposes of this paper, it will be necessary to proceed by piecemeal definition, unsatisfactory as that is. So far, then, any proper name is a Term and so is any quantifying expression, that is, any hypostatization of a quantifying connection.

This is enough to enable us to consider the phonetic projection of a network containing quantifying connections: The hypostatizations are written beside the networks.

(6a)

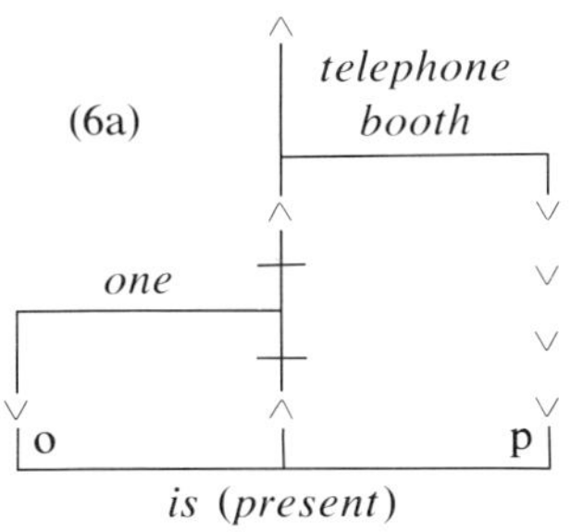

HYP:
every telephone booth

someone

← o-1 op-*is* p-2
p-1 po-*is* o-2 →

(6b)

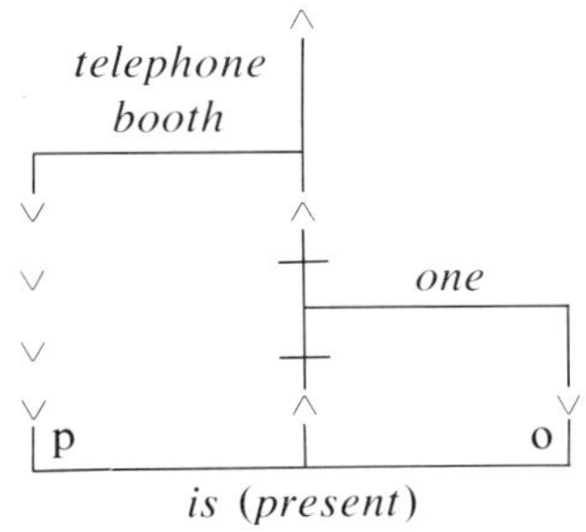

These networks will be projected onto the second of the English phonetic structures in the series above, and we can set out the linear insertion procedure for network (6a) as follows:

(P1)	T_1	V	T_2
1.	o-1	op-*is* (*present*)	p-2
2.	o-*someone*	op-*is* (*present*)	p-(*every telephone booth*)

(7) *Someone* *is* (*present*) *in every telephone booth.*

It is to be observed that each quantifying expression is inserted into the position of the schema corresponding to the name input of the first level connection that is joined to the name output of the quantifying connection from which that quantifying expression is derived. Thus linear insertion is guided by the **name pathways** of the network.

In effecting the linear insertion, the ordering of the quantifying connections in the vertical dimension of the network—their scope ordering—has been lost. This is not surprising because two dimensions have been assimilated into one, with the horizontal dimension of the network taking precedence. In this example, however, it does not matter. No one is going to suppose that the linear ordering of the quantifying expressions in the phonetic structure is the same as their scope ordering, and that the sentence means that there is at least one person such that he or she is present in every telephone booth, because we know that one person can only be in one place at a time and assume that the speaker would not have said *every telephone booth* if, in fact, there were only one booth. Other features of the meaning of the sentence thus compensate for the loss of information resulting from a linearization in which no distinction is drawn between proper names and quantifying expressions.

We are not always so fortunate. The result is then either that the sentence is ambiguous, or that we construe it upon the assumption that the linear ordering of the quantifying expressions is the same as their scope ordering. If it were really important to be exact, we should probably try to preserve the scope ordering. This can usually be done by exploiting the voices of the Verb, which allow us to alter the linear positions of the Terms. Thus, the positions of the name inputs in the horizontal dimension have been reversed in network (6b). Bearing in mind what was said earlier about the loss of a dimension in representing first level connections, networks (6a) and (6b) are representations of the same semantic structure. If either network were a mobile, suspended from the final propositional output, the other would simply represent what we should see by viewing it from the other side (assuming that the words attached to it were turned around). So a phonetic projection from network (6b) will be a semantically equivalent alternative to the projection from (6a), that is, a paraphrase of that part of the meaning represented by either network.

Network (6b) will project as follows:

(P1)	T_1	V	T_2
1.	p-1	po-*is* (*present*)	o-2
2.	p-(*every telephone booth*)	po-*is* (*present*)	o-*someone*
(8)	*In every telephone booth*	*is* (*present*)	*someone*.

This is awkward in English, because *is* (*present*) does not have a proper passive voice. We can improve it by saying *In every telephone booth, someone is present* instead, but most people would probably prefer the projection from network (6a). Consequently, there is a certain tension between phonetic and semantic considerations. Certain constructions will be preferable to others if it is paramount to convey our meaning exactly, but the order of preference will be changed if style and euphony are especially important. I do not suppose that one would have much difficulty in finding counterexamples to almost every rule of phonetic structure which has been proposed by grammarians if one were allowed to quarry poetry. This means that we cannot say that some phonetic projections are perfectly legitimate and others totally illegitimate, without further qualification. In between, there will often be projections that will be preferable to others according to the point of view, semantic or phonetic. To give a full account of the interplay between semantic and phonetic structure would be an enormously complex task. "Everyday speech is a part of the human organism and is no less complicated than it" (Wittgenstein, 1922, 4.002). At the beginning we must simplify; so I shall assume here that the main consideration is to express our meaning exactly and that no help is available either from the context of an example or from any feature of its meaning that is not represented in the network under consideration.

A name connection is hypostatized as a pronoun, and a pronoun is also a Term. Pronouns take many different phonetic forms in English, so I shall use PRO for this purpose, in the first instance. Where both name pathways leading from the name connection go to the same first level connection, the pronoun is always reflexive. We can indicate this in the hypostatization by the prefix R-. The gender of the pronoun is determined by the input to the name connection. Where this input is a proper name, the pronoun has the same gender as the proper name; where it is a quantifying connection, it has the same gender as that of the count noun written over the horizontal stroke of the connection. We can indicate the gender in the hypostatization by one of the prefixes m- (masculine), f- (feminine) or n- (neuter). Thus, for networks (9) and (10), we have the following hypostatizations:

(9)

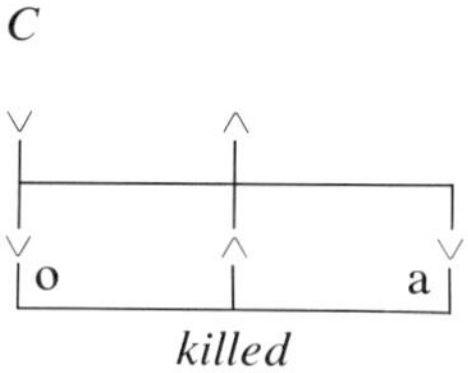

HYP:
(*Canning*)

mR-PRO

o-1 oa-*killed* a-2

(10)

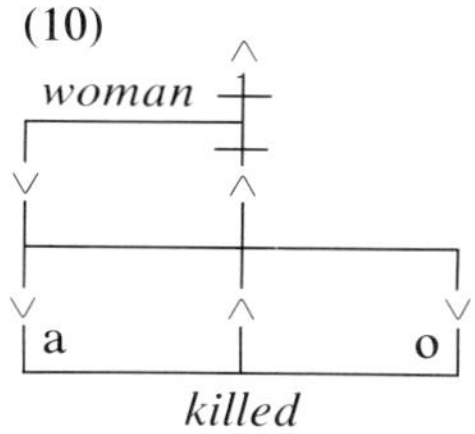

HYP:
some woman

fR-PRO

a-1 ao-*killed* o-2

The basic linear insertion rule for pronouns is that they are inserted into the linear position of the first level schema which corresponds to the name input joined to the **rightmost** of the two name outputs from the name connection. The proper name or quantifying connection is inserted into the position corresponding to the name input joined to the **leftmost** of the two name outputs from the name connection to which it is the name input. This is the preferred method of projection, but there are exceptions to it, though not for reflexive pronouns.

The projection of network (9) goes as follows:

(P1)	T_1	V	T_2
1.	o-1	oa-*killed*	a-2
2.	o-*Canning*	oa-*killed*	amR-PRO
(11)	*Canning*	*was killed*	*by himself.*

We can assume that the phonetic form of the pronoun is determined from the prefix of the hypostatization by consulting tables drawn up for the purpose. Some linguists might still object to a sentence like *Canning was killed by himself,* but I think that the ground for their objection is no more than our general preference for the active voice of a verb when no special purpose is served by the passive. One can quite well imagine, though, that if someone were asked, "By whom was Canning killed?" he might reply, "I'm afraid that Canning was killed by himself" (cf. the idiom: *died by his own hand*). From network (10), we obtain:

(P1)	T_1	V	T_2
1.	a-T_1	ao-*killed*	o-T_2
2.	a-(*some woman*)	ao-*killed*	ofR-PRO
(12)	*Some woman*	*killed*	*herself.*

4. PERSONAL PRONOUNS

There is, in my view, no difference in meaning between personal and reflexive pronouns. Both have essentially the same function, but a name connection is phonetically projected as a personal pronoun in some circumstances and as a reflexive pronoun in others. The relevant circumstances are these: If a propositional connection occurs within the scope of a name connection, it is projected as a personal pronoun, but otherwise as a reflexive pronoun. It will be necessary to qualify this slightly later, but the complications are best hidden until some familiarity has been gained with the normal application of the rule. The utility of having two different phonetic forms is that it enables us to "recover" the right name connections in sentences containing several pronouns where, if we had to rely solely on the clues afforded by gender, more than one possibility would remain open. We have here an instance of the typical difference between phonetic and semantic structures: Phonetic structures, lacking the multiplicity of semantic structures, compensate by introducing a much wider range of constituents, and the differences between them are used to represent differences of semantic interrelationships among the other constituents.

We can represent the hypostatizations of name connections as personal pronouns by the same method as their hypostatizations as reflexive pronouns, with P- as a prefix instead of R-. The gender and case of a personal pronoun is determined in exactly the same way as that of a reflexive pronoun. A simple example of a network whose phonetic projection will call for a personal pronoun is:

(13)

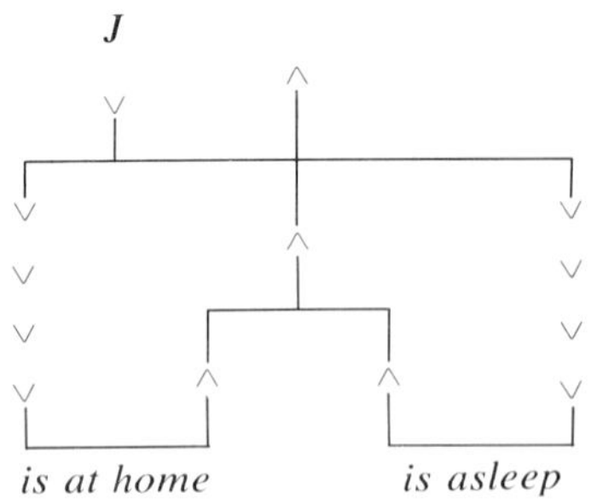

HYP: $I_{xy}KFxGya$

mP-PRO

both 1 and 2

3 is at home, 4 is asleep

English offers us two alternatives of linear insertion for networks of this type; we can either organize the phonetic structure around the propositional connective, or around one of the Verbs. The next section will be devoted to the second alternative; in this section, we shall explore the first.

The hypostatization of a propositional connection is best regarded as having two parts, after each of which an insertion has to be made, for example, **both 1 and 2** for conjunction. The first part has the role of an opening bracket and can be used, in cases where ambiguity might otherwise arise, to tell us how far back in the sentence the scope of the connective extends. This is necessary because the other part of the connective is written **between** the remaining constituents of the sentence, each insertion into it consisting of a Verb together with its associated Terms. Thus phonetic structures, too, make use of scope; but scope in a phonetic structure does not always coincide with scope in a semantic structure, so I shall call it **phonetic scope** in order to distinguish it. Sometimes it runs from left to right, but also sometimes from right to left, as here; it can usually be indicated by bracketing, or by a tree, though not if left-to-right scope becomes too intertwined with right-to-left scope. In representations of phonetic structures, I shall use subscripts where necessary, since these are more versatile than brackets. The two expressions inserted into a propositional connective are both within its phonetic scope, and each Term of a Verb falls within the Verb's phonetic scope. Phonetic scope determines the stages by which a phonetic structure is built up, and so can be used to guide linear insertion.

As a schematic symbol for which any propositional connective may be substituted, I shall use $C_1 1 C_2 2$. The first half of a propositional connective is usually omitted, only receiving a phonetic instantiation when necessary, for example when it acts as an opening bracket to indicate scope, or a phonetic function like style or euphony. If it is omitted, I shall regard the omission as equivalent to a null substitution for C_1, indicated by the sign $\varnothing$, in the hypostatization of the propositional connection. In English, it is often placed immediately before the main part of the Verb of the first insertion, so that it comes after the Verb's first Term. The same treatment is also accorded to *not*. I shall ignore this complication here; since the Terms of a Verb lie within its phonetic scope, it is evident that, for example *both* can still fulfill its bracketing function even if it occurs after the first Term, provided that its relation to the Verb remains clear.

In order to show which Terms lie within the phonetic scope of each Verb in a structure containing more than one, I shall add double subscripts to the schematic symbols for which Terms are substituted, the first number being that of the Verb within whose scope the Term lies, and the second indicating whether it is the first, second, third, etc., Term of that Verb. Thus a phonetic structure onto which network (13) can be projected may be represented as:

(P2) $\quad C_1 \quad T_{11} \qquad V_1 \qquad C_2 \quad T_{21} \qquad V_2$

Taking the hypostatization of the propositional connection with C_1 null, the first step in the linear insertion is to substitute it in (P2):

1. ∅ T_{11} V_1 *and* T_{21} V_2

The hypostatizations of the two first level connections are inserted next:

2. ∅ T_{11} *is at home* *and* T_{21} *is asleep*

This yields a schema containing two schematic symbols and, hence, of category 2PNN, exactly the same as that of *1 killed 2*, used in the examples of the last section. Wherever a propositional connection is used to combine two first level connections, the first having *m* name inputs and the second having *n* name inputs, the result will be a first level connection having $m + n$ name inputs; and similarly for schemas. This is quite clearly shown in the two-dimensional notation, for example, by joining the propositional connection in network (13) to the two first level connections without any intervening arrowheads:

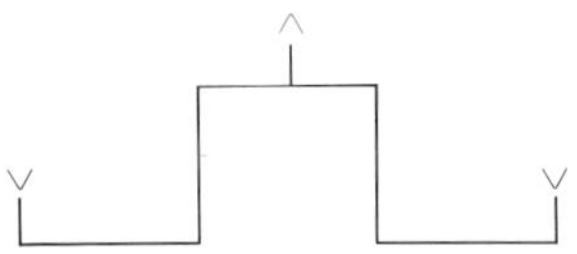

is externally equivalent to:

The final insertions can then be effected in exactly the same way as with reflexive pronouns:

3. ∅ *John* *is at home* *and* mP-PRO *is asleep*

which, by eliminating the null sign and looking up mP-PRO in the tables, yields:

(14) *John* *is at home* *and* *he* *is asleep.*

We could omit the pronoun from sentence (14) and also the second occurrence of *is*, but these two omissions are not, I think, on a par. The second, to extend a phrase of Geach's, is an omission of laziness, to avoid repetition. The pronoun, however, is not a repetition of anything, and its omission is equivalent to a null hypostatization of the name connection. It can be omitted from sentence (14) because we know from the meaning of *is asleep* that it is a Verb with which one Term is associated. If the Term is a pronoun, we look for another Term with which it establishes a connection, having due regard to the form of the pronoun; but here, there is only one candidate for the second Term, so the pronoun can safely be omitted.

Network (13) is symmetrical about its vertical axis (the position of *J* in the horizontal dimension is arbitrary) so that, seen from the other side, it would look exactly the same except that the order of the two first level connections would be reversed. The back view of the network can thus be phonetically projected in exactly the same way as the front view, but because the order of the two first level connections is reversed, it will yield *John is asleep and (he) (is) at home*. This would also apply if the propositional connection in network (13) had been disjunction instead of conjunction, and the symmetricality of both signs in our notation represents the property of commutativity. An implication, by contrast, has a **direction:** We have to distinguish, in order to explain its meaning, between its antecedent and its consequent. Accordingly, our sign for implication is asymmetrical.

If we were thinking of meaning, it would be natural, when expressing an implication in linear form, to put the antecedent first, as in the logician's schema *if p, then q*. But from the phonetic point of view, the arrangement with the consequent first, *1 if 2,* is the analogue of *both 1 and 2* and *either 1 or 2*. *Then,* like *both* and *either,* plays the role of a bracket, though here the closing bracket of the antecedent, so that it is always omitted from the transposed version. Thus, in order to obtain a network with implication which is as close as possible to network (13), we shall have to rotate the sign for implication (given in Section 2) on its vertical axis. As it is not a symmetrical sign, there can be no danger of confusion in doing this. Otherwise keeping to the same example, we then obtain:

(15)

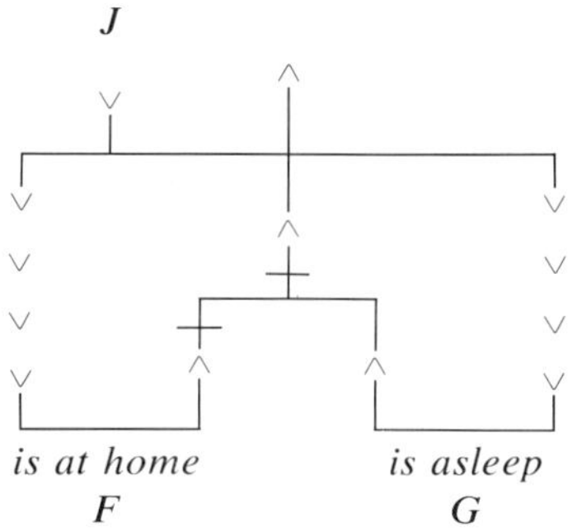

HYP: $I_{xy}CGxFya$

mP-PRO

∅ *1 if 2*

3 is at home, 4 is asleep

The linear insertion procedure from this network is then exactly the same as that from network (13), and will yield:

(16) *John is at home if he is asleep.*

Moreover, we can also omit the pronoun if we also omit the repeated *is*. *John is at home if asleep* may not be in the best style, but there are plenty of occasions when we do omit the pronoun quite naturally from sentences of this phonetic type, so the fault, if any, lies in the particular example, perhaps in its simplicity.

We can also have

(17) *If John is asleep, then he is at home.*

from which the pronoun may not be omitted. There are two ways in which we could regard this as being projected from the network. If it were projected from the back view of the network, then the two first level connections would already be in the correct order in the horizontal dimension and the linear insertion would be exactly the same as with conjunction above.

But in the hypostatization of the propositional connection, *if* would be substituted for C_1 and (*then*) for C_2. Alternatively, we could project from the front view of network (15), but introduce a transposition into the linearization procedure. The second alternative sounds more complicated than the first, but it will pay off later to keep the substitution into $C_1 1 C_2 2$ regular, that is, always substituting the part of the connective which has the bracketing function for C_1. There is also theoretical support for this decision: The effect of rotating a network containing a conjunction or disjunction is that of commutation, that is, of moving from *p and q* to *q and p*; whereas *if p, then q* is not a commutation of *q if p* but a reversal of the direction of the implication, accompanied by a reversal of the order of its arguments, that is, a transposition. The stage in the projection at which the transposition occurs is, of course, only a matter of expository convenience. We could equally regard sentence (17) as a phonetic projection of the back view of network (15), but with the hypostatization of the propositional connection being a substitution in the phonetic schema $C_2 2 C_1 1$, which would be a transposition of (and distinct from) $C_1 1 C_2 2$. It is not to be supposed, therefore, that the transpositions introduced in the course of linear insertion have any psychological analogues.

I shall say, then, that a network containing a sign of implication has an English phonetic projection only if the negated input to the sign occurs to the left of the unnegated part. If the network is to be projected without transposition, then the propositional connection will be hypostatized as ∅ *1 if 2*, but if it is projected with transposition, as *then 1 if 2;* the latter hypostatization will then act as a trigger for subsequent transposition. The transposition consists in reversing the linear order of whatever is substituted for 1, together with the *then* preceding it, and whatever is substituted for *2*, together with the *if* preceding it. A null substitution for the pronoun is excluded by *then 1 if 2;* apart from this, the linear insertion is exactly the same as before, the transposition being effected either immediately before or immediately after the insertion of the Terms. If the former option is taken, we get:

(P2)	C_1	T_1	V_1	C_2	T_2	V_2
1.	*then*	T_1	V_1	*if*	T_2	V_2
2.	*then*	T_1	*is at home*	*if*	T_2	*is asleep*
Trs.	C_2	T_2	V_2	C_1	T_1	V_1
3.	*if*	T_2	*is asleep*	*then*	T_1	*is at home*
4.	*if*	*John*	*is asleep*	*then*	mP-PRO	*is at home*
(17)	*If*	*John*	*is asleep*	*then*	*he*	*is at home.*

If the transposition is delayed until after step (3), the result will be: *If he is asleep, then John is at home*. Although sentence (17) is usually preferable to this, we do sometimes allow a pronoun to occur in the antecedent of an implication when the proper name or quantifying expression which it connects is in the consequent. This cannot occur with conjunction or disjunction, another piece of evidence that we have a transposition and not a commutation here, even at the phonetic level. When we hear a pronoun in the antecedent of an implication (which has already been signalled by the *if*; there is no comparable signal with conjunction and disjunction, because their commutativity renders a distinction between antecedent and consequent unnecessary), we wait for the associated proper name or quantifying expression just as one waits for the verb at the end of a German sentence.

The following network sheds further light upon the circumstances in which a name connection may receive a null hypostatization:

(18)

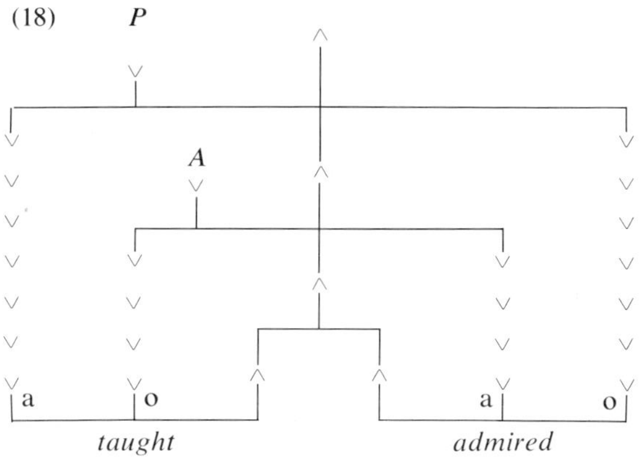

HYP: $I_{xy}I_{zw}KFxzGwyba$

mP-PRO$_1$

mP-PRO$_2$

∅ *1 and 2*

a-3 ao-*taught* o-4, a-5 ao-*admired* o-6

The standard hypostatizations are set out beside the network; if we use these and follow the normal insertion procedure, the result, taking P = *Plato* and A = *Aristotle*, will be *Plato taught Aristotle and he admired him*. Since the two pronouns are of the same gender, we cannot tell from their form which one establishes a connection with which proper name, as we could with *Plato taught Sappho and she admired him*. So we assume that the pronoun which is the first Term of the second Verb effects a connection with the first Term of the first Verb, and that the pronoun which is the second Term of the second Verb effects a connection with the second Term of the first Verb. Unfortunately, these are not the connections shown in the network, so the linear insertion procedure will have to allow for such supplementary conventions when more than one pronoun is of the same gender. In this case, the requisite pattern can be preserved by linearizing the second first level connection in the passive voice: *Plato taught Aristotle and he was admired by him*. Better still, we can also omit the first pronoun: *Plato taught Aristotle and was admired by him*. This is parallel to the omission of the pronoun from sentence (14), except that there is an additional motive for the omission here, to avoid having two pronouns of the same gender in the sentence. Given the convention that, in case of doubt, the *n*th Term of Verb 2 is to be assumed connected to the *n*th Term of Verb 1, this motive is, however, phonetic rather than semantic.

Notice that although the pronoun substituted for the first Term of Verb 2 may be omitted, that substituted for its second Term may not. In general, it seems that no pronoun following the second Verb may be omitted, except, perhaps, as part of an omission of laziness in which some or all of the Verb is omitted too. But the most elegant way of expressing the thought represented by the network is *Plato taught and was admired by Aristotle*. This is an exception to the standard insertion rule, which only applies when the name connection to which a proper name or quantifying connection is input is connected to the last name input (in the linearization) of the last first level connection. It illustrates the general rule that phonetic representations are to be kept as simple as possible, and the reason for this rule is that the more complicated they become, the greater the likelihood that the conventions which allow us to grasp their semantic structures will break down. "So-and-so has a very involved style", we say, "and it is difficult to follow the train of his thoughts." The use of pronouns, accordingly, is to be minimized, so name connections receive a null hypostatization wherever possible. If, then, a name connection which is to receive a null hypostatization is connected to what is to become the last Term of the phonetic structure, and if the last Term can never be omitted, the standard insertion rule must yield.

One reason we prefer *Plato taught and was admired by Aristotle* to the other alternatives is that it appears, at first sight, to have the phonetic structure T_1VT_2, with a complex Verb, and that is a much simpler structure than $C_1T_{11}V_1T_{12}C_2T_{21}V_2T_{22}$. But this cannot be, for each of the conjoined Verbs has two associated Terms, and two plus two make four; moreover, how could we then explain the possibility of the alternatives containing pronouns?

The phonetic projection of networks containing quantifying connections brings to light still further conventions to compensate for the loss of structural complexity. If we replace the proper name in network (15) by a universal quantifying connection and, for the sake of a little variety, change the two first level connections, we get:

(19)

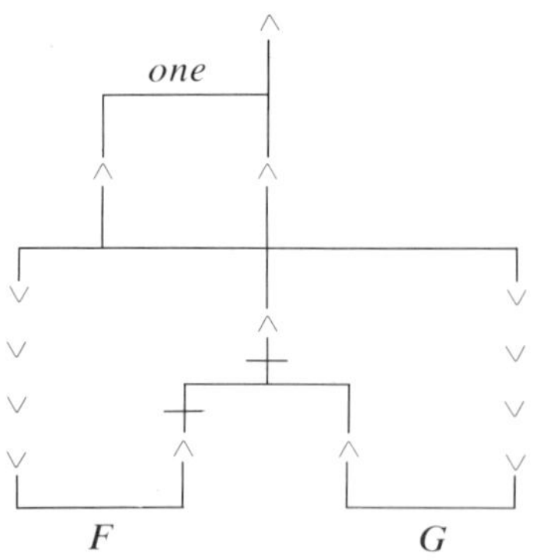

HYP: $\Pi_x I_{yz} CGyFzx$

everyone

mP-PRO

then 1 if 2

3 will get a first, 4 works hard

In the hypostatization of the propositional connection, which is implication,

C_1 is not null, so a transposition is signaled. Transposing before inserting the Terms, we shall obtain **If everyone works hard, then he will get a first.* This is not right, and the reason is obvious: At the point at which the Terms are inserted, *everyone* is brought within the phonetic scope of *if*, whereas, in the network, the propositional connection is within the semantic scope of the quantifying connection.

If the semantic scope were the same as the phonetic scope, we could not construct a network, because the antecedent of the implication would be the proposition *everyone will get a first* and this would exclude the possibility of any subsequent name connection:

(*19) $*I_{yz}C\Pi_x GxFy\zeta$

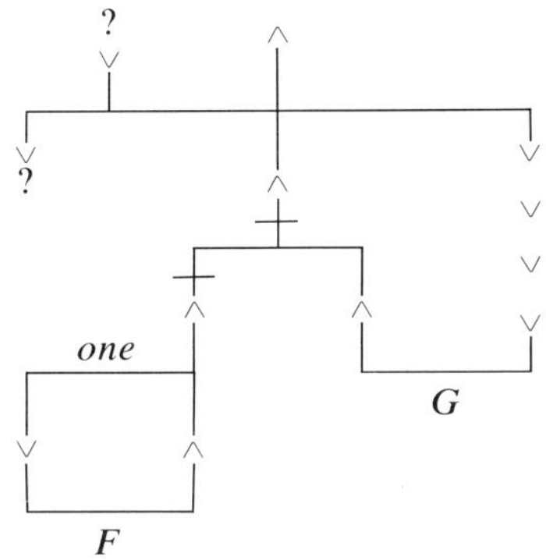

There is no name input to which we can join the first name output from the name connection, nor any proper name or quantifying connection to serve as input to the name connection; so the putative network cannot be completed. Similarly, there is no place in the linear formula for a second z, nor any filling for the argument place marked by ζ.

English offers us three solutions to this difficulty. The first is not to transpose the connective, so that we get *Everyone will get a first if he works hard.* The second is to project onto a phonetic structure in which the phonetic scope is the same as the semantic scope *Everyone, if he works hard, will get a first.* Discussion of the phonetic structure involved here will be deferred to the next section. Finally, there is a phonetic device available: We may retain the original transposed projection if we hypostatize the quantifying connection as *anyone* instead of *everyone: If anyone works hard, he will get a first.* Complication of the phonetic structure is thus avoided by a different hypostatization of the quantifying connection, which acts as a signal that the phonetic scope of *If anyone* . . . is the reverse of its semantic scope. This is not, of course, a comprehensive account of the difference between *any* and *every,* but only of the combination *if any* . . . . It is very important to observe that a single phonetic device may often serve several *different* semantic purposes.

There are comparable problems with existential quantification when phonetic scope gets out of step with semantic scope. We can investigate these with the aid of the following networks:

(20a)

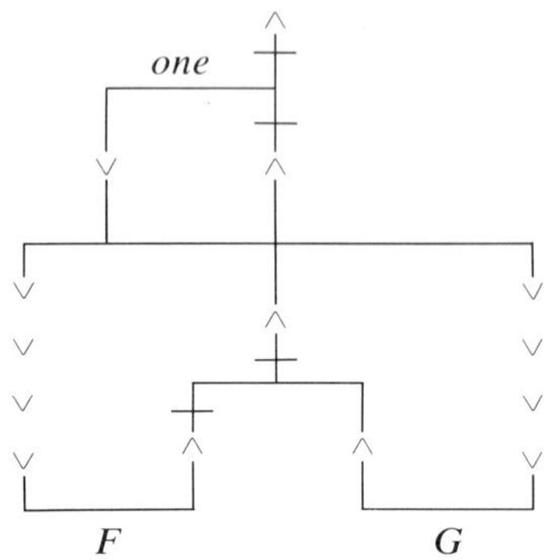

$\Sigma_x I_{yz} CGyFzx$

(20b)

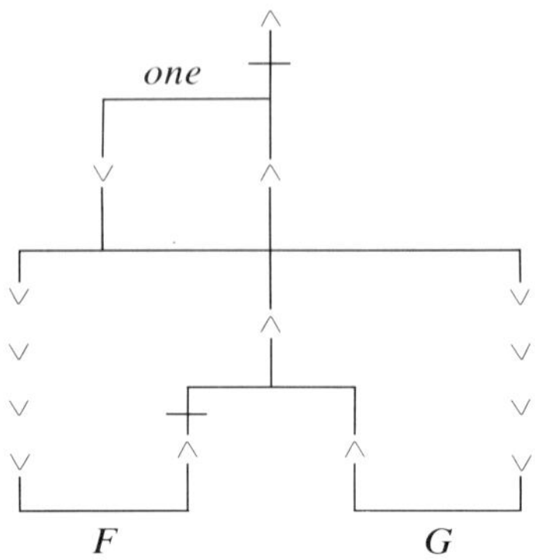

$N\Pi_x I_{yz} KGyNFx$

These networks are equivalent, given the classical notion of negation. Two *nots* cancel out when one lies within the **immediate** scope of the other (that is, where no horizontal stroke or intermediate sign comes between them). So far as this rule is concerned, intermediate name connections may also be ignored, so two of the negation strokes in network (20a) can be cancelled. It is a defect of the two-dimensional, as of the linear notation, that these alternative representations of the same semantic structure are available within it. We could eliminate alternatives by only allowing a negation stroke to occur **above** the horizontal stroke of a connection, but, while theoretically more satisfactory, that would make the projection procedure more complex.

The uppermost negation stroke, however, must remain in both networks; so, in the projection of either network onto phonetic structure (P2), it will be brought within the phonetic scope of the propositional connection. Now, there is a difference between the truth conditions of *not: p if q,* in which the whole implication is negated, and of *not-p if q,* in which only the consequent is negated. The former is true just in case *p* is false and *q* true, the latter false just in case both *p* and *q* are true (it means the same as *not both p and q*). So we can expect trouble from this projection.

The meaning of both networks is that there is at least one person who will get a first if he works hard. Projection of network (20a) without transposition yields *Someone will get a first if he works hard,* which is acceptable. With transposition, however, the result is *If someone works hard, then he will get a first,* which it is most natural to understand as meaning the same as *If **anyone** works hard, then he will get a first* (cf. *If someone were to say . . . , If someone would only help me . . .*). We can still fall back on the alternative construction and say *Someone, if he works hard, will get a first* but, this time, there is no phonetic device available by way of an alternative hypostatization of the quantifying connection. The indefinite article, which is often interchangeable with *some,* fares no better: *If a man works hard, then he will* *get a first;* for here it is interchangeable with *any* and not with *some*. The reason why the phonetic device is available in the one case but not in the other is, presumably, that the combination of universal quantification with implication is much more common than the combination of existential quantification with implication; existential quantifiers are more frequently followed by conjunction.

In hypostatizing the propositional connection of network (20b), we have to use conjunction because there is no longer a negation stroke on the upper part of the sign. Projection will then yield:

(P2)	C_1	T_1	V_1	C_2	T_2	V_2
	both not	*not everyone*	*will get a first*	*and*	∅	*will work hard*

The second *not* is now within the immediate phonetic scope of the first, although neither negation stroke is within the immediate semantic scope of the other in the network, so that, if we cancel them out here, we shall certainly change the meaning. I remarked earlier, however, that the bracketing part of a connective, as well as *not,* can (and often must) be placed immediately before the main part of the Verb in English. The semantic purpose which this serves comes out very clearly in the present example: It keeps the phonetic scope in line with the semantic scope, while still allowing the Term–Verb arrangement when the Term is a quantifying expression: *Not everyone will both not get a first and will work hard.* The form of expression may be clumsy because, in trying to keep phonetic structures as simple as possible, we avoid multiplying occurrences of *not*. Yet the meaning is clear and is that of the network. This device is not available when the connective is transposed, but transposition only occurs with implication; so we shall have to require that when an implication lies within the immediate scope (ignoring intermediate name connections) of a negated quantifying connection, the former must be hypostatized as *∅ 1 if 2,* so that transposition is blocked.

No abbreviates *not some* and there are, consequently, comparable problems when projection brings *no* within the phonetic scope of a propositional connective which, in the semantic structure, lies within the scope of the quantifying connection. I do not have the space to discuss these here, but there is a temptation which is especially strong when faced with unacceptable expressions beginning with *no* and which should be resisted. In order to explain why an example like *"If you passed no one on the road, then he walks slower than you" is unacceptable (cf. Carroll, 1872, Ch. 7), we are inclined to say that, *ex hypothesi,* there is no one for the *he* to refer back to. But, if that explanation were correct, a pronoun could *never* effect a connection with a quantifying expression beginning with *no*. Yet there are plenty of counterexamples to that suggestion, for example *No one will get a first unless he works hard*. The correct explanation lies, rather, in the conflict between semantic and phonetic scope.

It has been necessary to concentrate initially on some rather simple networks in order to clarify the method of projecting them onto phonetic structures. Actually, the networks never become very complex in practice; if they did so, it would probably be psychologically impossible for us to take in at a single hearing the meanings of the sentences onto which they could be projected. Here, now, is a network which may be compared with network (13). The two first level connections of network (13) are each replaced by a combination of a quantifying connection with a first level connection having two name inputs, and the proper name by a third quantifying connection:

(21)

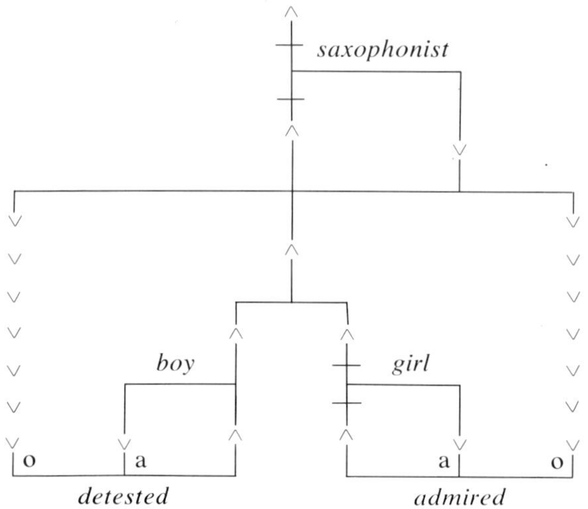

HYP:
some saxophonist

∅

∅ *1 and 2*

every boy, some girl

a-3 ao-*detested* o-4,
a-5 ao-*admired* o-6

$$\Sigma_x I_{yz} K \Pi_v F v y \Sigma_w G w z x$$

We can project this onto the phonetic structure:

(P3) $C_1 \quad T_{11} \quad V_1 \quad T_{12} \quad C_2 \quad T_{21} \quad V_2 \quad T_{22}$

the null hypostatization of the name connection overriding the normal insertion rule, as in the projection of network (18):

1.	∅	T_{11}	V_1	T_{12}	*and*	T_{21}	V_2	T_{22}
2.	∅	a-T_{11}	ao-*detested*	o-T_{12}	*and*	a-T_{21}	ao-*admired*	o-T_{22}
3.	∅	a-(*every boy*)	ao-*detested*	o-∅	*and*	a-(*some girl*)	ao-*admired*	o-(*some saxophonist*)
4.		*Every boy*	*detested*		*and*	*some girl*	*admired*	*some saxophonist*

If we assume that the context presupposes restricted groups of boys, girls, and saxophonists, we could hypostatize the quantifying connections in an alternative form to yield:

(22) *All of the boys detested, but some of the girls admired, one of the saxophonists.*

There is, perhaps, a slight change of meaning in that *some of the girls* suggests at least two, whereas our network says only *at least one*. *∅ 1 but 2* is an alternative hypostatization of conjunction; the exact circumstances which make *but* preferable to *and* turn out to be extraordinarily difficult to specify (cf. Dummett, 1973, pp. 2–3, 84–88). More interestingly, neither of the first two quantifying expressions is within the phonetic scope of the third, and we **could** interpret (22) in a sense that did not require that it was the **same** saxophonist which every boy detested and some of the girls admired. So, if the exact meaning were important, it would be preferable to passivize both of the first level schemas and obtain *One of the saxophonists was detested by all of the boys but ∅ was admired by some of the girls.*

Before leaving the topic of personal pronouns, we ought to look at an example that contains both a personal and a reflexive pronoun, especially as it is easy to think of apparent counterexamples to the rule for distinguishing the two types of pronoun proposed at the beginning of this section. In representing a semantic structure for a sentence like

(23) *If John is bored, then he ought to find some way of amusing himself.*

there must be a name connection within whose scope a propositional connection falls. How do we get the reflexive pronoun at the end of the sentence? Consider the following network:

(24)

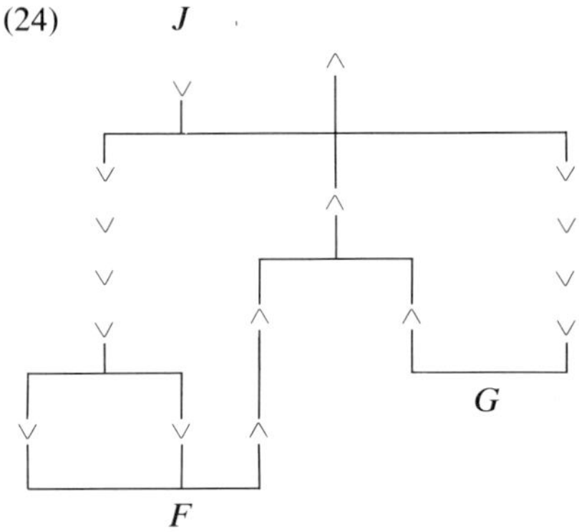

HYP: $I_{xy}CFxI_{vw}Gvwya$

mP-PRO

then 1 if 2

mR-PRO, *3 is bored*

4 ought to find some way of amusing 5

As there are two pronouns in (23), we need two name connections and, as the network shows, it is possible to place one of these so that no propositional connection falls within its scope, with the result that it is hypostatized as a reflexive pronoun. All the same, it is also possible to construct a network for (23) in which the propositional connective lies within the scope of *both* name connections:

(25) $I_{xy}I_{vw}GFxGvwya$

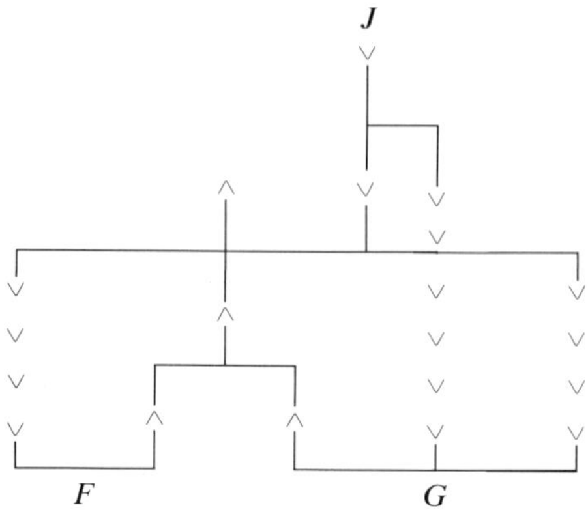

This is an analogue, for the *I*-operator, of prenex normal form for quantifiers, so that network (25) is an equivalent alternative to network (24). The rule for determining whether a name connection is to be hypostatized as a reflexive or as a personal pronoun must therefore be applied only to networks in which each name connection is given the **minimum** scope.

The question may be raised, here, whether name connections should always be represented as **binary.** The right answer to this, I think, is that, since name connections are not constituents of semantic structures, there is no distinction between, for example, a trinary connection and two binary connections of which one output of one is joined to the input of the other. If we eliminate the intervening arrowheads, this combination of two binary connections amounts, considering its inputs and outputs, to a trinary connection:

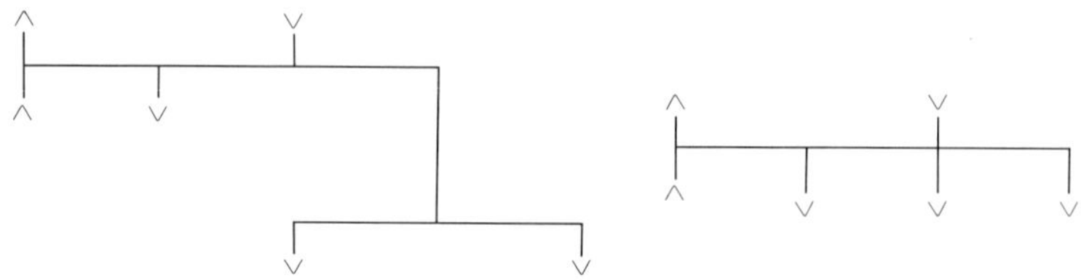

As regards the **representation** of name connections, however, it is more convenient to limit ourselves to signs for binary connections; otherwise, the hypostatization of name connections would be a much more complicated procedure. By limiting ourselves to binary signs, we shall have one such sign for each pronoun that is needed. This means, of course, that our representations already include certain features which are phonetic rather than semantic in character and, by the same token, that these representations allow extra equivalent alternatives. Thus, given trinary signs whose use was obligatory whenever it was possible, instead of networks (24) and (25) we should have only one network, obtainable from (25) by combining its two name connections into a single sign. Another disadvantage of trinary signs is that they would make the hypostatization rule much more complicated. We pay a price for the convenience of having only binary connections in an increased degree of phonetic "contamination" of our representations of semantic structures, but, as there is no way of avoiding concessions to phonetics altogether, in any method of representation, we may as well settle for what is most suited to the purpose at hand. But we should be careful not to attribute to semantic structures every feature of our representations of them, which always rest upon a compromise.

In order to obtain sentence (23), network (24) will have to be projected onto the phonetic structure:

(P4)	C_1	T_{11}	V_1	T_{12}	C_2	T_{21}	V_2	
1.	*then*	T_{11}	V_1	T_{12}	*if*	T_{21}	V_2	
2.	*then*	T_{11}	*ought to find . . .*	T_{12}	*if*	T_{21}	*is bored*	
Trs.	C_2	T_{21}	V_2		C_1	T_{11}	V_1	T_{12}
3.	*if*	T_{21}	*is bored*		*then*	T_{11}	*ought to find . . .*	T_{12}

For the first time in the examples discussed so far, the *order* in which the Terms are inserted becomes important. In order to get them right, we must work upward from the bottom of the network, taking the name connection with the least scope first, and not inserting proper names or quantifying expressions until we come to them. In the present case, the hypostatization of the name connection of least scope is mR-PRO, so that is inserted first:

4. *if* T_{21} *is bored then* T_{11} *ought to find some way of amusing* mR-PRO

The other name connection comes next, and, as T_{12} has already been replaced, this only leaves T_{11}:

5. *if* T_{21} *is bored then* mP-PRO *ought to find some way of amusing* mR-PRO

The proper name goes in last to yield sentence (23), after the right form of the pronouns has been found in the tables.

From my treatment of this example, of the last alternative from network (18), and of sentence (22), the reader will see that examples like

(26) *John dressed and shaved himself.*

are not exceptions to the hypostatization rule either. Here we have three name connections, two of which receive the null interpretation, one of which is the only one of the three within whose scope the propositional connection *must* fall. Some trickier cases arise, however, with sentences containing verbs of perception or psychological verbs followed by a *that* clause (including verbs like *said that*). The reader will have noticed that when I discussed William James's example, I treated *he himself* in *Baldy knew that he himself had fallen out of the wagonette* as a reflexive pronoun without further comment, although *himself* could be omitted from this sentence and *he himself* is not a straightforward reflexive pronoun. One reason for so doing follows. Some psychological verbs are followed by a *that* clause in English, others by an accusative-and-infinitive clause, others by either. *Know* can take either form, and, if we opt for the accusative-and-infinitive construction, the pronoun **must** be reflexive, for example, *Baldy knew himself to have fallen out of the wagonette*. It seems, therefore, that we should regard such pronouns as essentially reflexive, with some qualification to their form being effected by a *that* clause. We also seem to have a choice with verbs of perception, for example, *John saw a snake near him,* but also *John saw a snake near himself.* It is not, therefore, absolutely clear that these are exceptions to the hypostatization rule and, pending an analysis of the meanings of perceptory and psychological complement verbs, we shall do best to suspend judgment as to whether or not they are genuine exceptions.

5. RELATIVE PRONOUNS AND RELATIVE CLAUSES

It will hardly have escaped the reader's attention that the thoughts expressed by many of the examples in the last section could also have been expressed in sentences containing relative pronouns: *John, who is at home, is asleep; Plato, who taught Aristotle, was admired by him, Anyone who works hard will get a first; All of the boys detested one of the saxophonists, whom some of the girls admired;* and so on. In each case, the relative pronoun replaces two phonetic constituents, a personal pronoun and half of a propositional connective (the half that cannot receive a null hypostatization). Moreover, the choice of a relative pronoun in preference to a personal pronoun and propositional connective is often just a matter of style determined by phonetic and not by semantic considerations, although one type of construction may be avoided sometimes because of possible ambiguity. Finally, there is one case in which we do not have the option of a relative pronoun: a transposed implication lying within the scope of a name connection whose name input is a proper name, for example *If John is at home, then he is asleep*.

To the extent that relative constructions are phonetic variations upon (personal) pronominal constructions, both should be obtainable, by phonetic projection, from the same semantic structures. If our representations of semantic structures are properly adapted to the present task, therefore, no new signs should be needed in order to extend their application to relative constructions. Further, since it seems that where we have the option between a pronominal and a relative construction, the relative pronoun in the latter replaces a propositional connective plus a personal pronoun, the obvious inference is that the corresponding propositional and name connections in the semantic structure are together receiving a single hypostatization. The condition for this to be possible is that the propositional connection should lie within the **immediate** scope of the name connection, that is, that the propositional output of the propositional connection should be joined, without any intermediary, to the propositional input of the name connection. Since neither connection is a constituent of the semantic structure, but an interrelationship between its constituents, we are always at liberty to regard the two together as forming a single interrelationship. This will have two propositional inputs and one propositional output, but one name input and two name outputs, as we can easily see by combining the two signs (taking conjunction as the example of a propositional connection):

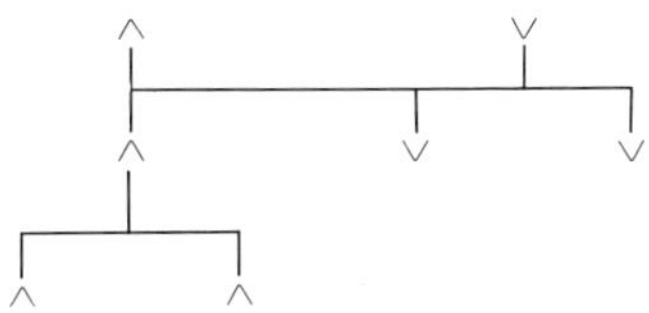

joins into

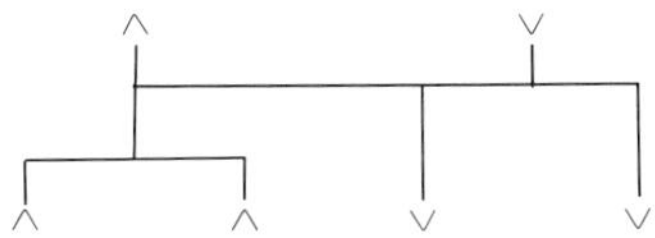

or

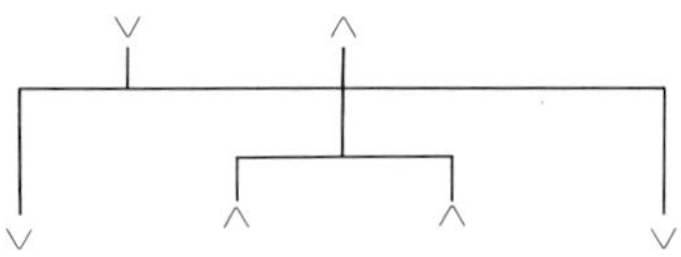

In combining the two signs, the horizontal stroke of the propositional connection must be kept **below** that of the name connection, in order to keep the scope of each, in relation to that of the other, unchanged. It will be convenient, in representing semantic structures, to combine the two signs in this way and to call the combination a **relative connection,** but no new interrelationship is being introduced and a corresponding representation in which they were distinct would always be an equivalent alternative.

The semantic category of relative pronouns can be read off from the sign for a relative connection. The simplest way of using it to form a propositional sign is as follows:

(27)

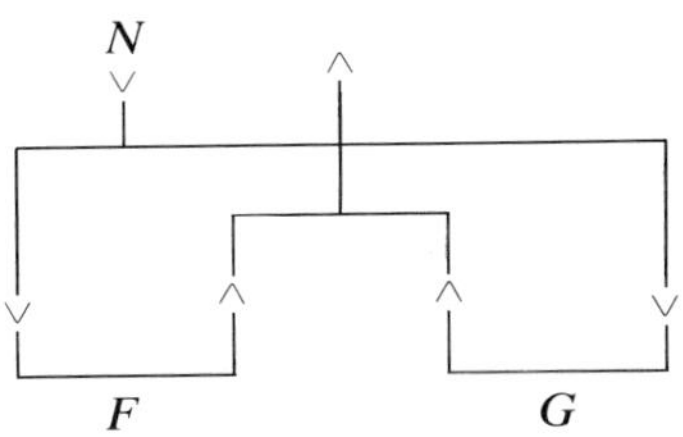

It will form a proposition from one proper name and two first level connections each having one name input. Its category is therefore 3PN1PN1PN. The linear notation yields the same result; within the immediate scope of $I_{xy}\varphi xy\zeta$ lies a conjunction, disjunction, or implication, so, combining the sign for conjunction with the *I*-operator, we get $I_{xy}K\varphi x\psi y$, and corresponding signs for the other two propositional connectives. Any of these could be substituted for $W_{xy}\varphi x\psi y\zeta$, which is of category 3P1PN1PNN, that is, 3PN1PN1PN, as the order of the βs in a category name does not matter. A proof is given in Appendix I that 3PN1PN1PN will split into 2PN2PNN 2PPP, the respective categories of a (personal) pronoun and a propositional connective. If, then, we can give a correct account of English relative constructions based on these assumptions, this will lend additional support to the hypothesis that pronouns are not semantic constitutents of sentences.

The network in (27) is an alternative to (13), and we projected (13) onto sentence (14), *John is at home and he is asleep.* It is clear that we cannot transform this into a relative construction just by replacing *and he* by a relative pronoun: **John is at home who is asleep.* Instead, we need:

(28) *John, who is asleep, is at home.*

which would correspond to *John—and he is asleep—is at home.* Now this is reminiscent of a transposition, but it is not parallel to the transposition of implication discussed in the last section, because it separates the first Term from its Verb. Our account of the phonetic structure of sentences containing relative clauses will be much simpler if we do not regard them as formed by a transposition but, rather, as being organized, in the first instance, around their main Verb instead of around the connective. That is to say, the basic structure of sentence (28) will simply be T_1V_1. The Term must then be the whole sentence apart from the main Verb, that is, *John who is asleep.* We are considering phonetic structure here, so let us not make the error of being guided by the meaning. *John who is asleep* does *not* form a unit with respect to the meaning in sentence (14), is not even a restrictive relative clause; but that is quite irrelevant at this point, because the meaning of the sentence depends upon what *semantic* structure we see it as having.

If *John is asleep* is a Term in sentence (14), however, it is certainly a complex Term, composed of a Term, a relative pronoun, and a Verb. It seems, then, that a relative pronoun forms a Term from a Term and a Verb. Therefore it is not itself a Term, although it replaces the first Term of a Verb (If the Verb in question has more than one Term associated with it, then these must be supplied in order to form the complex Term). The complex Term is organized around the relative pronoun, and the Term and Verb which it relates will therefore fall within its phonetic scope. I shall indicate this by writing T W-PRO V for the hypostatization of a relative pronoun, and by enclosing complex Terms in brackets in representations of phonetic structures. The subscript number of the complex Term will be that of the Term lying within the scope of the relative pronoun, and so it need not be shown separately. A relative pronoun receives its gender from its Term, but its case from its Verb, another indication that its role is to combine them.

If this is correct, then, as a phonetic structure for sentence (14), we shall have:

(P5) $\varnothing$ (T_{11} W-PRO V_2) V_1

where the null sign represents the hypostatization of C_1. Since the structure as a whole is organized around the main Verb, the first step in linearization will be to select one of the hypostatizations of the first level connections in the network and to insert it. This must always be the one occurring to the left of the propositional connection. Assigning the same interpretation to network (27) as to network (13), this yields:

1. $\varnothing$ (T_{11} W-PRO V_2) *is at home*

The insertion of the Verb of the complex Term comes next. This is determined from the network, by looking at the propositional connection which forms part of the relative connection, and finding the first level connection it joins to that connection whose hypostatization constitutes the main Verb. In this example, there is only one other first level connection, so we insert its hypostatization, *4 is asleep,* for V_2:

2. $\varnothing$ (T_{11} W-PRO *is asleep*) *is at home*

Finally, the name input to the relative connection supplies us with the substitution for T_{11}:

3. $\varnothing$ (*John* mW-PRO *is asleep*) *is at home*

This will yield sentence (14) when mW-PRO has been found in the tables. We can also project from the back view of network (27) to obtain *John, who is at home, is asleep.*

We cannot tell, from the phonetic form of an English relative pronoun, *which* propositional connection is incorporated into its meaning. Consequently, there must be restrictions upon the hypostatization of a name connection and a propositional connection within its immediate scope as a single morpheme, in order that we may be able to recover the meaning of a sentence from its phonetic representation. The restrictions are quite simple. First, if the input to the name connection is a proper name, the propositional connection must be conjunction. Thus, we cannot have a relative construction instead of *John, if he is at home, is asleep.* All the same, this construction is most simply accommodated by regarding it as a variant upon (P5), in which the general organization of the structure is the same, but the connective and the personal pronoun are not allowed to coalesce. As it is restricted to implication, we may simply write:

(P6) $\varnothing$ (T_{11} *if* T_{21} V_2) V_1

If the input to the name connection is a quantifying connection, the permissible combinations depend upon which quantifier is present. When the quantifier is universal, a name connection having implication within its immediate scope may be combined with it into a relative connection. When the quantifier is existential, the combination requires that the propositional connection may be conjunction. As these are the most frequent combinations, English provides a special phonetic device in the form of relative pronouns to keep its phonetic structures as simple as possible. Other types of quantifiers require separate consideration, which I cannot undertake here. The following network illustrates the combination of universal quantification with implication:

(29)

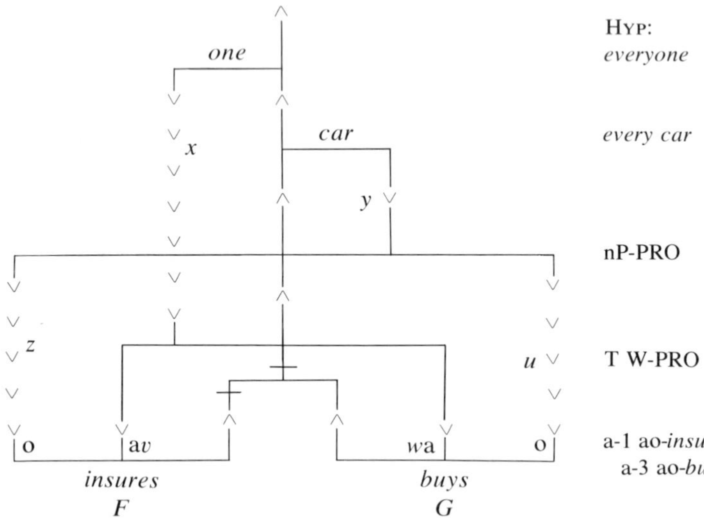

$\Pi_x \Pi_y I_{zu} W_{vw} GvzFwuyx$

As the linear formulas are now becoming rather complicated, it may be helpful to the reader to describe a simple method for constructing them from a network. We assign an index letter to each name pathway in the network (this has been done above) and then write out the second level operators, each with its index letters (read off from the name pathways leading away from each operator), working down from the top of the network along the propositional pathway until we come to a propositional connection. The matrix can then be read off from the propositional connectives and the first level connections, the second occurrences of the index letters being written out according to the order of the name inputs to the first level connections from their name pathways. Working back from right to left along the operators, we find the index letter of the name pathway leading to each and add it in the final step.

In this example, I have linearized the initial first level connection in the opposite order to that of its name inputs in the horizontal dimension of the network. That network could quite easily have been drawn with its name inputs in the same order as in the schema, but the arrowheads leading to it from the name connection would then have had to cross over the upper horizontal stroke of the relative connection. It seemed preferable to keep the network as uncluttered as possible, and I shall continue to allow myself freedom in hypostatizing first level connections to put the schematic numbers into any order; the voice prefix on the verb will, of course, be adjusted accordingly. I have also simplified this example by ignoring such questions as *From whom was the car bought?, And for how much?* etc., which are raised by the meanings of the first level connections.

This network may be projected onto the phonetic structure:

(P7) $\varnothing$ (T_{11} W-PRO V_2 T_{22}) V_1 T_{12}

Insertion of the Verbs yields:

1. $\varnothing$ (T_{11} aW-PRO ao-*buys* o-T_{22}) ao-*insures* o-T_{12}

The case of T_{11} is determined by V_1 in the normal manner. Working up from the bottom of the network to insert the Terms, we come first to the relative connection, whose input is *everyone,* which thus replaces T_{11}:

2. $\varnothing$ (a-*everyone* amW-PRO ao-*buys* o-T_{22}) ao-*insures* o-T_{12}

Next we come to the name connection. The normal insertion rule would substitute *every car* for T_{22}, but this would eventually yield the ill-formed expression, **Everyone who buys every car insures it.* And although we can have, for example, *Everyone who buys every car must be a multimillionaire,* this is *not* the meaning (mutatis mutandis for *F*) expressed by network (29), which may be paraphrased in logician's style as: *For everyone and for every car, if he buys it, then he insures it.* In order to express this sense in ordinary English, the universal quantifying expression which has been hypostatized as *every car* at the side of network (29) must be hypostatized upon insertion as *a car* or, at least, as *any car* instead. The meanings of the English determiners *every, each, any, a* and *some* are partly, at any rate, context dependent; I do not think that anyone has yet succeeded in giving a satisfactory account of the way in which the context determines or restricts the selection of an appropriate item, given that a certain sense is to be expressed. I cannot offer a rule here stipulating when universal quantification is to be expressed by the indefinite article. In any case quantification is not my present topic, so I proceed on the assumption that hypostatization and insertion for T_{22} yields:

3. $\varnothing$ (a-*everyone* amW-PRO ao-*buys* o-(a *car*)) ao-*insures* onP-PRO

By inflecting and looking up the pronouns in the tables, this yields:

(30) *Everyone who buys a car insures it.*

Everyone is interchangeable with *anyone* here.

As an example of existential quantification combined with conjunction, I take the well-known Bach–Peters sentence:

(31) *A boy who was only fooling her kissed a girl who really loved him.*

One version of a network which represents a semantic structure for this sentence is:

(32)

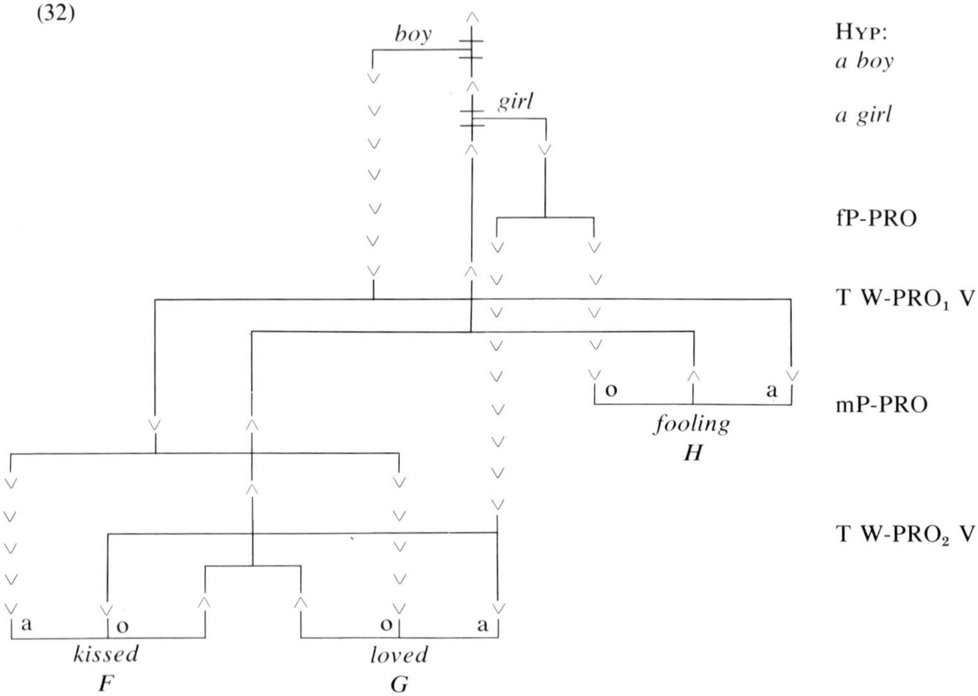

$\Sigma_x\Sigma_y I_{zw} W_{uv} I_{st} W_{qr} KFsqKGrtHvwzuyx$

I have omitted the hypostatizations of the first level connections and, from now on, will also omit their voice prefixes. The phonetic structure that we want here will be:

(P8) ∅ (T_{11} W-PRO_1 V_3 T_{32}) V_1 ∅ (T_{12} W-PRO_2 V_2 T_{22})

1. ∅ (a-T_{11} aW-PRO_1 *was only fooling* o-T_{32}) *kissed* ∅ (o-T_{12} W-PRO_2 *really loved* o-T_{22})

2. ∅ (a-(*a boy*) amW-PRO *was only fooling* o-T_{32}) *kissed* ∅ (o-(*a girl*) afW-PRO *really loved* o-T_{22})

3. ∅ (*a boy* amW-PRO *was only fooling* ofP-PRO) *kissed* ∅ (*a girl* afW-PRO *really loved* omP-PRO)

It is to be observed that no special rule is needed to ensure that *her* precedes *a girl*.

I can hardly conclude this section without an example of nested complex Terms, for nesting is one of the principal facilities afforded by the relative construction, as A. A. Milne brings out so well in his poem "The Engineer" (1932, pp. 150–151).

(33) *Anyone who hurts anyone who hurts him hurts himself.*

This sentence contains, in addition to a reflexive pronoun, one occurrence of *anyone* which cannot be changed to *everyone*. The following network will serve to represent a semantic structure for it:

(34)

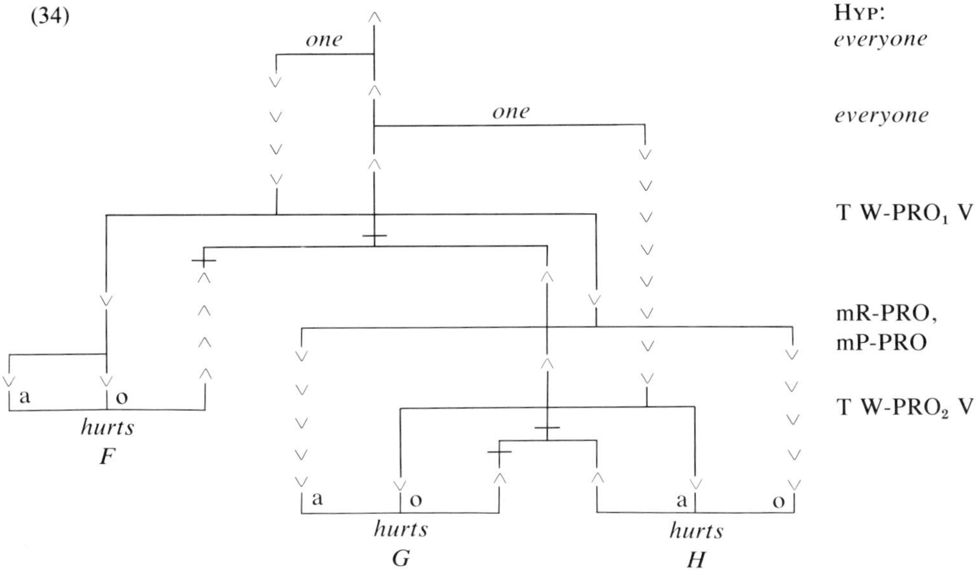

$\Pi_x \Pi_y W_{zu} I_{vw} Fvwz I_{st} W_{qr} Gsq Hrtuyx$

The phonetic structure and first part of the linear insertion are:

(P9) ∅ (T_{11} W-PRO V_2 ∅ (T_{22} W-PRO V_3 T_{32})) V_1 T_{12}

1. ∅ (a-T_{11} aW-PRO *hurts* ∅ (o-T_{22} aW-PRO *hurts* o-T_{32})) *hurts* o-T_{12}

Working up from the bottom of the network, we find that the input to the relative connection is the lower *everyone*. By inserting this for T_{22}, however, we are bringing it within the phonetic scope of the relative pronoun of the outer complex Term, which contains, as part of its meaning, an implication within whose semantic scope the quantifying connection does not lie. Hence, as in the projection of network (19), *every* must be changed to *any:*

2. ∅ (a-T_{11} aW-PRO *hurts* ∅ (o-*anyone* amW-PRO *hurts* o-T_{32})) *hurts* o-T_{12}

We come next to the name connection, which links T_{11} with T_{32}; thus, by the usual insertion rule for pronouns, mP-PRO will be substituted for T_{32}. The input to the name connection, however, comes from the relative connection above it, so the insertion of T_{11} will have to be deferred:

3. ∅ (a-T_{11} aW-PRO *hurts* ∅ (*anyone* amW-PRO *hurts* omP-PRO)) *hurts* o-T_{12}

The propositional pathway leading out of the name connection goes to the relative connection above it, but that lies above the name connection on the far left which must, accordingly, be tackled first. Here, by the normal insertion rule, mR-PRO will replace T_{12}:

4. ∅ (a-T_{11} aW-PRO *hurts* ∅ (*anyone* amW-PRO *hurts* omP-PRO)) *hurts* omR-PRO

Finally, we reach the upper relative connection. Its right hand output has already been dealt with, so only its input remains to be considered. This is the upper *everyone,* and the only unfilled position is that occupied by T_{11}; this does not come within the phonetic scope of an implication (the phonetic scope of the C_2 part of a relative pronoun extends only to the right), so we end with:

5. ∅ (*everyone* amW-PRO *hurts* ∅ (*anyone* amW-PRO *hurts* omP-PRO)) *hurts* omR-PRO

with the option of changing the first *every* to *any*. If we do not change it, we shall get: *Everyone who hurts anyone who hurts him hurts himself,* which still has the same meaning as sentence (33).

It would be necessary to consider a much wider range of examples than has been possible here before one could be confident that the method of projection and the system of phonetic structures which have been described above are correct in every detail, or that they provide us with the most economical account of English sentences containing pronouns. That is one reason why I have not set out a system of formal projection rules. Another reason is that it appears to me much more important to understand **why** a language obeys certain rules, and what functions they serve, than merely to have a set of unexplained rules which yield correct results. Indeed, if we do not try to explain the purposes which each rule serves, I very much doubt whether we shall succeed in obtaining satisfactory ones, in the long run. In any case, the most important point of doing formal linguistics is surely to obtain a better understanding of **why** general-purpose languages take the forms they do; we cannot stop short at merely finding out **how** they work, difficult enough as that is.

If the present approach to pronouns turns out to be sound, the major lesson to be learned from it is that formal linguistics has hitherto been crippled by its failure to distinguish between semantic and phonetic syntax. Trying to make do with a single syntax, it constantly oversimplifies semantic structures and overcomplicates phonetic ones. Nor can it develop a clear notion of grammaticality, because it is seldom clear whether an unacceptable expression is to be rejected on semantic or on phonetic grounds. Our judgments rest on the quicksand of intuition, a useful guide but a poor master: As Frege and Wittgenstein saw, the task of philosophy is to replace intuitions by arguments. Semantic syntax is the custodian of meaning, phonetic syntax of grammaticality, and it is possible for an expression to violate the requirements of either without violating those of the other. If we keep them distinct, we have a means of diagnosing faults; if we assimilate them, the amalgam will serve only to confuse both semantics and phonetics.

6. POSSESSIVE PRONOUNS

If we could assume that the genitive case had a single meaning, a separate section on possessive pronouns would be unnecessary. They would be formed in the course of phonetic projection, by replacing the corresponding case role indicator by the genitive inflection. This applies equally to reflexive, personal, and relative pronouns, although in grammatical usage they rate as a special category because a double case inflection is necessary, for example, *in his car*. Unfortunately, however, the genitive case can make many different contributions to the meaning of a sentence, so that, if our representations of semantic structures are to show these contributions, their phonetic projections will be much more complex.

There are many occasions from which a genitive originates (with respect to meaning) from a relative clause; for example, *John's daughter* is a girl whom he begat, and *his car* is a (or perhaps the one and only) car which he owns. In both of these cases, our representation of a semantic structure will have the form:

(35)

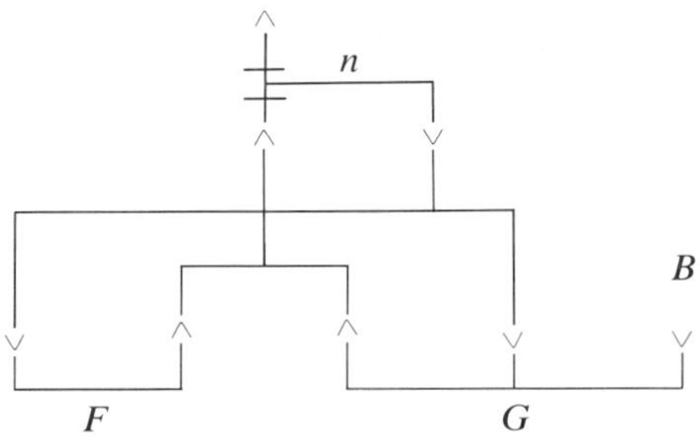

This network could be projected phonetically as: *An n, which F B, G;* but for some substitutions for *F*, it can be projected as *B's m G*. For example, instead of *A man who employs John walked past,* we can (and normally would) say *John's employer walked past.* In the first place, however, I do not know what is the restriction upon substitutions for *F* here; thus, the same substitution cannot be applied to *A girl, who was even taller than John, walked past.* And, in the second place, the substitution for *m* may be totally unrelated, phonetically, to those for *n* and *F*.

Perhaps the best that can be done here is to specify one fairly regular type of projection. Many verbs allow us to form a corresponding count noun ending in *-er*; the substitution for *F* in these cases is usually *man*. In the example above, we have *employer* from *employ*. So, where *n* = *man* in network (35), we shall, given a suitable substitution for *F*, be able to project the entire network onto *B's F-er G*. The name input to the main schema will, of course, have a case role indicator attached to it, and this will attach to *B's F-er* as a whole; hence the double case inflection, although the genitive here is not really a case *role* at all—it does not indicate anything comparable to an agent, a source, a goal, etc., but, rather, a projected relative connection.

Possessive pronouns are also used in English for quite different purposes. *John has a pain in his foot* does not have a semantic structure comparable to that of *John has a car in his garage:* That assimilation leads to the pseudo-problem of other minds and private languages. Linguists sometimes try to mark this difference by distinguishing alienable from inalienable possessions, but, as the latter include kinsfolk as well as parts of the body, it does not seem to be a very profitable distinction. For kinsfolk are encompassed by network (35). Moreover, "inalienable" suggests possessions that cannot be lost; yet we can lose not only our kinsfolk but also some parts of our bodies. In fact, it is curious to regard kinsfolk as possessions of **any** kind: *You treat me as if you owned me* is surely a rebuke, and is there no objection to slavery? What we need here, perhaps, is a distinction between what can and what cannot exist independently of us (or of some other specified thing).

These remarks are primarily addressed to problems concerning the genitive case. If those were solved, there seems no reason to suppose that possessive pronouns present any new or special difficulty.

7. PRO-VERBS

Pro-verbs are to schemas of first level what pronouns are to proper names. That is to say they are, hypostatizations of connections which identify two pairs, each consisting of one propositional input and the same number of name outputs, so that one first level schema can serve as input to two second level schemas, for example, to two quantifying connections. The two second level schemas must be joined by a propositional connection. The topic of pro-verbs is a large one, so I can only touch briefly upon it here. But, as they are the third-level analogues of pronouns and are, indeed, often phonetically indistinguishable from pronouns, some mention must be made of them.

An example of a pro-verb which is phonetically different from any pronoun is *so* as it occurs in:

(36) *Most undergraduates are impecunious and* ***so*** *are some graduates.*

This means the same as *Most undergraduates are impecunious and* (*also*) *some graduates are impecunious;* hence, *so* is here a pro-verb of laziness. In the expanded version, each subordinate sentence can be split into a quantifying expression (*Most undergraduates, some graduates*) and a first level schema having one name input (*I is impecunious*). In sentence (36), the two quantifying connections are first conjoined, so that, in categorial notation, we have:

$$2PP^1P^2\ 1P^21P^3N_4\ 1P^11P^5N_6$$

A proof will be found in Appendix 1 that this combination of category names can be derived from 2P1PN1PN, which is accordingly, the category of the whole conjunction of quantifying expressions. The pro-verb turns this into an expression of category 1P1PN, which can then work upon a single expression of category 1PN in order to form a sentence. Hence the category of the pro-verb will be 1(1P1PN)(2P1PN1PN), that is, 2P1PN2P1PN1PN.

This category name does not show how the internal propositional pathways run in the connection which it effects; in order to show this, a two-dimensional notation is again clearer. The structures involved are exact analogues of those for second level pronouns, everything being raised by one level. The additional complexity concerns only the pathways; each name pathway in the representations of second level structures now becomes a double pathway, a propositional pathway running alongside the original name pathway. The directions of the two are, however, **reversed** by comparison with the second level networks: In this case the name pathway runs upward, the propositional pathway downward. The reason for this is that, as a first level schema replacing a proper name at the top of the network, the sign for it has to be inverted. At the bottom of the network come the quantifying connections, which have, therefore, to be "bent back" upon themselves.

Previously, propositional pathways were distinguished from name pathways by their direction; now, it will be necessary to distinguish them typographically. For this purpose, I shall use double lines for propositional pathways within a connection and a different style of arrow for external pathways between the connections. The new signs for a quantificational connection and the sign for a pro-verb are then to be as follows:

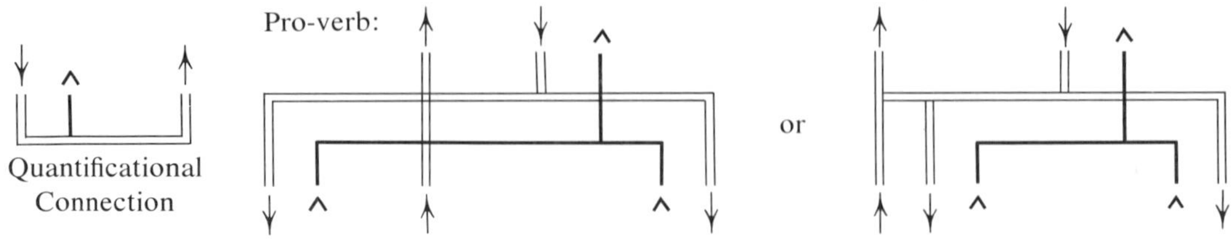

These complications could be avoided if three dimensions were available; they thus lend support to the view that any one notation will only give us a clearer view of semantic structure in one area of language at a time. The linguist is like an architect living among a people who can only see two-dimensionally, so that they cannot even read perspective drawings. The only idea he can give them of buildings is then through plans, elevations, and sections; though, even then, he would be much better off than the linguist, because his drawings are lacking in only one dimension.

Using these new conventions, we can represent a semantic structure for (36) as follows.

(37)

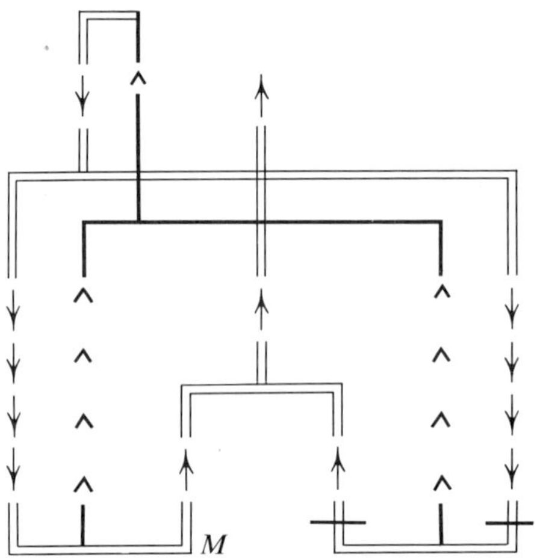

The reader is invited to compare network (37) with network (13): They are exact analogues. I have not yet been able to consider the method of projection for networks containing pro-verb connections.

It may be wondered why I began with an example containing quantifying expressions instead of one with proper names, like Chomsky's well-known sentence:

(38) *Bill has a chance to live and so has John.*

which appears to be much simpler. The reason is that examples like (38) raise an awkward problem. We cannot have a conjunction of proper names as we can of quantifying expressions, for 2PPP N N will not combine. Phrases like *Bill and John* may form units of phonetic structures, but anyone who thinks that they can form units of semantic structures will have much to explain. He will have to give an account of the meaning of *and* that will work not only in this but in more standard contexts as well, and show how the meaning of, *Bill and John* is determined by the meanings of *and,* of *Bill,* and of *John.* Faced with this difficulty, Frege said that, in an example like (38), the two proper names occur not as expressions of category N, but as a special type of quantifying expression, and thus are here of category 1P1PN (1893, §22). On this account, network (37) could also represent a semantic structure for (38), simply by changing the interpretations of the two quantifying connections and of the first level schemas.

The difficulty with this move is that it forces us to conclude that the proper names in (38) have different meanings from those they have when occurring as expressions of category N, for the whole point of the system of semantic categories is that the meaning of any expression is partly determined by its category. Hence, different category, different meaning. Now there are many cases of the same phonetic constituent having two meanings, for example, *move* as an intransitive verb, and *move* as a transitive verb which means, roughly, *cause x to move* in the intransitive sense. In such cases, we say that the expression in question is ambiguous. But there is absolutely no independent reason for supposing that *John* and *Bill* in (38) have different meanings from, for example, *John* in *John smokes a pipe* and *Bill* in *Bill plays golf.* Indeed, the supposition would lead to obvious difficulties.

Of course, it would be possible to say that proper names are **always** expressions of category 1P1PN; but, as I have argued elsewhere (Potts, 1976a), we have then either to find some candidates to count as expressions of category N, or else to give a convincing explanation of why there are none, although the category must be posited as one of the basic categories of the system. To resolve this difficulty would, I suspect, lead us to query Frege's assumption that proper names (or some other alternative, like "entity expressions") should be a basic category in the theory of meaning. This is, clearly, a trail to be followed upon another occasion.

As an example of a pro-verb which is phonetically indistinguishable from a pronoun, a relative pronoun will do:

(39) *John threw a tomato at the vice-chancellor, which was very unwise.*

A paraphrase brings out the relation with *so* as a pro-verb: *John threw a tomato at the vice-chancellor and was very unwise to do so.* This suggests that *which* as a pro-verb stands to *and so,* just as *which* as a pronoun stands to *and it.* In other words, it is a single hypostatization of the third level name connection and a propositional connection combined. But this needs further investigation.

8. PRONOUNS IN INTENTIONAL CONTEXTS

With the exception of the previous section, I have been concerned entirely with pronouns of second level. Even then, the investigation is only begun: Sentences containing pronouns together with negative universal quantifiers, *only* as part of a quantifying expression, quantifiers like *most* and *few,* truth functional connectives like *unless* or *if and only if* and non-truth-functional connectives like *because,* have not been touched. I have also steered clear of the definite article. There can be little doubt that personal, relative, and perhaps even reflexive pronouns are used as hypostatizations of many different kinds of connection, of which the pro-verbs are just one example. The nonrelative pro-verbs, indeed, seem to have a different phonetic hypostatization from their second level analogues, the personal pronouns, though the personal pronoun in *The man who deserves it will get the price he wants* may well be a disguised pronoun of third level.

What the reader will probably have missed above all, however, is any serious attempt to treat examples of pronouns occurring in intentional contexts, like: *Barbara talked for an hour and it surprised me* or *Hob thinks that a witch has blighted Hob's mare, and Nob wonders whether she (the same witch) killed Cob's sow.* The first example takes us into the thicket of quantification over propositions and the problem of propositional identity; the second would (on my account) wrongly commit anyone giving such a report to the existence of the witch concerned. Pronouns, identity, existential commitment, and the meaning of proper names are all closely interrelated topics: You cannot tinker with one without affecting your account of the others. A great virtue of Frege's philosophy of language is that it is consistent on all four of these topics, and minor modifications of it have usually come to grief. I said at the beginning of this chapter that I should work within Frege's assumption that the basic semantic category is that of proper names; that assumption is the basis of quantification theory, and, even if it has ultimately to be rejected, any alternative theory which hopes to win acceptance must, at least, incorporate what Frege achieved with his quantifier notation. To go substantially beyond Frege's philosophy of language will require a radical innovation and, as I am not in a position to offer a new theory on that scale, it seems reasonable to adopt the more modest aim of working out a companion, for pronominal constructions, to quantification theory in the hope that it too can be assimilated into a larger theory when the time comes.

APPENDIX 1. PROOFS IN FREGEAN CATEGORIAL NOTATION

Rules

(R1) $\Gamma\ \alpha\ \Delta$

$\Gamma\ 1\alpha\beta^n\ \beta^n\ \Delta$

where any category names may be substituted for α and for β, Γ being the string of category names occurring to the left of α, and Δ the string occurring to the right of α. The substitutions are to include any super- or subscripts, and contiguous numbers in $1\alpha\beta$ are to be added; n must be a number not already occurring in $\Gamma\ \alpha\ \Delta$. Both Γ and Δ may be empty or consist of just one category name.

(R2)

$$\begin{array}{l} \Gamma\ 1\alpha\beta\ \Delta \\ \quad \alpha \\ \qquad \vdots \\ \quad \gamma\ \delta\ \ \Sigma \\ \Gamma\ \ \gamma\ 1\delta\beta\ \ \Sigma\Delta \end{array}$$

(R3)

$$\begin{array}{l} \Gamma\ 1\alpha\beta\gamma\ \ \Delta \\ \quad 1\alpha\beta \\ \qquad \vdots \\ \quad 1\delta\epsilon\ \zeta\Sigma \\ \Gamma\ 1\delta 1\epsilon\gamma\ \ \zeta\Sigma\Delta \end{array}$$

Proof that a relative connection can be split into a name connection and a propositional connection:

0 $3PN_1 1P^2N_3 1P^4N_5$

1 $3PN_1 1P^2N_3P^4$

2 $2PN_1 1P^2N_3$

3 $2PN_1P^2$

4 $1PN_1$

5 $2PN_1P^6\ P^6$ **(R1)**

6 $2PN_1P^6\ 1P^6P^2$ **(R2)**

7 $2PN_1 1P^6N_3\ 1P^6P^2$ **(R3)**

8 $2PN_1 1P^6N_3\ 2P^6P^2P^4$ **(R2)**

9 $2PN_1 2P^6N_3N_5\ 2P^6P^2P^4$ **(R3)**

Proof that 2P1PN1PN can be split into one propositional and two quantifying connections:

0 $2P1P^1N_2 1P^3N_4$

1 $1P1P^1N_2$

2 P

3 $1PP^5\ P^5$ **(R1)**

4 $1PP^5\ 1P^5 1P^3N_4$ **(R2)**

5 $2PP^5P^6\ P^6\ 1P^5 1P^3N_4$ **(R1)**

6 $2PP^5P^6\ 1P^6 1P^1N_2\ 1P^5 1P^3N_4$ **(R2)**

APPENDIX 2. THE RELATION OF THIS ACCOUNT TO THE LATE PROFESSOR R. MONTAGUE'S TREATMENT OF PRONOUNS IN ENGLISH

Montague never wrote specifically about pronouns, but both personal and relative pronouns are included in the coverage of his paper "The Proper Treatment of Quantification in Ordinary English" (PTQ) and the logician's *such that,* which is akin to a relative pronoun, is discussed in "English as a Formal Language" (EFL) (both in Montague, 1974).

Montague's treatment is based, like mine, on a categorial grammar, but he does not argue for his category assignments and, in general, they correspond to categories of traditional grammar. Indeed his categorial grammar is, in a sense, idling, because it does not determine the possible combinations of expressions; this is specified by a whole series of supplementary syntactic rules which attempt to combine semantic and English phonetic syntax, so that the categorial grammar only comes into play in the semantic rules, which mirror the categories.

In PTQ, personal pronouns are assigned, together with proper names and quantifying expressions, to the category of **Terms,** in my notation 1P1PN and not to be confused with my Terms. Although he assimilates personal pronouns to proper names, it would be misleading, in view of his assignment of proper names to category 1P1PN, to call it a two-name theory. Relative pronouns are not, as such, categorized at all.

Montague's syntactic rules are based, so far as personal pronouns are concerned, on Frege's first order logic as interpreted by Quine; that is, these pronouns correspond to the so-called "bound variables" of the notation and the syntactic rules correspond to the formation rules of first order logic. This does not accord with Frege's own explanation of his notation, according to which the bound variables are integral parts of the quantifier (this point is pursued in more detail in Potts, 1973 and 1976b). Moreover, formulas containing only pairs of index letters, for example, $\Sigma_x Fx$, do not represent the structures of propositions containing pronouns. Thus, the formula cited could be used (assuming restricted quantification) to represent a structure for *A dog barked*. Some explanation must be given of why the last x in $\Sigma_x Fxx$ should represent a pronoun but the others not. This criticism applies more to Quine than to Montague, however, since, for Montague, personal pronouns are expressions of second level, of the same category as the quantifiers (Though there seems to be some tension between his category assignment and his syntactic rules on this point). In that case, it should be possible to combine a pronoun with a one-place predicate; yet this would not yield a "truth-value expression," Montague's term for what I have called a proposition.

Montague's account of the meaning of a personal pronoun needs to be spelled out in stages. First, the meaning of a truth-value expression is a proposition, and this is a function mapping possible worlds onto truth values. The meaning of a one-place predicate is a function mapping possible individuals onto propositions; so the meaning of a Term (and, hence, of a personal pronoun) is a function mapping a function of the latter kind onto a function of the former kind. In other words the meaning of a personal pronoun is a function which maps a function which maps possible individuals onto a function which maps possible worlds onto truth values onto a function which maps possible worlds onto truth values. Moreover, a function, for Montague, is an object in Frege's sense (his use of the lambda calculus shows this), so that every kind of expression, of whatever category, is a name and the function in which its meaning consists is the bearer of that name. This aspect of Montague's theory is pre-Fregean, but the motivation for the complexity of the functions involved is to give a theory of inference for sentences containing pronouns in intentional contexts. Even if it were successful from that point of view, however, I hope that I have established at least a *prima facie* case for holding that, in extensional contexts, there is no necessity to explain their meanings in this devious way.

Although Montague does not assign a category to relative pronouns, an expression of the form *A such that B,* where *A* is a Common noun phrase (category 1PN), and *B* is a Term (category 1P1PN) is a Common noun phrase. The implication here is that the category of *such that* is 2P1P1PN1PN, which would make it an expression of *third* level. I do not think that this can be substantiated.

Practitioners of Montague's grammar are now finding examples of acceptable sentences containing pronouns to which his theory fails to assign a meaning (for example, sentence (30)), and are therefore seeking to patch up his rules. But counterexamples are also being proffered to these amendments. Logicians with a taste for the history of their subject will be reminded of the similar fate which met medieval theories of *suppositio* until logic went into decline and, eventually, the theory was superseded by Frege's quantification theory. Apart from this straw in the wind, Montague's theory is extremely inflexible in two respects. First, it provides for only a limited number of "canonical" constructions; one might say that, given a certain thought, there are far less ways of expressing it in Montague's English than in ordinary English. It would be an interesting exercise, which I have not attempted, to take two sentences having the same meaning, one of which used a propositional connective and a personal pronoun, the other a relative pronoun, and see whether they are also assigned the same meaning by Montague's theory.

The other limitation is that, because his semantic rules mirror his categories, it is doubtful whether one could adapt his account to another language simply by changing the syntactic rules, especially if the other language were not Indo-European; my guess is that the semantic rules would also need substantial modification. Indeed, I think it would be fair to say that he rejected the distinction between two types of structure, semantic and phonetic, with the result that difficulties arise over paraphrase and translation. His motivation for this may well have been his contempt for transformational grammar but, although I think he did well to reject the latter's computational account of the relationship between the two types of structure, to abandon the very distinction itself was, in my view, to throw the baby out with the bath water.

REFERENCES

Ajdukiewicz, K. (1935). Die syntaktische Konnexität. *Studia philosophica* **1,** 1–27. [Part I translated from the Polish text by P. T. Geach (1967), On Syntactical Coherence, *The Review of Metaphysics* **20,** 635–647.]

Carroll, Lewis (1872). *Through the looking-glass and what Alice found there*. [Reprinted (1939) in *The complete works of Lewis Carroll*. London: The Nonesuch Press.]

Dummett, M. A. E. (1973). *Frege: Philosophy of language*. London: Duckworth.

Fillmore, C. J. (1968). The case for case. In E. Bach and R. Harms (Eds.), *Universals in linguistic theory*. New York: Holt, Rinehart and Winston, 1–88.

Frege, G. (1879). *Begriffsschrift*. Halle: L. Nebert. [Reprinted (1964), Hildesheim: Georg Olms.]

Frege, G. (1893). *Grundgesetze der Arithmetik, I. Band*. Jena: H. Pohle. [Reprinted (1962), Hildesheim: Georg Olms.]

Geach, P. T. (1972). *Logic matters*. Oxford: Basil Blackwell.

Gentzen, G. (1935). Untersuchungen über das logische Schliessen. *Mathematische Zeitschrift* **39,** 176–221, 405–431. [Translated by M. E. Szabo (1964, 1965), *American Philosophical Quarterly* **1,** 288–306; **2,** 204–218.]

James, W. (1890). *The Principles of Psychology,* 2 vols. New York: Henry Holt and Co. [Reprinted (1950), New York: Dover Publications.]

McCawley, J. (1971). Where do noun phrases come from? In D. Steinberg and L. Jakobovits (Eds.), *Semantics*. Cambridge: Cambridge University Press. Pp. 217–231.

Milne, A. A. (1932). *The Christopher Robin verses*. London: Methuen.

Montague, R. (1974). *Formal philosophy: Selected Papers of Richard Montague*. (R. Thomason, Ed.). New Haven: Yale University Press.

Potts, T. C. (1973). Fregean Categorial Grammar. In Bogdan and Niiniluoto (Eds.), *Logic, probability and language*. Dordrecht: Reidel. [Revised version with simplified rules in *Studia Logica*. **37** (1978), 7–26.]

Potts, T. C. (1976a). Montague's Semiotic: A Syllabus of Errors, *Theoretical Linguistics* **3,** 191–208.

Potts, T. C. (1976b). Reply to Jaakko Hintikka, Quantifiers in Logic and Quantifiers in Natural Languages. In Körner (Ed.), *Philosophy of logic*. Oxford: Basil Blackwell.

Potts, T. C. (1976). The Place of Structure in Communication. In Vesey (Ed.), *Communication and Understanding. Royal Institute of Philosophy Lectures,* **10** (1975–1976).

Wittgenstein, L. (1922). *Tractatus logico-philosophicus*. London: Routledge and Kegan Paul.

PROBLEMS WITH PRONOUNS IN TRANSFORMATIONAL GRAMMAR[1]

THOMAS WASOW

1. INTRODUCTION

There are at least three sorts of questions which transformational studies of pronouns are relevant to. I will use the labels "syntactic," "semantic," and "anaphoric" in discussing them. The main point of this paper is to argue that answers to one type of question are not, in general, answers to the other types of questions. Hence, it is important to distinguish among these types of questions. I will argue that a major shortcoming of transformational investigations of pronouns is that they have, in general, failed to make the necessary distinctions among what I am calling syntax, semantics, and anaphora. In the course of making this argument, I will summarize what I consider to be the main accomplishments of transformational analyses of pronouns.

[1] I am very grateful to G. Harman and F. Heny for correcting certain mistakes in an earlier version of this paper. They are not, of course, responsible for any that remain.

SYNTAX AND SEMANTICS
VOLUME 10

ISBN: 0-12-613510-X

1.1. Syntax

Syntax, as I am using the term, is concerned with distinguishing sentences from arbitrary strings of words. It deals primarily with distributional properties of classes of words and with systematic structural relationships between sentence types. Syntactic analysis of pronouns includes at least an account of their distribution and case marking.[2]

A very basic syntactic question about pronouns which I want to consider briefly is "What is a pronoun?" That is, I want to consider how to explicate the traditional grammatical category of pronouns.[3] I believe this to be a worthwhile enterprise for at least two reasons. First of all, I have discovered that traditional grammatical categories are usually very useful ones. Second, some interesting universal claims about pronouns have been made,[4] which suggest that an adequate linguistic theory (that is, an adequate characterization of the notion "possible human language") should include a specification of what a pronoun is.

The problem with trying to define "pronoun" cross-linguistically is that pronominal systems are so different from language to language. For example, pronouns in some languages (like English) behave very much like ordinary noun phrases, whereas in others (for example, Romance languages) some pronouns are attached to the verb ("cliticized"). In still other languages (for example, Luiseño), a free pronoun and the corresponding clitic can co-occur, and in some languages (for example, Classical Aztec), clitics are obligatory even when full noun phrases are present. In short, there is a broad range of types of pronominal systems, with pronouns at one end of the spectrum behaving very much like some sort of nouns, and those at the other end apparently functioning like agreement markers of some kind.

Further complicating matters is the fact that several of the properties which seem to characterize pronouns in many languages are shared by the auxiliary verbal systems of various languages. These properties include cliticization, lack of stress, a tendency to appear in second position in the sentence, and diachronic instability.[5]

[2] Curiously, it is the syntax of pronouns which has received the least attention from transformationalists. The only relevant articles I know of are Postal, 1966; Delorme and Dougherty, 1972; and Sommerstein, 1972.

[3] I call this a syntactic question mainly because I think distributional criteria can come closer than any others to capturing this concept. I will say more on this subject later.

[4] For example, Greenberg's (1963) universals 25, 42, and 45:

> 25. If the pronominal object follows the verb, so does the nominal object . . .
> 42. All languages have pronominal categories involving at least three persons and two numbers . . .
> 45. If there are any gender distinctions in the plural of the pronoun, there are some gender distinctions in the singular also. . . . [pp. 112–113].

[5] I am grateful to Sue Steele for informing me of the cross-linguistic facts cited here.

Even if we restrict our attention to English, the task of defining pronouns is by no means simple. I have been able to discover several sufficient conditions for calling an English word a pronoun, and I can think of a couple of plausible candidates for necessary conditions, but I have been unable to find any set of conditions that is both necessary and sufficient. For example, although it is true that only pronouns may serve as subjects of tags in tag questions, relative pronouns may not appear in tags, since tags are possible only on main clauses, whereas relative pronouns are possible only in subordinate clauses. Similarly, although only pronouns exhibit gender and case marking in English, the second person pronoun *you* exhibits neither. Conversely, it seems that pronouns are generally stressless (aside from contrastive stress), but so are some nonpronouns, for example, *people* in the normal pronunciation of (1)

(1) *Only a fool trusts people.*

(cf. *Only a fool trusts politicians,* in which *politicians* normally gets stressed). The possibility of having either an anaphoric or a deictic use may also be a necessary condition for pronouns, but, of course, many other kinds of elements (for example, expressions like *the former, do so,* and *now*) also satisfy this criterion. In short, though grammarians tend to make ready use of the category pronoun (alternatively, of the feature [pro]), the problem of explicating the notion precisely—either across languages or within a language—is presently unsolved.

As with the problem of definition, other aspects of the syntax of pronouns have largely been ignored in recent work. I will return to some other syntactic questions later.

1.2. Semantics

The salient semantic questions about pronouns are all connected with the problem of determining truth conditions for sentences containing pronouns. The question which has dominated the literature on this is whether pronouns should be treated as bound variables in assigning interpretations to sentences containing them. Among the examples which have been most widely discussed in this connection are those in (2).

(2) a. *If anyone owns a donkey, he beats it.*
b. *The pilot who shot at it hit the MIG that chased him.*
c. *The man who gave his paycheck to his wife was wiser than the man who gave it to his mistress.*

Sentence (2a) is of interest because of difficulties that arise in assigning proper scope to the quantifiers when the pronouns are treated as bound variables (cf. Geach, 1962, and Carlson and Hintikka, 1975). Karttunen's (1971) discovery that Bach–Peters sentences [for example, (2b)] are ambiguous has shown that the problem of assigning interpretations to pronouns may interact in complex ways with the uniqueness conditions associated with definite descriptions. Finally, (2c) shows that the interpretation of a pronoun may differ from that of its antecedent in ways which are not readily captured in a bound variable treatment. These important and difficult problems are discussed by other contributors to this volume; I will have little to say about them.

1.3. Anaphora

A pronoun's interpretation is, typically, a function of its association with something else in the (linguistic or extralinguistic) context, usually a noun phrase or an object in the environment.[6] Pronouns that are so associated with linguistic expressions are called "anaphoric," and the associated expressions are called their "antecedents." The fundamental problem of anaphora (which I will refer to as "the anaphora problem") is to determine which features of discourse govern the association of anaphoric pronouns with the appropriate antecedents.

Most of the transformational literature on pronouns is concerned primarily with the anaphora problem, though the discussion of it has generally been phrased as though it were syntactic ("pronominalization") or semantic ("coreference"). I will argue that existing attempts to assimilate anaphora to either syntax or semantics are inadequate. Nevertheless, I believe that transformationalists have made considerable progress toward solving the anaphora problem, and this is what I now wish to discuss.

A possible solution to the anaphora problem (which has only very recently been seriously challenged) was discovered, apparently independently, by a number of investigators (among them Langacker, Postal, Lakoff, and Ross, and Matthews and Gross) in the mid-1960s (cf. Ross, 1967). I state it here under the name "Precede–Command Condition" (PCC).

(3) **Precede–Command Condition:** A noun phrase A may serve as the antecedent for a pronoun B (which agrees with A in the relevant features, including person, number, and gender) if and only if either
  (a) B follows A in the discourse, or
  (b) A and B are in the same sentence, and B does not command A.

[6] I say "typically" because, as Lasnik (1976) observes, if sufficiently many assumptions are shared by speaker and hearer, then neither a linguistic antecedent nor a referent need be present. If, for example, I stay home with a sick daughter while my wife is at work, my wife's first words on returning home can be "How is she?" with no problem of vagueness or ambiguity regarding the interpretation of the pronoun.

A node X is said to command a node Y if every S dominating X dominates Y.[7] PCC has the effect of allowing "backward" (that is, pronoun first) anaphora only when the pronoun is in a subordinate clause not containing the noun phrase. This is illustrated by the contrast in (4): *John* may be the antecedent of *he* in (4a) but not in (4b), since *he* commands *John* only in (4b).

(4) a. *When he woke up, John felt better.*
b. *He felt better when John woke up.*

PCC has enjoyed unusually widespread acceptance by transformational grammarians, during a period when they could agree on little else. Recently, however, some significant (and, to my mind, well-motivated) changes have been suggested. Making precise my vague suggestion (Wasow, 1972) that the notion of "command" should be generalized to "relative depth of embedding," Reinhart (1975, 1976) and Culicover (1976) independently arrived at a formulation that replaces "command" with what Reinhart calls "C-command," defined as (5a) and illustrated in (5b).

(5) a. X C-commands Y if every branching node dominating X dominates Y.[8]

b.

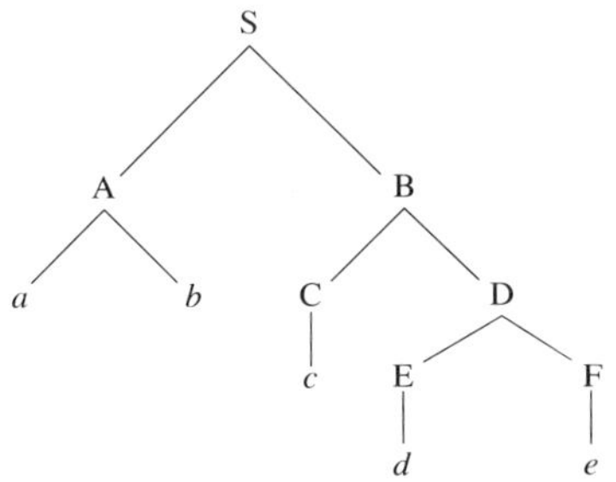

c C-commands both *d* and *e*, but neither *d* nor *e* C-commands *c*; *a* and *b* C-command each other, but nothing else.

Reinhart and Culicover argue that left-to-right order plays no role in the determination of anaphoric relations within sentences; that is, they effectively posit two anaphora rules: one operating within sentences and the other between sentences. Aside from agreement, the former requires only that the pronoun *not* C-command the antecedent, whereas the latter requires the pronoun to follow the antecedent. The Reinhart–Culicover intrasentential rule is given in (6).

[7] To avoid further complicating the statement of PCC, I stipulate that items in different conjuncts of a coordinate structure are not "in the same sentence."

[8] This notion is equivalent to Klima's (1964) concept of "in construction with"; I use Reinhart's terminology because it reflects the asymmetry of the relation far better.

(6) **C-Command Condition** (CCC): A noun phrase A may serve as the antecedent for a pronoun B in the same sentence if and only if they agree in person, number, and gender, and B does not C-command A.

In a great many cases, CCC agrees with PCC in its predictions. There are some differences, however, and CCC seems to handle these somewhat better. The sentences in (7) give some examples in which CCC allows anaphora between *him* (or *his*) and *John* but PCC does not. The parenthetical symbol following each sentence indicates the subsentential branching node that dominates the pronoun but not the antecedent and hence by the definition of C-command and CCC, permits coreference.

(7) a. *His mother loves John.* (NP)
b. *In his apartment, John smokes pot.* (NP)
c. *Near him, John saw a snake.* (PP)
d. *People admire him, in John's home town.* (VP)
e. *The chairman hit him on the head before John had a chance to say a word.* (VP)

Sentences in (8) give examples in which CCC forbids anaphora between *John* and *he*, but PCC allows it.

(8) a. *Near John, he saw a snake.*
b. *In John's apartment, he smokes pot.*
c. *John's wife and he have split up.*

The predictions of CCC strike me as more accurate than those of PCC in both (7) and (8). CCC is not, of course, without its problems (for example, accounting for apparent dialect differences regarding examples like (7a) and (8c)), but it seems to be a significant advance over PCC.

2. GENERALIZING THE PROBLEM OF ANAPHORA

Anaphora is clearly intimately related to both syntax and semantics: On the one hand, proposed solutions to the problems connected with assigning referents to pronouns generally assume that the appropriate pairing of pronouns and antecedents is given; on the other hand, both PCC and CCC are formulated in terms of syntactic properties. It is my contention, however, that the problem of anaphora is best treated as being neither a syntactic nor a semantic one.

One reason I take this position is that the notion of anaphora can be generalized in ways that cut across both syntactic and semantic distinctions. There are elements other than pronouns whose reference is a function of their association with antecedents, for example, epithets like *the fool* or expressions like *someone else*. Further, there are cases in which it is not determination of reference, but rather, interpretation of sense which depends on the association of some element with an antecedent. This is especially clear where the antecedent is not a noun phrase. Thus, the notion of anaphora can be generalized to designate any case in which an element's interpretation is a function of its being paired with something else in the linguistic context. Some examples of different types of anaphora are given in (9), with anaphoric elements and antecedents indicated.

(9) a. *After* ***John*** *pushed me in the pool,* ***the bastard*** *ate my sandwich.*
b. *Unless* ***John*** *is alone,* ***someone else*** *will come.*
c. *The man who said he is* ***bringing the wine*** *will_____.*
d. *People who* ***see that movie*** *regret* ***doing*** $\left\{ \begin{matrix} \textit{so} \\ \textit{it} \end{matrix} \right\}$.
e. *First John* ***fell in the lake,*** *and then* ***it*** *happened to me.*
f. *The people who understand how* ***John killed himself*** *don't know why_____.*
g. *Even the people who believe* ***that John is a Communist*** *publicly deny* ***it.***

These examples show that anaphoric elements need not be pronouns, noun phrases, or even words. They also show that the relationship between the interpretations of the anaphoric element and its antecedent need not have anything to do with reference, nor need it be a relationship of identity. Yet there are a number of properties which are common to these various types of anaphora. Foremost among these is that they all obey something like PCC or CCC[9] [see Wasow (to appear) for others].

(10) a. *After the bastard pushed me in the pool, John ate my sandwich.*
b. *Unless someone else comes, John will be alone.*
c. *The man who said he will is bringing the wine.*
d. *People who do* $\left\{ \begin{matrix} \textit{so} \\ \textit{it} \end{matrix} \right\}$ *regret seeing that movie.*
e. *After it happened to John, I fell in the lake.*
f. *The people who understand how don't know why John killed himself.*
g. *Even the people who believe it deny publicly that John is a Communist.*

[9] The examples given in (10) and (11) do not distinguish between CCC and PCC; to my knowledge, no one has tried to test the difference between the two conditions for any kind of anaphora besides simple pronominal anaphora.

(11) a. *The bastard ate my sandwich after John pushed me in the pool.*
b. *Someone else will come unless John is alone.*
c. *The man will who said he is bringing the wine.*
d. *Doing* $\begin{Bmatrix} so \\ it \end{Bmatrix}$ *upsets people who have seen that movie.*
e. *It happened to me after John fell in the lake.*
f. *Why isn't known by the people who understood how John killed himself.*
g. *It is denied publicly by even the people who believe that John a Communist.*

The anaphors in (10) may be associated with the same antecedents as in (9); those in (11) may not. This is just what the generalization of PCC or CCC to these other kinds of anaphora predicts. Thus, the class of relationships that obey CCC is heterogeneous, both syntactically and semantically. This suggests that it would be inadvisable to try to use formal machinery developed for either syntax or semantics when dealing with anaphora.

3. SYNTACTIC ANALYSES OF ANAPHORA

In the remainder of this paper, I will survey some of the attempts that have been made to handle anaphora with syntactic and/or semantic formalisms, and I will show some of the problems they encounter. In this section, I will deal with three attempts to handle anaphora by means of syntactic transformations.

3.1. "Pronominalization"

The first analysis of pronouns by transformational grammarians postulated a transformation of "Pronominalization," which turned full noun phrases into pronouns under appropriate conditions of identity (see especially Lees and Klima, 1963; Langacker, 1966; and Ross, 1967). It was widely accepted at that time that pronouns and their antecedents had to have the same referents; this was subsumed under the identity condition on the transformation by attaching "referential indices" to all noun phrases in deep structure and assigning identical referents to pairs of noun phrases if and only if they had identical indices [see Chomsky, 1965 (pp. 145–146)]. The transformation was written with two different "structural changes," one for "forward" and one for "backward" Pronominalization. PCC was made a condition on the applicability of Pronominalization. Thus, this analysis used syntactic machinery to answer all three kinds of questions: The distribution of pronouns, the association with antecedents, and the assignment of reference were all supposed to be accounted for by the transformation of Pronominalization, together with the indices. Not surprisingly, this theory encountered difficulties in all three domains.

A number of syntactic problems with Pronominalization have been noted in the literature [see Wasow (to appear) for a summary of these]. Instead of repeating them here, I will point out one which has not previously been discussed, so far as I know.

The generalization given here as (12) appears to hold in English.

(12) The set of environments in which full noun phrases may appear is a proper subset of the set of environments in which pronouns may appear.

That is, any position which allows a full noun phrase will also allow some pronoun, but not conversely.[10] If (12) is correct, it is surely a linguistically significant generalization, and hence should be captured by an adequate theory of pronouns.

The Pronominalization theory does not capture (12). Rather, it predicts that the inclusion relation should be just the opposite, that is, all pronoun positions should be possible noun phrase positions.[11] The sentences in (13) shows that this is not the case.

(13) a. *John left, didn't* $\left\{\begin{array}{l}\textit{he}\\ \textit{*the man}\end{array}\right\}$?

b. *Mary lost* $\left\{\begin{array}{l}\textit{her}\\ \textit{*Sue's}\end{array}\right\}$ *temper.*

c. *I know a person* $\left\{\begin{array}{l}\textit{whose}\\ \textit{*the person's}\end{array}\right\}$ *mother is 106 years old.*

[10] There are several apparent counterexamples. The clearest is the position immediately following a verbal particle.

(i) *John put away* $\left\{\begin{array}{l}\textit{the book}\\ \textit{*it}\end{array}\right\}$.

Another, pointed out by Colin Biggs (personal communication) is after *just* in subject position.

(ii) *Just* $\left\{\begin{array}{l}\textit{the women}\\ \textit{??they}\end{array}\right\}$ *left early.*

I suspect that these are just instances of constructions requiring stressed constituents, so the normal stresslessness of pronouns makes them sound bad. The examples are much improved, in my opinion, if put in contexts in which the pronouns get contrastively stressed. In any event, even if (12) is a bit too strong as given, I think there is a significant generalization to be captured here. Notice, by the way, that my proposals in Section 4 (p. 218) do not rule out the possibility of constraints against the appearance of pronouns in certain environments. However, they do entail that exceptions to (12) will require complicating the grammar by adding such constraints, thus predicting that there will be few, if any, exceptions.

[11] Even if Pronominalization is obligatory, pronoun positions would be filled by full noun phrases on the surface when the identity condition is not met.

Such examples were dealt with by advocates of Pronominalization by means of transformations that obligatorily created pronouns and whose conditions for applicability had to be satisfied, under certain circumstances [for example, those characterizing (13b) and (13c)]. Though the individual cases might be handled in this way, the Pronominalization theory offers no principled reason why (12) should be the case. Further, it says nothing at all about the distribution of nonanaphoric pronouns, which, it turns out, may appear in precisely the same environments as full noun phrases.

The Pronominalization theory is also inadequate on semantic grounds. This is basically because it treats all pronouns as "pronouns of laziness."[12] As a consequence, it can offer no plausible source for pronouns which are interpreted as variables. Consider (14).

(14) *Only Jerry believes he can win.*

One interpretation of (14) is that Jerry is the only individual who believes that he himself can win; the salient features of this reading can be represented very roughly as (15).

(15) $-(\exists x)(x \neq \textit{Jerry} \ \& \ x \textit{ believes } x \textit{ can win})$

But the Pronominalization theory has only two sources for (14), deriving *he* either from *Jerry* or from *only Jerry* under identity with (a part of) the subject NP, and neither of them would mean anything that could be analyzed as (15). The sources for (15) available to the Pronominalization theory are given as (16a) and (17a), with the corresponding interpretations crudely schematized as (16b) and (17b).

(16) a. *Only Jerry believes Jerry can win.*
b. $-(\exists x)(x \neq \textit{Jerry} \ \& \ x \textit{ believes Jerry can win})$

(17) a. *Only Jerry believes only Jerry can win.*
b. $-(\exists x)(x \neq \textit{Jerry} \ \& \ x \textit{ believes } -(\exists y)(y \neq \textit{Jerry} \ \& \ y \textit{ can win}))$

Thus, Pronominalization does not suffice as an account of the semantics of pronouns.

Finally, Pronominalization even fails as an account of the distribution of anaphoric relations (quite apart from any problems with PCC). Consider the underlying structure abbreviated here as (18).

(18) *A man* $[_S$*who* $\textit{John}_i$ *respected*$]$ *told* $\textit{John}_i$ $[_S$*that* $\textit{John}_i$ *was a fool*$]$.

Since there are three occurrences of $\textit{John}_i$ in (18), Pronominalization will be applicable twice. Suppose the first application pronominalizes the second occurrence of $\textit{John}_i$ under identity with the first, and the second application

[12] This term is originally due to Geach (1962), though it has come, over the years, to be used to cover a much wider range of cases than Geach had suggested.

pronominalizes the first occurrence under identity with the third. The resulting sentence will be (19), which does not permit an interpretation on which *he, him,* and *John* are all anaphorically related (that is, in this case, coreferential).

(19) *A man who he respected told him that John was a fool.*

There is apparently no way to block this derivation under a Pronominalization analysis. The underlying structure (18) cannot be ruled out, for it also underlies (20), which **does** allow the anaphoric reading.

(20) *A man he respected told John that he was a fool.*

Nor can either of the applications of Pronominalization which led to (19) be ruled out: Completely parallel applications would be needed in order to derive the anaphoric readings of (21a) and (21b).

(21) a. *A man John respected told him that I was a fool.*
b. *A man he respected told me that John was a fool.*

Thus, the Pronominalization theory cannot account for the impossibility of an anaphoric interpretation of (19).

3.2. The Bound Variable Theory

One alternative to Pronominalization which has attained considerable currency is what I have referred to elsewhere (Wasow, 1975) as the **Bound Variable Theory** (BVT). The essential feature of all variants of BVT is that each noun phrase position in underlying structure is occupied by a variable, bound by some sort of specification of the noun phrase which is later to occupy that position. For example, McCawley (1970) sketches an analysis which would derive (22a) from (22b).

(22) a. *John said that he meditates.*
b.

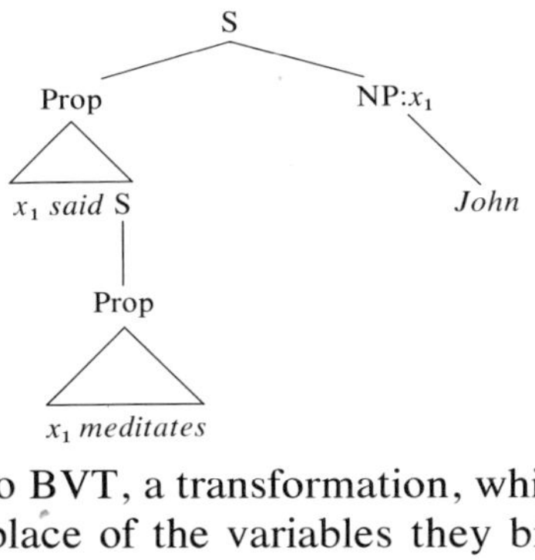

According to BVT, a transformation, which I call Substitution, inserts noun phrases in place of the variables they bind. When one noun phrase binds more than one variable, as in (22b), it is substituted for only one of them; the remaining ones become the appropriate pronouns. Most versions of BVT assume that something like PCC is a condition on Substitution, so that not every occurrence of a bound variable is a possible Substitution site (but see my discussion of Harman's version of BVT in the next section).

Assuming (as most adherents of BVT have) that a different source can be provided for "pronouns of laziness," I can find no major **semantic** problems with BVT as such (though McCawley's version used here is undoubtedly too simple). Although the semantic questions concerning pronouns have not, by any means, been completely answered, there seems to be a fairly solid consensus among people working on them that the answers will involve representing at least some pronouns as bound variables. I see no reason to quarrel with this conclusion.

However, BVT asserts more than this. According to BVT, surface pronouns are **transformationally derived** from underlying bound variables. Consequently, BVT makes claims not just about semantics, but about syntax, too. I have argued elsewhere in some detail that these claims are incorrect (Wasow, 1975), so here I shall present only a short sketch of my syntactic objections to BVT.

All of my arguments against BVT were generalizations of an argument against Pronominalization due to Bresnan 1970. The essence of the argument is the following: If full noun phrases and pronouns are derived from the same syntactic source, the rule which creates the difference between them (Pronominalization or Substitution, as the case may be) must apply before any rule which treats pronouns and full noun phrases differently; this leads to ordering paradoxes.

For example, consider (23c), which is a rough approximation of the underlying structure from which (23a) and (23b) would be derived in McCawley's version of BVT (with *they* and *them* interpreted anaphorically).[13]

(23) a. *Crooks assassinated a candidate who found that they were on the payroll.*

b. *A candidate who found that there were crooks on the payroll was assassinated by them.*

c.

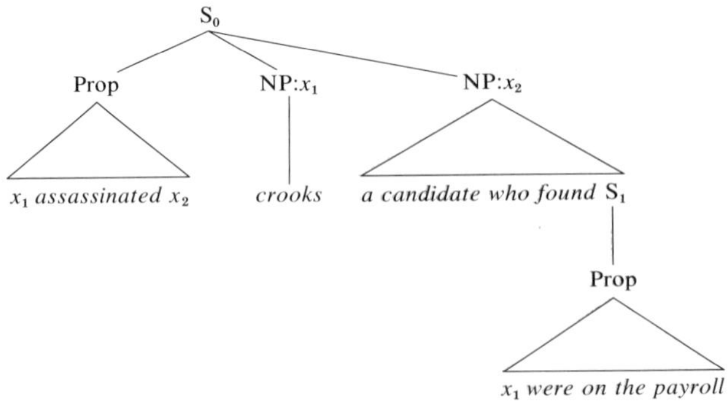

The derivation of (23b) involves the application of two transformations which play a crucial role in my argument, namely Passive and *There*-Inser-

[13] This argument also holds for other versions of BVT that I am familiar with.

tion. I will assume without argument that these rules exist and apply cyclically, that is, one clause at a time, starting at the bottom of the tree and working up.[14] It follows from the cyclicity of these rules that *There*-Insertion applies before Passive in the derivation of (23b), for *there* is inserted into S_1, but Passive is applied to S_0. I now wish to consider the question of when Substitution for x_1 applies. It cannot apply before Passive, for whether Passive applies may determine which occurrence of x_1 is replaced by *crooks*. If Passive does apply, *crooks* may be substituted for the occurrence of x_1 in S_1, yielding (24a), which permits an anaphoric interpretation of the pronoun; if Passive does not apply, *crooks* may not be substituted for the occurrence of x_1 in S_1, as evidenced by the fact that (24b) does not permit an anaphoric interpretation of the pronoun.

(24) a. *A candidate who found that crooks were on the payroll was assassinated by them.*
b. *They assassinated a candidate who found that crooks were on the payroll.*

In short, how Substitution applies depends on whether Passive applies. We may conclude that Passive precedes Substitution for x_1 in derivations from (23c). Now, notice that Substitution for x_1 must precede *There*-Insertion in S_1, for *There*-Insertion may not apply if *crooks* is substituted for the main clause occurrence of x_1, as (25) shows.

(25) **Crooks assassinated a candidate who found that there were they on the payroll.*

In other words, whether *There*-Insertion may apply depends on how Substitution applies. We may conclude that Substitution for x_1 precedes *There*-Insertion in derivations from (23c). So we have shown that Passive must apply before Substitution for x_1, which must apply before *There*-Insertion, which (by the transformational cycle) must apply before Passive. This implies that no derivation from (23c) can involve applying all three transformations. However, (23b) shows that this conclusion is incorrect. Hence, we have an ordering paradox.

The point is that certain syntactic processes differentiate between full noun phrases and their corresponding pronouns. Since BVT derives both from underlying variables, it requires that Substitution must apply before these processes can take place. My claim is that this is impossible without considerable ad hoc modification of BVT.

I believe that BVT also provides an unsatisfactory answer to what I am calling the problem of anaphora. If CCC (or something like it) is simply a condition on the applicability of Substitution, then there is no reason to expect the same condition to obtain elsewhere. As noted in Section 1.3, however, PCC or CCC seems to hold for all anaphoric relations, not just those

[14] These assumptions would be widely (but not universally) accepted by transformationalists; for arguments supporting them see Akmajian and Heny, 1975; and Wasow, 1975.

involving pronouns (with "bound variable" interpretations). It would be no trivial matter to generalize BVT in such a way that all varieties of anaphora would involve Substitution. Doing so would require introducing verb phrase (or, possibly, predicate) variables, sentence (or proposition) variables, and perhaps other types of variables as well, in order to accommodate anaphora involving these different kinds of antecedents. Although it may well be possible to construct a theory of this sort, the burden of proof clearly lies with its advocates.

Like Pronominalization, BVT is an attempt to use the transformational machinery developed for syntax to answer the semantic and anaphoric questions about pronouns. BVT is semantically far more sophisticated than the Pronominalization theory, but, once again, failure to distinguish among the types of questions leads to problems with the syntax and anaphora.

3.3. Harman's Variant of BVT

Harman (to appear, and 1976) has suggested a variant of BVT which is not subject to the above criticisms, at least in the form given here. Harman's position is that Substitution is not subject to any special condition such as PCC or CCC. Rather, he argues (i) that Substitution replaces only the leftmost occurrence of any variable by the corresponding noun phrase, and (ii) that Substitution "is not part of any transformational cycle, and can apply freely in any order in relation to many transformations that are often taken to be part of the cycle [1976, p. 78]." Thus, under Harman's theory, cases of "backward" pronominal anaphora will always be cases in which Substitution has applied prior to some reordering rule (in some cases, other applications of Substitution itself).[15] As Harman (1976, observes (numbering changed):

> Within this framework, certain movement transformations must be carefully reformulated. For example, Passive must not be allowed to transform (26) into (27)
>
> (26) *The woman John loved kissed him.*
>
> (27) *He was kissed by the woman John loved.*
>
> [Harman's] proposal, therefore, is that such transformations should be explicitly stated so as to move only variables. In that case, Passive could not apply to (26) [p. 78].

Rather, it would apply to a pre-Substitution structure of the form *x kissed y*.

In an earlier paper (Wasow, 1975), I raised several objections to this proposal, to which Harman (1976) has replied. At this point I would like to consider some aspects of that reply and to raise some further objections.

One of my arguments was that Passive could not be restricted to applying

[15] Notice that this requires that examples like (i) involve a rule preposing the adverbial clause, which, in this instance, applies after Substitution.

(i) *After he left, John drove home.*

only to variables, since it could front the existential *there,* which is not represented by a variable at any stage of a derivation. This is illustrated by (28).

(28) *There is believed to be a monster in Loch Ness.*

Harman's reply to this objection is the following: "We can easily accommodate this point . . . by means of a slight modification of the original proposal. We need only say that variable moving rules can also move such *there's* [p. 79]." It seems to me that this reply leaves my original criticism largely intact. Any finite number of exceptions to a generalization can be accommodated by stipulating that they are exceptions; however, a competing generalization which does not have to treat these cases as exceptional is naturally preferable. Harman has proposed the generalization that Passive moves only variables, but he must stipulate *there* as an exception: The usual formulation of Passive as moving noun phrases needs no such stipulation. Notice, further, that *there* is not the only exception to Harman's generalization. Idiom chunks and the so-called "weather" *it*, which are presumably not underlying variables, can also passivize.

(29) a. *Advantage has been taken of John by unscrupulous operators.*
b. *Tabs are being kept on all known radicals.*
c. *It is believed to be raining in the affected areas.*

In support of the contention that the fronted noun phrases in (29) are not derived from variables, notice that they may not be antecedents to pronouns. Thus, the examples of (30) are ungrammatical, at least on the anaphoric interpretations.

(30) a. ?? *John took advantage of me, and then Mary took it of him.*
b. ?? *The FBI keeps tabs on radicals, and the CIA keeps them on diplomats.*
c. * *It was raining, but nobody knew where it was.*

It seems as though Harman would have to expand his list of exceptions, so I think my objection has considerable force.

A second objection had to do with the English double-object construction.[16] It is well-known that the second object in this construction cannot be a definite pronoun.[17]

[16] In my earlier paper, I formulated this in terms of a transformation of Dative Movement. Since I no longer believe in the rule (thanks mostly to the influence of Richard Oehrle—see Oehrle, 1975), I will formulate the argument in more neutral terms here. I no longer think my (1975) formulation of the Dative Movement argument against BVT goes through, but I am still convinced that double object constructions provide evidence that pronouns and full noun phrases must be differentiated earlier in a derivation than BVT allows.

[17] The indefinite pronoun *one* may be the second object, so this is no counterexample to (12).

(i) *John wouldn't send the children one.*

(31) *John was* {*sending* / *telling*} *the children* {*a story* / *stories* / **it* / **them*}.

Since many of the same verbs may take single pronominal objects (as in (32)), and since passivization of the first object (as in (33)) does not improve examples like (31), it is reasonable to conclude that the constraint which rules out (31) must apply prior to Passive; otherwise (32) and (33) would be very hard to distinguish.

(32) *The children were* {*sending* / *telling*} {*a story* / *stories* / *it* / *them*}.

(33) *The children were* {*sent* / *told*} {*a story* / *stories* / **it* / **them*}.

In Wasow 1975, I concluded from this that Substitution would have to apply before Passive in examples like (33), and I incorrectly claimed that this conflicted with Harman's analysis of Passive as a variable moving rule. As Harman (1976) points out, these observations only require that Substitution must apply to the second object before Passive applies. Since Passive fronts the first object, there is no inconsistency.[18] There is, however, a slightly different problem. There are speakers who allow the second object of a double object construction to be fronted by Passive, in case the first object is a pronoun. Thus, many speakers [who agree with the judgments given in (31)–(33)] find examples like (34) quite acceptable.

(34) *This ring was given me by my mother.*

Such examples, if grammatical, pose a serious problem for Harman. They

[18] In fact, Harman (1976, p. 80) tries to turn these and related facts into an argument for his analysis. Since *Tough* Movement can alter anaphoric relations, Harman must treat it as a variable movement rule as well. Hence, though *Tough* Movement normally can front any noun phrase following the complement verb, Harman's analysis predicts that it cannot front the second object in a double object construction (since that must already be a full noun phrase). Harman gives (i) and (ii), starring (ii), in support of his prediction.

(i) *John is easy to give presents.*
(ii) *Presents are easy to give John.*

My intuitions on these and related examples are not clear; I certainly do not find in them the clear contrast Harman claims exists. Indeed, the existing literature on the operation of *Tough* Movement in double object constructions claims that only the second object can be fronted (see Culicover and Wexler, 1973).

require Passive to apply to the second noun phrase of the double object construction, which, on Harman's theory, must be a full noun phrase when Passive applies; this contradicts Harman's hypothesis that Passive only moves variables.[19]

Though the arguments are a bit trickier to construct, I think I have shown that Harman's version of BVT is subject to the same sort of syntactic objections as other versions. Similarly, I think that Harman's analysis is an inadequate theory of anaphora, though, once again, the general argument I presented in the last section requires some modification if it is to apply to Harman's version of the theory.

My argument was, in essence, that the generality of PCC (or CCC) cannot adequately be captured by BVT. Since Harman's variant of BVT does away with any condition like PCC, this objection clearly requires modification. What replaces PCC in Harman's theory is the freedom from ordering constraints accorded Substitution; backward anaphora results from permutations following Substitution. There is no reason, however, to think that Substitution would be the only unordered rule in the grammar of English. And if there are other unordered rules, one might expect that they, too, would appear to obey a constraint like PCC. In short, Harman's theory leads one to expect that PCC would not be restricted to pronominal anaphora.

Let me illustrate this with a concrete example, namely, the rule often referred to as VP Deletion. Under Harman's theory, this rule could be stated very simply as (35).

(35)

X –	VP –	Y –	VP –	Z	
1	2	3	4	5	⇒
1	2	3	∅	5	

where 2 = 4

Sentences like (36a) and (36b) would both be derived from a structure like (37), but in the derivation of (36a), Substitution would apply before VP deletion and in (36b), the order would be reversed.

(36) a. *A woman who wants to run for office should.*
b. *A woman who wants to should run for office.*

[19] Note also that passivization of the second object is possible only if the first one is a pronoun, as (i) illustrates.

(i) ??*This ring was given my wife by my mother.*

Hence, it must be possible to distinguish pronouns from full noun phrases in the first object at the point in the derivation where Passive applies. This appears to violate Harman's assumption that the first noun phrase position must be occupied by a variable when Passive applies. In fact, however, Harman's proposal only requires that the items which Passive moves be variables; since it is the second object which gets fronted in (34), the first object need not be a variable.

(37)

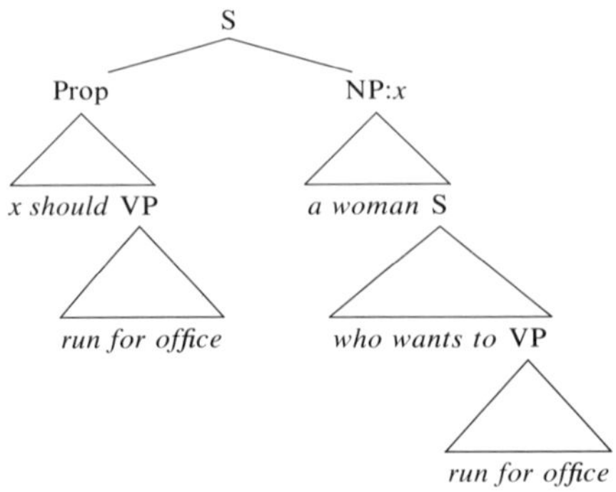

Thus, the fact that VP deletion obeys a constraint like PCC can be expressed very naturally in Harman's analysis. This is a very appealing feature of the analysis.

There are, however, complications. As was the case with Substitution, VP deletion must be blocked from applying before certain transformations. For example, if VP deletion is applied to (38a), yielding (38b), we cannot subsequently passivize to get (38c).[20]

(38) a. *After John saw me, Mary saw me.*
b. *After John saw me, Mary did.*
c. *After I was seen by John, Mary did.*

Some sort of ordering (presumably intrinsic) between VP deletion and Passive must be established to prevent the latter from applying to a clause whose verb phrase has served as the second term in some application of (35).

Similarly, the rule of Sluicing (see Ross, 1969) cannot be allowed to apply before Passive, or (39a) will result in (39c).

(39) a. *The man who slapped me did not explain why he slapped me.*
b. *The man who slapped me did not explain why.*
c. *Why was not explained by the man who slapped me.* (cf. *Why he slapped me was not explained by the man who slapped me.*)

Indeed, it can be shown that no anaphoric relation can be allowed to be established before passivization.[21] Hence, Harman would have to construct his grammar in such a way that Passive always applies before any of these otherwise unordered rules.

It is not clear how this could be done, short of extrinsic ordering. The device which Harman used for pronominal anaphora, namely, formulating Passive in terms of variables only, will not suffice for the other cases. Perhaps it

[20] Harman explicitly rejects the principle of the transformational cycle, so it cannot be invoked in (38c).

[21] See Wasow (to appear) for the relevant evidence (though presented in another context).

can be extended by introducing verb phrase and sentence variables, but, as I noted before, the burden of proof is on the person who claims that such a solution is feasible. Although extrinsic ordering could, in principle, be invoked, I assume Harman would not accept such a solution; for allowing the same rule to be extrinsically ordered with respect to some rules, and unordered with respect to others seems quite ad hoc. Further, it leaves unexplained the fact that Passive must precede all anaphora rules; extrinsic ordering would permit it to apply in the midst of them.

In short, whereas some aspects of Harman's variant of BVT generalize elegantly to nonpronominal anaphora, others do not. Hence, it is subject to much the same charge of failing to capture generalizations about anaphora as are other versions of BVT.

One final problem with Harman's proposal is that it cannot readily accommodate some of the examples which seem to prefer CCC over PCC. Other analyses of anaphora discussed here can be modified simply by replacing PCC with CCC; since Harman's analysis rejects PCC in favor of free ordering of Substitution, it is not clear how the corresponding modification can be made. More specifically, Harman's analysis, as presented, would not allow anaphoric interpretations of (7d) and (7e), repeated here for convenience.

(7) d. *People admire him, in John's home town.*
e. *The chairman hit him on the head before John had a chance to say a word.*

If Reinhart and Culicover are right in claiming that these sentences do allow an anaphoric interpretation of *him,* then Harman's theory must be revised in some way which is not obvious.

4. THE INTERPRETIVE THEORY

The leading alternative to BVT within transformational grammar takes as its main premise that the syntax and semantics of pronouns are independent problems. Under this theory, pronouns are generally assumed to be base-generated, and their meanings are determined by "interpretive" rules operating on derived structure. Most of the interpretivist literature emphasizes the autonomy of syntax from semantics, with little, if any, attention paid to the distinction between semantic questions and the problem of anaphora.[22]

In the remainder of this paper I want to sketch a version of an interpretivist theory of pronouns which distinguishes among the following: syntactic rules of the familiar types used in transformational grammar, a rule or set of

[22] Jackendoff (1972) and Lasnik (1976) make no distinction between these at all. Wasow (1973, to appear) and Chomsky (1976) do make the distinction, but give it little attention.

rules for pairing anaphoric elements with antecedents (on the basis of syntactic information), and rules for assigning interpretations to anaphoric elements (on the basis, at least in part, of semantic properties of their antecedents). In particular, I will attempt to show how some of the problems I have discussed in this paper could be handled in such a theory.

First of all, I will suppose that one base expansion for the category noun phrase is the lexical category "pronoun," and that pronouns are inserted into such positions in the base. It follows that any (deep structure) position allowing full noun phrases will allow pronouns.

Second, I claim that pronoun positions which do not permit full noun phrases result from the application of transformations of a certain kind, specifically, those which mention "identity" between two noun phrases. More precisely, I wish to propose as a universal constraint on the formulation of transformations that structural conditions and structural changes of transformations never mention true identity between terms [as, say, Peters and Ritchie's formalization (1973) would allow them to]. Instead (at least in the case of noun phrases), they may specify that one term must be a pronoun which is anaphorically bound to another term. This allows transformations to make obligatorily anaphoric pronominal copies of existing noun phrases, to move noun phrases, leaving behind bound pronominal copies. or to perform some operation (movement or deletion) to a noun phrase just in case it is a pronoun anaphorically related to some other noun phrase. The existence of these sorts of cases is widely (though not universally) accepted by transformationalists.

(40) a. ***John** is here, isn't **he**?* (copying)
b. ***He**'s a fool, **your candidate**.* (movement with a copy left behind)
c. ***The man** (**whom**) I saw was wearing a mask.* (movement or deletion subject to anaphora)

In contrast, there are no well-established cases of copying, movement, or deletion of noun phrases being contingent on true identity.[23] Thus, the transformational literature lends some plausibility to the proposal advanced here that transformations may mention anaphora, but not identity. If all nonpronominal noun phrases are base-generated, while pronouns can be either base-generated noun phrases or transformationally inserted, then the generalization given as (12) follows (*modulo* my remarks in Footnote 10). That is, any noun phrase slot may be filled by a pronoun, but some pronouns are inserted into positions from which full noun phrases are excluded,[24] so the set

[23] See Peters, 1973 for one rather unconvincing attempt to establish the need for noun-phrase deletion under true identity.

[24] This formulation needs some modification in the case of relative pronouns, which I am assuming are not transformationally inserted. Rather, I assume that any underlying structure with a full noun phrase in the place of the relative pronoun will be filtered out by the relativization transformation (see Chomsky, 1965).

of possible full noun phrase positions is a proper subset of the set of possible pronoun positions.

Third, I propose that only those anaphoric relations which are required by some transformational operation are obligatory. Anaphora rules may prohibit the establishment of anaphoric relations between certain pairs (cf. Lasnik, 1976), but they may not require two items to be anaphorically related.[25] Hence, except where transformations produce obligatory anaphora, anaphoric relations must be optional. Since we know that pronouns which are assigned antecedents transformationally are precisely those in environments which exclude full noun phrases, it follows that no pronoun in a position allowing full noun phrases may enter into obligatory anaphoric relations. This conclusion appears to be borne out by the facts. Indeed, Dougherty (1969) observed just this state of affairs, labeling it "The Anaporn Relation." By stipulating that all and only transformations that mention something like "identity" can establish obligatory anaphoric relations, we are able to deduce both Dougherty's Anaporn Relation and the generalization given as (12) above (but cf. Footnote 10).

Fourth, I propose that the following constraint is universal:

(41) No anaphoric element may C-command its antecedent (in surface structure).

Thus, even if different anaphoric relations may arise in rather different ways (as suggested by Hankamer and Sag, 1976), they will all obey CCC.[26] Under such a theory, the generality of CCC does not require that all rules which may establish anaphoric relations be formally alike,[27] so no analogue to this problem for BVT arises.

Finally, I propose that the relation "is anaphorically related to" is transitive, that is, if A is anaphorically related to B, and B is anaphorically related to C, then A is anaphorically related to C. I expect that this transitivity will be an automatic consequence of any adequate definition of anaphora, but I

[25] The reciprocal (*each other*) construction in English appears at first to be a counterexample, that is, a case in which obligatory anaphora obtains, but no transformation is involved. I would argue, however, that the anaphora is not obligatory, but, rather, that the anaphora rule for *each other* is quite specific and excludes most potential antecedents. Hence, there is usually only one possible antecedent for *each other*. This is not always the case, as the ambiguity of (i) illustrates.

(i) *The men told the women about each other.*

Stronger evidence for the nonobligatoriness of reciprocal anaphora is provided by the fact that *each other* need not have a linguistic antecedent; thus, as Emmon Bach has observed, (ii) is possible if addressed to two people, regardless of linguistic context.

(ii) *Here's a book to read to each other.*

See Lasnik and Fiengo, 1973 for some interesting discussion.

[26] Hankamer (1976) makes a similar point, though he formulates it very differently.

[27] Thus, my position here is not the same as it was in my earlier work, especially Wasow (to appear).

cannot really justify this expectation. In any event, if the anaphoric relation is transitive, the unwanted interpretation of (19), repeated here for convenience, is easily excluded.

(19) *A man he respected told him that John was a fool.*

If anaphoric relations are established between *he* and *him* and between *he* and *John,* then, by transitivity, we can conclude that an anaphoric relation must hold between *him* and *John*. Since *him* is a pronoun and *John* is not, *John* must be the antecedent of *him*. But this violates the general prohibition against anaphoric elements C-commanding their antecedents (that is, (41)). Thus, the problem which the Pronominalization theory encountered in connection with (19) can be easily and naturally avoided in an interpretive theory. The actual distribution of pronouns under this theory results, therefore, in part from base generation, and in part from the application of transformational deletion and copying rules, while the distribution of anaphoric relationships is controlled quite separately by the universal constraint (41). In this it contrasts with the earlier proposals within transformational grammar, Pronominalization and BVT, described in Section 3. Moreover, under the present proposal, no attempt is made to account for the semantics of pronouns by means of the same mechanism that accounts for their distribution or that which controls anaphora. The interpretation of pronouns, whether as bound variables or as pronouns of laziness is a separate question about which I shall say nothing on this occasion.

5. CONCLUSION

Much more could be said. I have not touched on many important ideas in the transformational literature on pronouns. I hope, however, that I have painted a fair picture of the linguistic research into pronouns and anaphora over the past decade and a half. There has been definite progress on some fronts (especially the anaphora problem), but, in my view, a good deal of energy has been expended chasing up blind alleys, often in pursuit of solutions to problems which were artifacts of the theories current at the time.

I have tried to argue that the genuine questions about pronouns which need to be answered fall into three general categories, and that fewer problems arise if we keep our answers to the different categories of questions distinct.

REFERENCES

Akmajian, A., and Heny, F. (1975). *An Introduction to the Principles of Transformational Syntax*. Cambridge, Mass: MIT Press.

Bresnan, J. (1970). An argument against pronominalization, *Linguistic Inquiry,* **1**(1), 122–123.

Carlson, L., and Hintikka, J. (1975). Conditionals, generic quantitifers, and other applications of subgames. Manuscript.
Chomsky, N. (1965). *Aspects of the Theory of Syntax.* Cambridge, Mass: MIT Press.
Chomsky, N. (1976). On wh-movement. In P. Culicover, T. Wasow, and A. Akmajian (Eds.), *Formal Syntax,* New York: Academic Press.
Culicover, P. (1976). A constraint on coreferentiality. *Foundations of Language,* **14**(1), 109–118.
Culicover, P., and Wexler, K. (1973). An application of the freezing principle to the dative in English. *Social Sciences Working Papers,* **39,** School of Social Sciences, University of California. Irvine, California.
Delorme, E., and Dougherty, R. (1972). Appositive NP constructions. *Foundations of Language,* **8**(1), 2–29.
Dougherty, R. (1969). An interpretive theory of pronominal reference, *Foundations of Language,* **5**(4), 488–519.
Geach, P. (1962). *Reference and Generality,* Ithaca, New York: Cornell University Press.
Greenberg, J. (1963). Some universals of grammar with particular reference to the order of meaning elements. In J. Greenberg (Ed.), *Universals of language.* Cambridge, Mass: MIT Press.
Hankamer, J. (1976). The semantic interpretation of anaphoric expressions. Manuscript.
Hankamer, J., and Sag, I. (1976). Deep and surface anaphora. *Linguistic Inquiry* **7,** 391–428.
Harman, G. (1976). Anaphoric pronouns as bound variables: Syntax or semantics? *Language,* **52**(1), 78–81.
Harman, G. (In press) Noun phrases derived from variable binding operators. In B. Freed, A. Marras, and P. Maynard (Eds.), *Forms of representation,* Amsterdam: North-Holland.
Jackendoff, R. (1972). *Semantic Interpretation in Generative Grammar,* Cambridge, Mass: MIT Press.
Karttunen, L. (1971). Definite descriptions with crossing coreference, *Foundations of Language,* **7**(2), 157–182.
Klima, E. (1964). Negation in English. In J. Fodor and J. Katz (Eds.), *The Structure of Language,* Englewood Cliffs, New Jersey: Prentice-Hall.
Langacker, R. (1966). Pronominalization and the chain of command. In D. Reibel and S. Schane (Eds.), *Modern Studies in English.* Englewood Cliffs, New Jersey: Prentice-Hall.
Lasnik, H. (1976). Some remarks on coreference. *Linguistic Analysis,* **2,** 1–22.
Lasnik, H., and Fiengo, R. (1973). The logical structure of reciprocal sentences in English, *Foundations of Language,* **9**(2), 447–468.
Lees, R., and Klima, E. (1963). Rules for English pronominalization, *Language,* **39**(1), 17–28.
McCawley, J. (1970). Where do noun phrases come from? In R. Jacobs and P. Rosenbaum (Eds.), *Readings in English Transformational Grammar.* Waltham, Mass: Ginn & Co.
Oehrle, R. (1975). *The Syntax and Semantics of the Dative Alternation in English.* Unpublished doctoral dissertation, MIT.
Peters, S. (1973). On restricting deletion transformations. In M. Gross, M. Halle, and M-P. Schützenberger (Eds.), *The Formal Analysis of Natural Languages.* The Hague: Mouton.
Peters, P., and Ritchie, R. (1973). On the generative power of transformational grammars. *Information and Control,* **6,** 49–83.
Postal, P. (1966). On so-called "Pronouns" in English. In D. Reibel and S. Schane (Eds.), *Modern Studies in English.* Englewood Cliffs, New Jersey: Prentice-Hall.
Reinhart, T. (1975). Syntax and coreference. In *Proceedings of the Fifth Annual Northeastern Linguistic Society Conference.* Harvard University Department of Linguistics.
Reinhart, T. (1976). *The Syntactic Domain of Anaphora.* Unpublished doctoral dissertation, MIT.
Ross, J. (1967). On the cyclic nature of English pronominalization. In D. Reibel and S. Schane (Eds.), *Modern Studies in English,* Englewood Cliffs, New Jersey: Prentice-Hall.
Ross, J. (1969). Guess who? In *Papers from the Fifth Annual Meeting of the Chicago Linguistic Society.* University of Chicago, Department of Linguistics.

Sommerstein, A. (1972). On the so-called definite article in English. *Linguistic Inquiry,* **3**(2), 197–209.
Wasow, T. (1972). *Anaphoric Relations in English.* Unpublished doctoral dissertation, MIT.
Wasow, T. (1973). More migs and pilots. *Foundations of Language* **9**(3), 297–305.
Wasow, T. (1975). Anaphoric pronouns and bound variables. *Language,* **51**(2).
Wasow, T. (In press). *Anaphora in Generative Grammar.* Ghent: Story-Scientia.

PART II

OTHER ANALYTICAL PROBLEMS

A MODEL-THEORETIC ACCOUNT OF THE CONCEPT OF CAUSALITY IN NATURAL LANGUAGE

JOACHIM BALLWEG

1. INTRODUCTION

The ideas I want to sketch in this paper are to be understood within the framework of a verbal grammar which is roughly of the following shape:

The base consists of a categorial syntax with S (for Sentence) and d (for Designator) as main categories and with lambda abstraction. The structures yielded are input to a semantic interpretation component which is stated below in some detail, and to a Lexicon II (as opposed to Lexicon I, which lists the items of our base language $^{\lambda}$L), which maps expressions of $^{\lambda}$L onto corresponding lexical items. For example, in λx[WALK(x)] *John*, λx[WALK(x)] is translated into the English verb *to walk;* by this procedure, we arrive at a shallow syntax I, which is still categorial. In the next step, surface case indices are introduced, and finally some morphological spelling-out rules, topological rules, etc. give us the surface structures. In this

SYNTAX AND SEMANTICS
VOLUME 10

ISBN: 0-12-613510-X

paper, however, I will be mainly concerned with the semantic aspects of the treatment of causative verbs; that is, their entries in the Lexicon II.[1]

The general intuitive basis of the concept of causality given here is that two events are causally linked just in case we can reproduce the second event by merely producing the first event—under circumstances not too dramatically different from the original ones. This last condition somehow reflects the notion of an experimental setting: In natural sciences, one is mainly interested in statements of the type *x causes y* under a certain set of given circumstances, (for example, atmospheric pressure, temperature and the like); in natural language and everyday speech, however, the circumstances are normally not mentioned explicitly, and if I say *Drinking five whiskies yesterday almost gave me the works,* it is normally understood that the drinking of such an amount of booze will bring about the deplorable effect of almost giving me the works if I am in a condition like I was yesterday, but that is not to say that every five whiskies I drink in a row will produce the same effect. To capture these observations, we must construct our semantical apparatus in such a way that we are able to give a formal treatment of the idea of similarity of circumstances.

2. MACHINERY

Let $\mathbf{M} = \langle w_0, \mathrm{S}, \mathbf{W}, \mathbf{D}, \langle T, <, \bar{T}, \bar{t}_0, \{\underset{A}{\geq}\}_{\bar{T}}\rangle, V\rangle$ be a model, where

w_0 is a state of the "actual world."
S is a function (defined below).
$\mathbf{W}$ is a set of "possible world states."
$\mathbf{D}$ is the individual domain.
$\langle T, <, \bar{T}, \bar{t}_0, \{\underset{A}{\geq}\}_{\bar{T}}\rangle$ is a temporal frame.
V is the evaluation function.

A possible world state w_x is determined

[1] The work presented here is part of a project "Erarbeitung einer Verbvalenzgrammatik mit semantischer Basis," which is carried out under the auspices of the Deutsche Forschungsgemeinschaft at the Institut für deutsche Sprache in Mannheim. My thanks go to everyone who works within the project—Angelika Ballweg-Schramm, Pierre Bourstin, Jacqueline Kubczak, Sabine Pape, and Helmut Schumacher—for many valuable contributions, as well as to all others who discussed these topics with me: Lennart Åqvist, Franz Günthner, Klaus Baumgärtner, and Manfred Pinkal from Stuttgart, Hannes Rieser and Janos Petöfi from Bielefeld, Leopold Auburger, Tohru Kaneko, and Gerhard Stickel from the institute in Mannheim, and finally, to all the people at the Groningen Round Table whose comments made me see flaws and problems in my presentation there.

I am especially indebted to Frank Heny, who corrected my English, and who made some valuable suggestions in the final version of this paper.

by its domain $\mathbf{D}_{w_x} \subseteq \mathbf{D}$
by its sortal structure $\mathbf{D}_{w_x} = \mathbf{D}_{w_x^1} \cup \mathbf{D}_{w_x^2} \cup \mathbf{D}_{w_x^3} \cup \cdots \cup \mathbf{D}_{w_x^n}$,
where a sort $\mathbf{D}_{w_x^n}$ is the value of a selection function f^*
applied to $\mathbf{D}_{w_x}$.

To express the degree of similarity between a given world state w_0 and any possible state w_x, we define a function from ordered pairs of world states to ordered pairs of numbers

$$S\langle w_x, w_0\rangle = \langle n, m\rangle$$

where n is the quotient of the intersection of the sets of functions f^* which have an extension in w_x and w_0, respectively, and the union of the same sets:

$$\frac{|\{f^*:\ f(\mathbf{D}_{w_x}) \neq \varnothing\} \cap \{f^*:\ f^*(\mathbf{D}_{w0}) \neq \varnothing\}|}{|\{f^*:\ f^*(\mathbf{D}_{w_x}) \neq \varnothing\} \cup \{f^*:\ f^*(\mathbf{D}_{w0}) \neq \varnothing\}|} = n \qquad (0 \leq n \leq 1)$$

The number n provides for the intuition that the worlds in question are structured in a more or less similar way. The number m ensures that the two given worlds contain more or less the same elements, and is simply the quotient of the intersection of $\mathbf{D}_{w_x}$ and $\mathbf{D}_{w0}$ and their union:

$$\frac{|\mathbf{D}_{w_x} \cap \mathbf{D}_{w0}|}{|\mathbf{D}_{w_x} \cup \mathbf{D}_{w0}|} = m \qquad (0 \leq m \leq 1)$$

Whereas n gives us a weak ordering of possible world states according to their being structured more or less similar to w_0, world states which are of the same rank according to n can again be ordered according to their having more or less the same elements, that is, according to m.

To give an example of what is meant by the different sorts, consider a certain position on a chessboard where there is a white king, bishop, and knight and a black king. To decide whether the black king suffers check from the white knight or not, you have to know what sort of agent a knight is in the game at stake, what kinds of moves he can make, etc. If you take such a chessboard position as a world w_0, and two other positions on a chessboard involving the same pieces as w_0, say, w_x and w_y, then you can compare the similarity of w_x and w_y with w_0 by seeing if the pieces are of the same sorts; for instance, w_y could be a fairytale chess position, where the piece which is normally a knight is a so-called nightrider which moves in a different way, and so w_y would be less similar to w_0 than w_x. For instance, no positions where the black king suffers check by the white knight in w_0 and w_x need be check positions with respect to the nightrider in w_y.

The temporal frame $\langle T, <, \bar{T}, \bar{t}_0, \{\underset{A}{\geq}\}_{\bar{T}}\rangle$ is a degenerated version of Åqvist and Günthner 1975 where

$T \neq \varnothing$ is a non-empty set of time points.
$< \subseteq T \times T$ is a strict, linear, and dense ordering of T to be understood as a relation of precedence; $<$ is such that for any t, t', t'' in T:

(a) $t \nless t$ (irreflexivity) in T of $<$)
(b) if $t < t'$ and $t' < t''$, then $t < t''$ (transitivity in T of $<$)
(c) $t < t'$, or $t' < t$, or $t = t'$ (connectedness in T of $<$)
(d) if $t < t'$, then there is a t'' in T such that $t < t'' < t'$ (density in T of $<$)
(e) for each t in T there is a t' in T such that $t < t'$
for each t in T there is a t' in T such that $t' < t$ (this guarantees that T is infinite in both directions.)

$\bar{T}$, the set of stretches of time or time intervals, is the subset of $\mathcal{P}T$, the power set on T, such that for all members $\bar{t}$ of $\bar{T}$, for all t, t', t'' in $\bar{t}$, $<$ is defined as above. (See Åqvist and Günthner, 1975, for a more sophisticated treatment.)
$\bar{t}_0$ is a designated member of $\bar{T}$, representing a time of speech.
$\{\underset{A}{\geq}\}_{\bar{T}}$ is a family of weak-preference orderings on $\bar{T}$.
$\bar{t} \underset{A}{\geq} \bar{t}'$ means that time $\bar{t}$ realizes the state of affairs A at least as much as time $\bar{t}'$.
$\underset{A}{\geq}$ is such that for all $\bar{t}$, $\bar{t}'$, $\bar{t}''$ in $\bar{T}$.

(a) $\bar{t} \underset{A}{\geq} \bar{t}$ (reflexivity in $\bar{T}$ of $\underset{A}{\geq}$)
(b) If $\bar{t} \underset{A}{\geq} \bar{t}'$, and $\bar{t}' \underset{A}{\geq} \bar{t}''$, then $\bar{t} \underset{A}{\geq} \bar{t}''$ (transitivity in $\bar{T}$ *of* $\underset{A}{\geq}$)
(c) either $\bar{t} \underset{A}{\geq} \bar{t}'$ or $\bar{t}' \underset{A}{\geq} \bar{t}$ or both (strong connectedness in $\bar{T}$ of $\underset{A}{\geq}$)
(d) if $\bar{t} \underset{A}{\geq} \bar{t}'$, then $\bar{t}' \underset{\neg A}{\geq} \bar{t}$
(e) if $\bar{t} \underset{A}{\geq} \bar{t}'$ and $\models_{\bar{t}'} A$, then $\models_{\bar{t}} A$
(f) if $\models_{\bar{t}} A$ and $\models_{\bar{t}'} \neg A$, then $\bar{t}' \underset{A}{\ngeq} \bar{t}$

Åqvist and Günthner do not give any further explanation of what it means for a time to realize a state to a degree, but it seems clear that one could construct the similarity between states in a parallel fashion to the similarity of worlds outlined here.

V is the evaluation function in the usual way.

Relative to the model just presented, we can now define two different "earlier than" relations between stretches of time, namely,

$$\bar{t} \lessdot \bar{t}' =_{df} \exists t_x: \quad t_x \in \bar{t}\ (\forall t_y: \quad t_y \in \bar{t}'\ (t_x < t_y)) \wedge \bar{t} \cap \bar{t}' \neq \varnothing$$

and

$$\bar{t} \lll \bar{t}' =_{df} \forall t_x: \quad t_x \in \bar{t}\ (\forall t_y: \quad t_y \in \bar{t}'\ (t_x < t_y))$$

and a generic relation which is an alternative of the inclusive $\lessdot$ and the exclusive $\lll$, symbolized by $<$.

So far, our model only contains possible worlds in a static sense; to remedy this shortcoming, we introduce the concept of a finite binary game tree.[2]

Let a finite, binary game tree with uniform path length k be a structure

$\mathrm{Tr} = \langle s_0, \mathcal{S}, \{\langle \boldsymbol{X} \rangle\}_k, k, \bar{ft}, \bar{T}_{\mathrm{Tr}} \rangle$ where

(i) $\mathcal{S}$ is a finite, nonempty set of situations.

(ii) s_0 is a designated member of $\mathcal{S}$, the starting situation.

(iii) $\{\langle \boldsymbol{X} \rangle\}_k$ is a set of power k of possible events $\langle \boldsymbol{X} \rangle$; $\langle \boldsymbol{X} \rangle$ is a function, such that

$$\forall s((\langle \boldsymbol{X} \rangle(s)) = \langle s'\colon \quad s' = X(s), s''\colon \quad s'' = \neg(X(s))\rangle)$$

(iv) k is a positive integer, the length of Tr.

(v) $\bar{ft}$ is a function which maps $\mathcal{S}$ into $\bar{T}_{\mathrm{Tr}}$ such that

(a) $\bar{ft}(s_0) = \bar{t}_x\colon \quad \bar{t}_x \subset \bar{T}_{\mathrm{Tr}} \wedge \neg\exists \bar{t}_y\colon \quad \bar{t}_y \subset \bar{T}_{\mathrm{Tr}} \wedge \bar{t}_y < d\, \bar{t}_x$

(b) $\forall s(\forall s'\colon \quad s' \in \{\langle \boldsymbol{X} \rangle(s)\}(\bar{ft}(s) < \bar{ft}(s')))$

(c) $\forall s(\bar{ft}(s) = \bar{t}_x\colon \quad \bar{t}_x \subset \bar{T}_{\mathrm{Tr}} \wedge \neg\exists \bar{t}_z\colon \quad \bar{t}_z \subset \bar{T}_{\mathrm{Tr}} \wedge \bar{t}_x < d\, \bar{t}_z$ iff $\langle \boldsymbol{X} \rangle(s) = \varnothing)$

(vi) $\bar{T}_{\mathrm{Tr}} \subset \bar{T}$.

If we define a move function mf, such that

$$\forall s(mf(\langle \boldsymbol{X} \rangle(s)) = s'\colon \quad s' \in \{\langle \boldsymbol{X} \rangle(s)\} \wedge \exists^{1!} s' = mf(\langle \boldsymbol{X} \rangle(s))$$

where

$$(\exists^{1!} x(F(x)) = {}_{df} \exists x(F(x) \wedge \forall y(F(y) \quad (y = x))),$$

we can define a path P in Tr as a sequency of values of mf, where mf is applied to S_0 to yield s_1, to s_1 to yield s_2, and so on to s_k.

A possible world in the dynamic sense of a possible world course through a certain stretch of time can now be regarded as such a path through a game tree as shown in the following diagram.

[2] See Åqvist, 1974 for the use of game trees in causality problems.

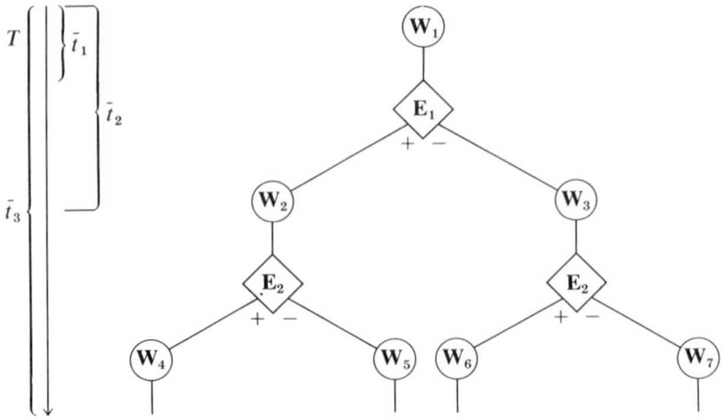

The pair $\langle w, \bar{t}_1 \rangle$ would be the path containing just w_1; $\langle w, \bar{t}_2 \rangle$ either the path $w_1 - w_2$ or $w_1 - w_3$; and $\langle w, \bar{t}_3 \rangle$ one of the paths $w_1 - w_2 - w_4$, $w_1 - w_2 - w_5$, $w_1 - w_3 - w_6$ or $w_1 - w_3 - w_7$. If any of them is selected as the course of the real world, then the other ones can be viewed as representing ''possible courses'' which the real world could have taken. S is restricted to the initial states of worlds in that sense; that is, $S\langle w_0, w_x \rangle$ means $S\langle w_{0_1}, w_{x_1} \rangle$, where both w_{0_1} and w_{x_1} are within the same time interval.

3. THE ANALYSIS OF CAUSALITY

3.1. Gradual Becoming

Within the model outlined, I will first give an analysis of gradual becoming, and then of verbs which can be described in terms of gradual becoming, like *to die, to fall asleep* etc.

For this purpose, I introduce an operator of the category $S/\langle S, S \rangle$ (that is, an operator which takes an ordered pair of sentences into a sentence), namely, the operator CHANGE, for which the truth conditions are:

$$\underset{\langle w, \bar{t} \rangle}{\overset{\mathbf{M}}{\models}} \text{CHANGE}\langle X, Y \rangle \text{iff } \exists \bar{t}_1 \colon \ \bar{t}_1 \subset \bar{t} \wedge \exists \bar{t}_2 \colon \ \bar{t}_2 \subset \bar{t} \wedge \bar{t}_1 \ll \bar{t}_2 \wedge$$

$$\underset{\langle w, \bar{t}_1 \rangle}{\overset{\mathbf{M}}{\models}} X \wedge \underset{\langle w, \bar{t}_2 \rangle}{\overset{\mathbf{M}}{\models}} Y \wedge$$

$$\forall \langle \bar{t}_x, \bar{t}_y \rangle \colon \ \bar{t}_1 < \bar{t}_x < \bar{t}_y < \bar{t}_2$$

$$(\bar{t}_x \underset{X}{\geq} \bar{t}_y \wedge \bar{t}_x \underset{Y}{\leq} \bar{t}_y)$$

$$\wedge \ \exists \langle \bar{t}_u, \bar{t}_v \rangle \colon \ \bar{t}_1 < \bar{t}_u < \bar{t}_v < \bar{t}_2$$

$$(\bar{t}_u \underset{X}{>} \bar{t}_v \wedge \bar{t}_u \underset{Y}{<} \bar{t}_v)$$

These truth conditions, which look a bit clumsy at first sight, simply say that CHANGE$\langle X, Y \rangle$ is true relative to the Model **M** and the world w at the stretch of time $\bar{t}$, if and only if there is a time $\bar{t}_1$ in $\bar{t}$ such that X is true relative to $\langle w, \bar{t}_1 \rangle$, and that there is a time $\bar{t}_2$ in $\bar{t}$ such that it is distinctly after $\bar{t}_1$ and Y is true at $\langle w, \bar{t}_2 \rangle$, and that for all ordered time pairs in between, X is realized at least as much at the earlier time as at the later, whereas Y is realized at the later time at least as much as at the earlier time; and for at least one such time pair, X is realized at the earlier time more than at the later, whereas Y is realized at the later time more than at the earlier. To take a metaphor from music, we have a decrescendo of X and a crescendo of Y, both possibly irregular.[3]

With the operator defined thus, we can give lexical entries for verbs of change, for example:

$$\lambda x[\lambda Y[\text{CHANGE}\langle \neg Y, Y\rangle](\text{SLEEP}(x)]$$

would be the lexical entry of *to fall asleep* or of the German verb *einschlafen*.

The construction of a complex function out of simpler ones by means of lambda abstraction is rather similar to the transformation of Predicate Raising in Generative Semantics, as Dowty (1976), has shown. But there are some differences which I consider important: First, that the lambda abstracts I use here are fully interpretable semantically, whereas the structures yielded by Predicate Raising are not, and that the method of lambda abstraction used here makes it possible to have structures which turn out to be true under exactly the same conditions, but have different compositional intensions (in the sense of Lewis, 1972). For example, compare our base structure of

John falls asleep.
$\lambda x[\lambda Y[\text{CHANGE}\langle \neg Y, Y\rangle](\text{SLEEP}(x))](John)$

with

It comes about that John sleeps.
$\lambda Y[\text{CHANGE}\langle \neg Y, Y\rangle](\lambda x[(\text{SLEEP}(x)](John))$

One brief remark: Dowty (1976) proposes truth conditions for a BECOME-operator, such that BECOME (S) is true iff S is true at some time point t, not-S is true at some time point t' before t and at all time points between t' and t. As opposed to the solutions outlined above, this approach seems to me to suffer from a rather serious flaw, namely, that Dowty's definition for BECOME is unable to give an account of gradual becoming, of becoming more and more . . . and the like, whereas my analysis, following the lines of Åqvist and Günthner, does.[4]

Dowty's proposal would, for instance, require that *John falls asleep* be

[3] This CHANGE-operator is parallel to Åqvist and Günthner's "It becomes more and more the case and then remains that *p*." See Åqvist and Günthner 1975 (pp. 13–17). The picture from music is not chosen arbitrarily; as Lennart Åqvist told me, he started to develop this conception in Åqvist 1973, "Music from a Set-Theoretical Point of View."

[4] Dowty's (1972) BECOME(X) would be rendered by $\lambda S[\text{CHANGE}\langle \neg S, S\rangle](X)$ in the system given here.

true at some t, and for some t' before t and all t'' in between, John is awake. This, of course, is empirically inadequate, as *to be awake* and *to sleep* are to be described as scalar notions, and falling asleep is not an event which takes place between two neighboring time points, but rather is to be analysed as going gradually down the "scale of awakeness" until one reaches a range of this scale which is labelled SLEEP, and this event surely takes more than two time points, as we see in sentences like

Last night, it took me an hour and five whiskies to fall asleep.

3.2. Conditionals

Within the model presented in the previous section, we can now give the semantics of a Conditional operator COND$S/\langle S, S\rangle$ in the following way:

$$\vdash^{\mathrm{M}}_{\langle w_x, \bar{t}_y\rangle} \mathrm{COND}\langle A, B\rangle \text{iff } \exists w_i\colon \vdash^{\mathrm{M}}_{\langle w_i, \bar{t}_y\rangle} (A \wedge B)(\forall w_j\colon \vdash^{\mathrm{M}}_{\langle w_j, \bar{t}_y\rangle} (A \wedge \neg B)([S\langle w_x, w_i\rangle] > [S\langle w_x, w_j\rangle]))$$

where $w_x \neq w_i$.

What this amounts to is that the conditional is true in the selected discourse world w_x if and only if there is at least one $(A \wedge B)$ world such that for all $(A \wedge \neg B)$ worlds, the latter are less similar to w_x than the former.

This definition, which is very much in the spirit of Lewis 1973, allows us to distinguish COND$\langle A, \neg B\rangle$ from $\neg$(COND$\langle A, B\rangle$). To show this, we first look at the truth conditions of COND$\langle A, \neg B\rangle$:

$$\vdash^{\mathrm{M}}_{\langle w_x, \bar{t}_y\rangle} \mathrm{COND}\langle A, \neg B\rangle \text{iff } \exists w_j\colon \vdash^{\mathrm{M}}_{\langle w_j, \bar{t}_y\rangle} (A \wedge \neg B)(\forall w_i\colon \vdash^{\mathrm{M}}_{\langle w_i, \bar{t}_y\rangle} (A \wedge B)([S\langle w_x, w_i\rangle] < [S\langle w_x, w_j\rangle]))$$

Now, the truth conditions for $\neg$(COND$\langle A,B\rangle$) are

$$\vdash^{\mathrm{M}}_{\langle w_x, \bar{t}_y\rangle} \neg\mathrm{COND}\langle A, B\rangle \text{iff } \neg(\exists w_i\colon \vdash^{\mathrm{M}}_{\langle w_i, \bar{t}_y\rangle} (A \wedge B)(\forall w_j\colon \vdash^{\mathrm{M}}_{\langle w_j, \bar{t}_y\rangle} (A \wedge \neg B)([S\langle w_x, w_i\rangle] > [S\langle w_x, w_j\rangle])))$$

By "driving in" the negation, we can transform the definiens. Through the equivalences $\exists x(Fx) \equiv \neg(\forall x(\neg Fx))$ and $\neg(\exists z(Fx)) \equiv \forall z(\neg Fx)$, we get

$$\forall w_i\colon \vdash^{\mathrm{M}}_{\langle w_i, \bar{t}_y\rangle} (A \wedge B)\neg(\forall w_j\colon \vdash^{\mathrm{M}}_{\langle w_j, \bar{t}_y\rangle} (A \wedge \neg B)([S\langle w_x, w_i\rangle] > [S\langle w_x, w_j\rangle]))$$

By $\neg(\forall x(Fx)) \equiv \exists x(\neg Fx)$, we get

$$\forall w_i\colon \vdash^{\mathrm{M}}_{\langle w_i, \bar{t}_y\rangle} (A \wedge B)(\exists w_j\colon \vdash^{\mathrm{M}}_{\langle w_j, \bar{t}_y\rangle} (A \wedge \neg B)(\neg([S\langle w_x, w_i\rangle] > [S\langle w_x, w_j\rangle])))$$

and as $\neg(a > b) \equiv (a \leq b)$, we finally arrive at

$$\forall w_i\colon \ \underset{\langle w_i, \bar{t}_y\rangle}{\overset{\mathbf{M}}{\models}} (A \wedge B)(\exists w_j\colon \ \underset{\langle w_j, \bar{t}_y\rangle}{\overset{\mathbf{M}}{\models}} (A \wedge \neg B)([S\langle w_x, w_i\rangle] \leq [S\langle w_x, w_j\rangle]))$$

If we compare this with the definition of COND$\langle A, \neg B\rangle$, we see that, by virtue of our definition, COND$\langle A, \neg B\rangle$ entails $\neg$(COND$\langle A, B\rangle$), but not vice versa.

This is necessary, as pointed out in Lewis 1973, in order not to mark as contradictions sequences like the well-known "It is not the case that if Bizet and Verdi were compatriots, Bizet would be Italian, and it is not the case that if Bizet and Verdi were compatriots, Bizet would not be Italian; . . . ."[5]

The difference between the Lewis-Conditional and the one defined here is that the former requires that both A and B be true in the sphere of worlds next to w_x and satisfying A, whereas the definition proposed here requires merely that there is at least one world where both A and B are true which is nearer to w_x than all worlds in which A and $\neg B$ are true. In other words, Lewis's definition has to check the nearest worlds, while mine has only to look for the nearest world which satisfies $A \wedge \neg B$, and then see if there is a nearer world satisfying $A \wedge B$. This difference takes care of the intuition that there might be many things which could make a world be the nearest to w_x, which have nothing to do with the question at hand. In our discussion of causality we will take up an example where this distinction turns out to be crucial.

However, there is one problem with the kind of approach given here, namely, that the truth conditions given for COND require either a finite number of possible world states to be compared with each other, or else something like a uniformedness assumption, which would be that for any two given worlds w_x and w_y, differing with respect to w_0, but not differing relative to the two conditionally linked propositions, all worlds in between behave alike with respect to these two propositions—an assumption which is as problematic as the spheres suggested by Lewis, But this assumption allows us to introduce a concept of a partial model, relative to which our definition of COND can be set to work. This seems natural, as speakers are not to be supposed to check infinitely many possible worlds or the like, if they are asked to decide the truth or falsity of a given conditional statement.[6]

We introduce a function P^*, which takes a model into a partial, finite model. P^* is such that for some w in $\mathbf{M}$, $P^*(w) = w'$ in the partial model $\mathbf{M}'$, and for some w in $\mathbf{M}$, $P^*(w)$ is 0 in $\mathbf{M}'$ and $P^*(w_0) = w'_0$. All sentences which are true for any w_n in $\mathbf{M}$ are also true relative to w'_n in $\mathbf{M}'$.

[5] Lewis 1973 (p. 80).

[6] When I presented my paper at the Third Groningen Round Table, I simply assumed a finite number of possible worlds, and hence was critized severely—and rightly so. Of course, I must admit that the uniformedness assumption and the derivation of partial models is by no means less of an ad hoc device than Lewis's (1973) spheres.

The truth conditions of COND, then, would be:

$$\vdash^{\mathbf{M}}_{\langle w_x, \bar{t}_y\rangle} \mathrm{COND}\langle A, B\rangle \text{iff } \exists \mathbf{M}': \quad \mathbf{M}' = P^*(\mathbf{M})(\exists w_j'(\vdash^{\mathbf{M}}_{\langle w_x, \bar{t}_y\rangle} (A \wedge \neg B) \wedge w_j' \in \mathbf{M}')$$
$$\wedge \exists w_i'(\vdash^{\mathbf{M}'}_{\langle w_i', \bar{t}_y\rangle} (A \wedge B) \wedge w_i' \in \mathbf{M}' \wedge \forall w_j'(\vdash^{\mathbf{M}'}_{\langle w_j', \bar{t}_y\rangle}$$
$$(A \wedge \neg B) \wedge w_j' \in \mathbf{M}' \rightarrow [S\langle w_x', w_i'\rangle] > [S\langle w_x', w_j'\rangle])))$$

This definition maintains the distinction, given above, between the stronger and weaker negation of the conditional. Its natural interpretation would be that a speaker, when defending a conditional statement, would give just the kind of argument which could be described as mentally setting up an experimental situation. The construction of the partial model can be viewed as parallel to the construction of an experiment insofar as it reduces the complexity of the original situation where the two propositions occurred.[7]

3.3. Causality

Armed with a Conditional as defined in Section 3.2, we propose the following definition of CAUSE$S/\langle S, S\rangle$

$$\vdash^{\mathbf{M}}_{\langle w_x, \bar{t}_y\rangle} \mathrm{CAUSE}\langle A, B\rangle \text{iff} \vdash^{\mathbf{M}}_{\langle w_x, \bar{t}_y\rangle} (A < B) \wedge \mathrm{COND}\langle A, B\rangle \wedge \neg(\mathrm{COND}\langle \neg A, B\rangle)$$

If we insert the truth conditions of COND, we get:

$$\vdash^{\mathbf{M}}_{\langle w_x, \bar{t}_y\rangle} \mathrm{CAUSE}\langle A, B\rangle \text{iff} \vdash^{\mathbf{M}}_{\langle w_x, \bar{t}_y\rangle} (A < B) \wedge \exists \mathbf{M}': \quad P^*(\mathbf{M}) = \mathbf{M}'$$
$$[(\exists w_j'(\vdash^{\mathbf{M}'}_{\langle w_j', \bar{t}_y\rangle} (A < \neg B) \wedge w_j' \in \mathbf{M}') \wedge \exists w_i'(\vdash^{\mathbf{M}'}_{\langle w_i', \bar{t}_y\rangle} (A < B) \wedge w_i' \in \mathbf{M}'$$
$$\wedge \forall w_j'(\vdash^{\mathbf{M}'}_{\langle w_j', \bar{t}_y\rangle} (A < \neg B) \wedge w_j' \in \mathbf{M}' \rightarrow ([S\langle w_x', w_i'\rangle] > [S\langle w_x', w_j'\rangle])))$$
$$\wedge \exists w_a'(\vdash^{\mathbf{M}'}_{\langle w_a', \bar{t}_y\rangle} (\neg A < B) \wedge w_a' \in \mathbf{M}') \wedge \forall w_a' (\vdash^{\mathbf{M}'}_{\langle w_a', \bar{t}_y\rangle} (\neg A < B) \wedge w_a' \in M' \rightarrow$$
$$\exists w_b'(\vdash^{\mathbf{M}'}_{\langle w_b', \bar{t}_y\rangle} (\neg A < \neg B) \wedge w_b' \in \mathbf{M}' \wedge ([S\langle w_x', w_a'\rangle] \leq [S\langle w_x', w_b'\rangle]))))]$$

For the moment, I restrict both A and B to sentences which describe events in order to discuss the causal relation in the limited sense of "cause-and-effect" relation.

[7] One could go further and reduce the set of sorts and of individuals relative to which the similarity is determined, thus getting from "possible worlds" to something like "possible experimental settings."

To show what this definition means, we look at an example:

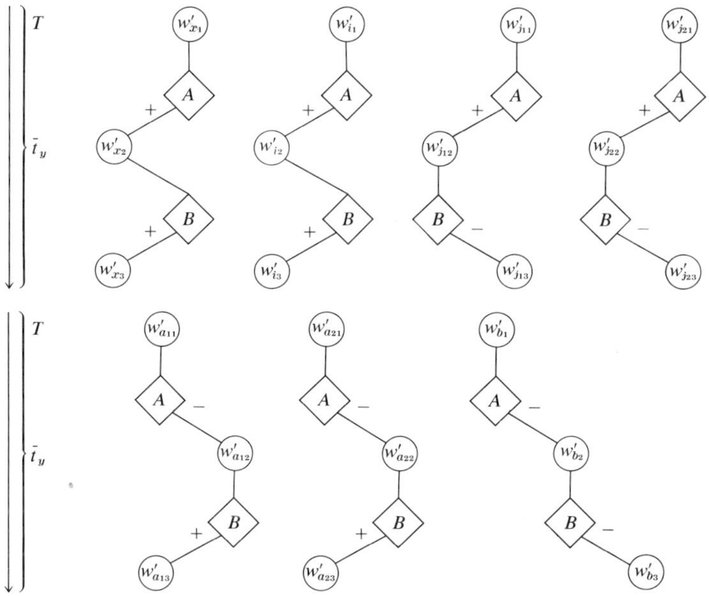

If this is a partial model, and the relations

$$[S\langle w'_x,\, w'_i\rangle] > [S\langle w'_x,\, w'_{j_{1/2}}\rangle]$$
$$[S\langle w'_x,\, w'_b\rangle] \geq [S\langle w'_x,\, w'_{a_{1/2}}\rangle]$$

hold relative to it, then CAUSE$\langle A, B\rangle$ is true relative to $\langle w_x, \bar{t}_y\rangle$.

People familiar with definitions of causality in terms of counterfactuals will immediately notice one difference, namely, that I make use of the outer —that is, weaker—negation.[8] I do this for the following reason: Imagine a situation where an event A and an event B take place simultaneously, and both A and B are sufficient to bring about C; in this case, I want both CAUSE$\langle A, C\rangle$ and CAUSE$\langle B, C\rangle$ to be true, and so I have to admit that there might be $\urcorner A$-and-C-worlds which are as near to the real world w_x as the $\urcorner A$-and-$\urcorner C$-worlds. As an example, consider a man sitting on a terrace and typing a paper on causality; on the same terrace, two young cats, Spotty and Zorro, are running around. Now let

Spotty runs around on the terrace.

be A,

Zorro runs around on the terrace.

be B, and

The man typing a paper on causality loses concentration.

be C. As anyone who knows young cats will agree one of them running

[8] Both Lewis (1973) and Dowty (1972) use the stronger negation.

around is just enough to cause difficulties with concentration; so I should be able to say truly, both

Spottie's running around causes the man to lose concentration.

and

Zorro's running around causes the man to lose concentration.

Consequently, I have to choose the weaker negation.

In Section 3.2, I mentioned that there was a case where the difference between the Lewis Conditional and mine would be crucial. Let

The hatchet falls down on the head of the delinquent.

be A,

The shadow of the hatchet falls down on the head of the delinquent.

be B, and finally

The delinquent's head falls down.

be C. Normally, we would say that A was the cause of both B and C, that A preceded B and B preceded C, that is, $\text{CAUSE}\langle A, B\rangle \wedge \text{CAUSE}\langle A, C\rangle \wedge B < C$ (that A precedes B and C is implicitly given within the truth conditions of CAUSE). If we now analyse this example with the Lewis Conditional, we would, in stating the truth conditions for $\text{CAUSE}\langle A, C\rangle$ look for the sphere of worlds closest to the real one where A holds, and if C holds there too, then $\text{COND}\langle A, C\rangle$ would be true. After that, we would look for the closest sphere of worlds where $\neg A$ holds, and if $\neg C$ holds there in at least one world, then $\neg(\text{COND}\langle\neg A, C\rangle)$ would be true; but unfortunately, it is clear that in the closest worlds where the hatchet falls, the shadow of the hatchet falls too, and in the closest worlds where the hatchet does not fall, its shadow does not fall. So, by using the Lewis Conditional for a counterfactual analysis of causality, we get entirely wrong causes if we look at chains of this type, as Lewis himself pointed out.[9]

In this case, $\text{CAUSE}\langle A, B\rangle \wedge \text{CAUSE}\langle A, C\rangle \wedge B < C$ would be sufficient to derive the truth of $\text{CAUSE}\langle B, C\rangle$.

Since I claimed earlier that my slightly refined definition does not lead to such results, I am now obliged to show that my proposal is not vacuous. To do this, I shall give a model relative to which $\text{CAUSE}\langle A, B\rangle$, $\text{CAUSE}\langle A, C\rangle$, and $B < C$ is true, and then show that $\text{CAUSE}\langle B, C\rangle$ is not necessarily true, that is, $\neg(\text{CAUSE}\langle B, C\rangle)$ is possible:

[9] See Abbott 1974 (p. 7).

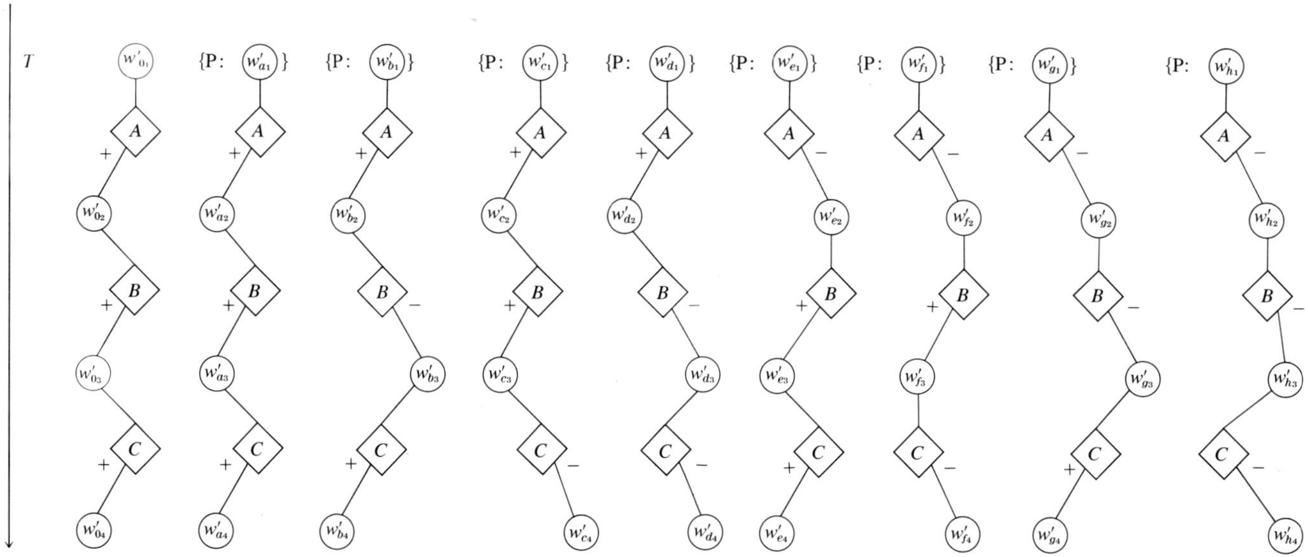
T
{P:
A
B
C
+
−

Relative to the above system of sets of paths or possible worlds, the following similarity conditions must hold to make CAUSE$\langle A, B\rangle$ and CAUSE$\langle A, C\rangle$ true:

(I shall use w'_{a_1/b_1} as a shorthand for w': $w' \in \{w'_{a_1}\} \cap \{w'_{b_1}\}$.)

$$\begin{aligned}
&(1)\quad \exists w'_{a_1/c_1}(\forall w'_{b_1/d_1}([S\langle w'_{0_1}, w'_{a_1/c_1}\rangle] > [S\langle w'_{0_1}, w'_{b_1/d_1}\rangle]))\\
&(2)\quad \forall w'_{e_1/f_1}(\exists w'_{g_1/h_1}([S\langle w'_{0_1}, w'_{e_1/f_1}\rangle] \leq [S\langle w'_{0_1}, w'_{g_1/h_1}\rangle]))\\
&(3)\quad \exists w'_{a_1/b_1}(\forall w'_{c_1/d_1}([S\langle w'_{0_1}, w'_{a_1/b_1}\rangle] > [S\langle w'_{0_1}, w'_{c_1/d_1}\rangle]))\\
&(4)\quad \forall w'_{e_1/g_1}(\exists w'_{f_1/h_1}([S\langle w'_{0_1}, w'_{e_1/g_1}\rangle] \leq [S\langle w'_{0_1}, w'_{f_1/h_1}\rangle]))
\end{aligned}$$

We consider first the truth conditions of CAUSE$\langle A, B\rangle$:

$A < B$ is satisfied by w_0 and, of course, by w'_0.

COND$\langle A, B\rangle$ is satisfied by $\begin{Bmatrix} w'_a \\ w'_c \end{Bmatrix} - \begin{Bmatrix} w'_b \\ w'_d \end{Bmatrix}$ and Condition 1.

$\neg$COND$\langle \neg A, B\rangle$ is satisfied by $\begin{Bmatrix} w'_e \\ w'_f \end{Bmatrix} - \begin{Bmatrix} w'_g \\ w'_h \end{Bmatrix}$ and Condition 2.

so CAUSE$\langle A, B\rangle$ is true relative to our model. Next, we consider CAUSE$\langle A, C\rangle$:

$A < C$ is satisfied by w_0.

COND$\langle A, C\rangle$ is satisfied by $\begin{Bmatrix} w'_a \\ w'_b \end{Bmatrix} - \begin{Bmatrix} w'_c \\ w'_d \end{Bmatrix}$ and Condition 3.

$\neg$COND$\langle \neg A, C\rangle$ is satisfied by $\begin{Bmatrix} w'_e \\ w'_g \end{Bmatrix} - \begin{Bmatrix} w'_f \\ w'_h \end{Bmatrix}$ and Condition 4.

So CAUSE$\langle A, C\rangle$ is also true relative to our model. Now we state the truth conditions for $\neg$(CAUSE$\langle B, C\rangle$):

$$\models^{M}_{\langle w_j, \bar{t}_y\rangle} \neg(\text{CAUSE}\langle B, C\rangle) \text{ iff } \models^{M}_{\langle w_0, \bar{t}_y\rangle} (\neg(B < C)) \vee (\neg(\text{COND}\langle B, C\rangle)) \vee (\neg(\neg\text{COND}\langle \neg B, C\rangle))$$

If any part of that formula is satisfiable without contradiction for CAUSE$\langle A, B\rangle \wedge$ CAUSE$\langle A, C\rangle \wedge (B < C)$, we have shown that we get the desired results.

The first part, $\neg(B < C)$ is, of course, not true. The second part, $\neg$(COND $\langle B, C\rangle$) would be true iff

$$\forall w'_{a/e}(\exists w'_{c/f}([S\langle w'_0, w'_{a/e}\rangle] \leq [S\langle w'_0, w'_{c/f}\rangle]))$$

This is not a contradiction to Conditions 1–4, and so we have shown that CAUSE$\langle A, B\rangle \wedge$ CAUSE$\langle A, C\rangle \wedge (B < C) \wedge \neg$(CAUSE$\langle B, C\rangle$) is, by virtue of our definitions, contingent, and so we do not have "overkill" in describing causal chains like the one discussed here.

With the CAUSE-operator defined, we can now formulate lexical entries for causative verbs, for example,

$$\lambda x(\lambda y[\text{CAUSE}\langle E \ldots x \ldots y \ldots, \lambda S[\text{CHANGE}\langle \neg S, S\rangle](\text{DEAD}(y))\rangle])$$

would be the lexical entry of "to kill," namely, in a rough paraphrase, "an event E, which involves both x and y caused a change of state such that it became the case that y is dead." By just "picking out" the individual variables, we change the sentence structure in the outer pair of square brackets into a two-place predicate over individuals. Using this simple device, we overcome the difficulty that, in natural languages like English or German, causative verbs are normally not constructed as operators over sentences describing events, but rather as two- or three-place predicates over individuals. Therefore, I disagree with Abbott's statement (1974) that ". . . any application of these [that is, model theoretic] accounts to natural language causing . . . will have to allow things other than events as both causes and effects."[10]

4. FINAL REMARKS

I have tried to develop a model within which we could give a treatment of the semantics of so-called causative verbs. My main concerns have been

to give an adequate account of the concept of change and the treatment of verbs of change
to develop a conditional logic
to combine these two in the analysis of causative verbs

Little emphasis has been placed on surface problems, and the procedure for deriving surface sentences from our proposed representations in $^{\lambda}$L has only been hinted at. But what should be clear even from this sketch is the extreme usefulness of the lambda operator, which enables us to have a very "flexible" deep syntax in which expressions with the same truth conditions can be rendered differently, according to their different compositional intensions, and which also plays a key role in our concept of lexical entries.[11] The main point has been to develop a semantics of causality which overcomes some of the difficulties of the Lewis and Dowty approaches. I have shown that it really does so for one tricky case. However, in the area of

[10] Abbott 1974 (p. 1).

[11] I just mentioned briefly the usefulness of lambda abstraction in describing sets of semantically and syntactically related lexical items ("*Wortfelder*"); for example,

$$\lambda x[\text{CHANGE}\langle \neg S, S\rangle](f(x))]$$

with the range of the predicate variable f specified as {SCHLAF, SCHLUMMER, DÖS} would constitute a part of the *Wortfeld* of *schlafen*, namely, the verbs of change *einschlafen, eindösen, einschlummern*. The formula captures all of the syntactic properties of the three verbs, and all they have in common semantically. The use of such schemes makes it possible to reduce the lexicon considerably. See Ballweg (1967a and 1967b) for a more detailed analysis.

causality, no one can ever be altogether sure not to be confronted tomorrow with an example which necessitates at least a refinement of the details.

REFERENCES

Abbott, B. (1974). Some problems of giving an adequate model-theoretic account of CAUSE. In C. Fillmore (Ed.), *Berkeley Studies in Syntax and Semantics,* Berkeley: University of California.

Åqvist, L. (1974). A new approach to the logical theory of actions and causality. In S. Stenlund (Ed.), *Logical Theory and Semantic Analysis.* Dordrecht: Reidel.

Åqvist, L. (1973). Music from a set-theoretical point of view, *Interface,* **2,** 1–22.

Åqvist, L., and Günthner, F. (1975). Fundamentals of a theory of verb aspect and events within the setting of an improved tense logic, Stuttgart, mimeographed. (1978) In F. Günthner and C. Rohrer (Eds.), *Studies in Formal Semantics,* Amsterdam: North-Holland.

Ballweg, J. (1976a). Semantische Grundlagen einer Theorie der deutschen kausativen Verben. Manuscript, Mannheim. (1977) *Forschungsbericht 38* des Instituts für deutsche Sprache. Tübingen: Narr.

Ballweg, J. (1976b). Versuch einer exemplarischen Teilanalyse eines Wortfeldes. "Wer schläft, $\begin{Bmatrix} s \\ k \end{Bmatrix}$ ündigt nicht." *Deutsche Sprache,* **4,** 305–312.

Dowty, D. (1972). On the syntax and semantics of the atomic predicate CAUSE. In *Papers from the Eighth Annual Meeting of the Chicago Linguistic Society.* Chicago: Department of Linguistics, Chicago University.

Lewis, D. (1970). Causation, *Journal of Philosophy,* **17,** 556–567.

Lewis, D. (1973). *Counterfactuals,* Cambridge: Harvard University Press.

Lewis, D. (1972). General semantics. In D. Davidson and G. Harman (Eds.) *Semantics of natural language,* Dordrecht. Pp 169–218.

NOTES FROM THE PRAGMATIC WASTEBASKET: ON A GRICEAN EXPLANATION OF THE PREFERRED INTERPRETATION OF NEGATIVE SENTENCES[1]

DAVID S. SCHWARZ

1. In her recent book on presupposition, Deirdre Wilson argues that in general what we identify as the presuppositions of a sentence are properly represented as ordinary **entailments** of that sentence, rather than as semantic presuppositions.[2] Central to this argument is an attempt to explain pragmatically why the presupposition-carrying interpretations of negative sentences are normally the preferred ones. Clearly, she needs such an account, since otherwise we can explain the phenomenon of preferred interpretation only by supposing that the presuppositions are truth conditions *both* of the negative sentences and of the sentences negated. From this, of course, it is a very short step to the position Wilson wishes to avoid, namely, that presupposi-

[1] An early draft of this paper was presented at the Western Division APA meetings in Chicago, April, 1975, under the title, "A Pragmatic Account for Presupposition." For help in developing that version into the present one, I would like to offer special thanks to Frank Thompson, Janet Fodor, Frank Heny, and G. Lee Bowie.

[2] Deirdre Wilson, *Presuppositions and Non-Truth-Conditional Semantics,* Academic Press (London), 1975.

SYNTAX AND SEMANTICS
VOLUME 10

ISBN: 0-12-613510-X

tions are conditions on both the truth and the falsity of sentences which exhibit them.[3]

To develop her pragmatic account, Wilson appeals to Gricean principles of conversation—the speaker's commitment to the presupposition in asserting the negative sentence then being a conversational implicature.[4] She develops this, in particular, for sentence (1), whose possible interpretations she identifies as (la)–(le) (1975, p. 99).[5]

(1) *John doesn't regret that Bill is ill.*
 a. *John does not exist.*
 b. *Bill does not exist.*
 c. *Bill exists but is not ill.*
 d. *John exists and Bill is ill, but John does not know that Bill is ill.*
 e. *John knows that Bill is ill, but isn't sorry about it.*

Of course, (1) is the negative of (2), and (2) has (3) as its presupposition.

(2) *John regrets that Bill is ill.*

(3) *John knows that Bill is ill.*

The task, then, is to give a Gricean explanation of why (1e), the presupposition-carrying interpretation of (1), is the **preferred** interpretation—and to do

[3] One *can* avoid taking this step, however, by appealing to Russell's distinction between internal and external negation. From the fact that not-A entails B, then, one need not conclude that false-A does as well. Instead—in view of the ambiguity of negation—one can say that whereas false-A is equivalent to the external reading of not-A, the reading of not-A which entails B is the internal one. For a discussion of this approach, and of its pitfalls, cf. Kempson 1975, (esp. p. 95 ff).

[4] Cf. Grice, 1974.

[5] Following Wilson, I rely here and throughout on an intuitive understanding of "interpretation of a sentence." Very roughly, I take this to be whatever would be a basis for asserting the sentence. Pressed to spell this out, I would confine myself to negative sentences, and I would begin by saying that an interpretation of a sentence not-S is another consistent sentence logically sufficient for not-S. Of course, this is not yet enough, since an infinite number of sentences are sufficient for not-S, whereas, clearly, not-S has only a finite number of interpretations. To get further than this, we would have to say something about the logical structure of S itself. Suppose that we determine a certain set of singular terms and predicate relation expressions to form the basic units in that structure. Such a determination may not be absolute, but may be relative to our current purposes, interests, knowledge, etc. At any rate, in (2) ((1) being not-(2)) we might determine *John, Bill, is ill, knows,* and *regrets* to be such units. Given these units, we may say that they form certain "sentential parts" of S: The singular terms combined with *exists,* and the combinations of these singular terms with predicate/relations expressions that both actually occur in S's structure and are entailed by S. In (2), given the units identified, the sentential parts are: *John exists, Bill exists, Bill is ill, John knows that Bill is ill,* and *John regrets that Bill is ill.* The interpretations of not-S are drawn from the possible truth functions of S's sentential parts and the existential generalizations of S's sentential parts. I think this does capture what we want as not-S's interpretations, as a function of what we count as S's logical structure. Exactly why the notion of "interpretation" works this way is, I think, a deep and complicated question.

this while supposing that (3), the presupposition itself, is simply an **entailment** of (2). Of course, this supposition entails that neither (1) nor (2) have semantic presuppositions, and hence, that these two sentences are contradictories.

With these understandings, then, Wilson's explanation runs roughly as follows (1975, p. 100): Interpretations (1a)–(1c) are eliminated by appealing to Grice's "Maxim of Quantity," which we state in (4).

(4) *Maxim of Quantity:* A speaker is to prefer the logically stronger to the logically weaker utterance where the extra information conveyed by the former is relevant, and where the logically stronger utterance does not violate other principles of cooperative conversation.[6]

Consider (1a), then. The interpretation in (1a) is logically stronger than (1)—entailing but not entailed by (1).[7] Moreover, (1a)'s extra logical strength is relevant: Where (1) is relevant, it is clearly relevant to know whether (1) is true by virtue of (1a). In addition, uttering (1a) rather than (1) does not appear to violate any other principles of cooperative conversation. Hence, were the speaker uttering (1) while in a position to utter (1a), he would be violating the Maxim of Quantity. Supposing the speaker cooperative, then, entails that he is not uttering (1) on the basis of (1a), and *mutatis mutandis* for (1b) and (1c). *Mutatis mutandis* for (1d) too, given that were the speaker uttering (1) on *this* basis, he could have said (1d′) instead.

(1d′) *John does not know that Bill is ill.*

Although logically weaker than (1d), (1d′) **is** stronger than, and no longer than, (1).

This leaves (1e), although one might wonder why the argument from the Maxim of Quantity does not apply here as well. The answer, apparently, is that where the speaker is uttering (1) on the basis of (1e), to say something relevantly stronger than (1) he must violate the principle expressed in (5).

(5) Avoid unnecessary prolixity—that is, don't waste words.

In this case, (1e) itself is the shortest expression the speaker can use to say something relevantly stronger than (1). Since (1e) is substantially longer than (1), then, it violates principle (5). Thus, the speaker uttering (1) while in a position to utter (1e) does *not* violate the Maxim of Quantity. Hence, the argument which eliminates (1a)–(1d) as interpretations does not eliminate (1e).

[6] In the interest of brevity, I will sometimes refer to **the saying of something logically stronger, where the extra information is relevant,** as saying something "relevantly stronger" or "being more relevantly informative."

[7] In general, A is logically stronger than B iff A entails B but B does not entail A. Interpretation (1a) is logically stronger than (1), given our supposition that (1) does not semantically presuppose (3), and that (3) is merely an **entailment** of (2). The information that (1a) adds to (1) is simply that it is by virtue of John's nonexistence that (1) is true.

And, of course, since (1e) is the only interpretation not thus eliminated, it remains as the **preferred** interpretation.

2. This, then, is Wilson's account. Although it has the ring of plausibility, I think it falls short of the mark in two ways. First, if it *does* leave (1e) as a preferred interpretation, it does not leave (1e) as the *only* preferred interpretation. The problem is that there are interpretations of (1) which Wilson fails to consider, for example, (1f) and (1g).

(1) f. *Someone regrets that Bill is ill, but not John.*
g. *John regrets something, but not that Bill is ill.*

These interpretations represent bases for the utterance of (1) distinct from (1a)–(1e), in that a speaker might be in a position to say (1f) or (1g) without being in a position to say any of (1a)–(1e) (cf. *infra,* Section 5). Moreover, given that interpretations (1a)–(1d) are already ruled out, a speaker uttering (1) on the basis of either (1f) or (1g) is also in the position of being unable to say something relevantly stronger than, but no longer than, (1). Hence, on Wilson's account, (1e) should share its status of preferred interpretation with (1f) and (1g). Given that (1e) does have this status all to itself, this is something which remains to be explained.

Second, and worse, Wilson's account is not really adequate to leave (1e) as even an **allowed** interpretation. The difficulty is the claim that the length of (1e) violates the principle in (5). Where (1) is true because of (1e), the difference in length between (1e) and (1) buys—and is *needed* to buy—a gain over (1) in relevant information. The additional length, then, does not seem to be a waste of words—unless providing any information at all, ever, is a waste of words. Hence, the length of (1e) cannot be taken to violate (5), and thus cannot be taken to exempt the speaker uttering (1) on the basis of (1e) from violating the Maxim of Quantity. So far as Wilson's account goes, then, the presumption that the speaker is cooperative ought to eliminate (1e) as an interpretation too. Of course, (1e) is not eliminated, but Wilson offers no real insight into why that is so.

In spite of these failings, I do think Wilson's account provides a promising beginning on which to build. In the remainder of this paper, I would like to do just that, hopefully constructing the sort of explanation at which Wilson was aiming. I will do this in three stages. First, I will show how we can get (1e) as an *allowed* interpretation, along with interpretations like (1f) and (1g). Second, I will explain how (1e) achieves the status of *the preferred* interpretation. Third, and finally, I will suggest how the account can be applied to other cases. Again, since the goal is a *pragmatic* explanation of preferred interpretations—with the presuppositions taken as just *entailments* of the positive sentences which exhibit them—throughout I shall suppose that the sentences I discuss do not have *semantic* presuppositions, and that therefore the negations of these sentences are their contradictories.

3. To start, let me say that I *do* take Wilson's argument from the Maxim of Quantity to succeed in eliminating (1a)–(1d) as interpretations of (1).

What we need to discover, then, is a feature of (1e)—not shared by (1a)–(1d)—which spares *it* from being eliminated by the same stroke. Of course, (1e) does differ from (1a)–(1d) in being the interpretation on which the speaker is unable to say something relevantly stronger, but not longer, than (1). As we have seen, this is not the feature of (1e) we are looking for. However, related to this is another, which, characterized in a *preliminary* way, is just that (1e) cannot be expressed without explicitly asserting the presupposition itself, *namely,* (3).

(3) *John knows that Bill is ill.*

That is, where the speaker is saying (1) on the basis of (1e), he cannot say something stronger than (1)—and still relevant to (1)—without uttering a version of (1e) itself, which includes an explicit assertion of (3). I shall put this by saying that (1e) "embeds" (3), explaining this in general with (6).

(6) For any two sentences S and A, and interpretation I of S, I **embeds** A if and only if where S is said on the basis of I, the speaker is unable to say something logically stronger than S—without adding information irrelevant to the question of S—unless he explicitly asserts A.

In these terms, it may be verified that (1c) does not embed *Bill exists,* since saying (1) on this basis, the speaker could have said (1c′) instead—and this is still both stronger than (1) and relevant, while omitting the explicit assertion of *Bill exists*.

(1c′) *Bill is not ill.*

Again, (1d) does not embed *John exists and Bill is ill,* since, as we have seen, the speaker could have said (1d′) instead.

At any rate, (1e) embeds (3), and, as we next note, (3) is interesting here because it is necessary for (2), the contradictory of (1). (Recall that (3), as (2)'s presupposition, is treated semantically here as just an ordinary entailment of (2).) Moreover, this is (3)'s **only** significant logical relation to (1): (3) is not **sufficient** for (2); nor is it either necessary or sufficient for (1) itself. Of course, if (3) is necessary for (2)—(1)'s contradictory—then the negation of (3) is sufficient for (1). Hence, (3) contradicts a proposition sufficient for (1), namely, the negation of (3), or, as I have written it, (1d′).

(1d′) *John does not know that Bill is ill.*

I will mark this by calling (3) an "assertion counter to (1)"—a term I define more generally with (7).

(7) For any two sentences S and A, an assertion of A is an **assertion counter to** S if and only if A is necessary for not-S (not-A is sufficient for S) but not sufficient for not-S, and neither necessary nor sufficient for S.

Combining terms, then, in embedding (3), (1e) embeds an assertion counter to (1). Again comparing (1e) with (1a)–(1d), (1c) and (1d) as written exhibit assertions counter to (1), namely *Bill exists* and *John exists and Bill is ill*, but (1c) and (1d) do not **embed** these assertions. However, (1a) and (1b) embed themselves, although these are not assertions **counter** to (1), each being sufficient for (1).

Interpretation (1e), then, is distinguished from (1a)–(1d) in embedding an assertion counter to (1), namely, (3). It can be argued that because of this, (1e) is exempted from Wilson's argument based on the Maxim of Quantity. The general idea is this: Since (1e) embeds an assertion counter to (1), to be more relevantly informative when saying (1) on the basis of (1e), the speaker has to explicitly utter an assertion counter to (1). Given that saying (1) is conversationally *in* order, uttering an assertion counter to (1) is *out* of order. Hence, the speaker uttering (1) on the basis of (1e) does not violate the Maxim of Quantity; he is being as informative as he can be without violating other conversational principles. What needs to be spelled out, now, is why an assertion counter to (1) is *out* of order where saying (1) is *in* order.

4. Suppose, then, that an assertion of A is an assertion counter to S, and that saying S is conversationally in order. It then follows that neither (8) nor (9) will be the case.

(8) The hearer believes not-A.

(9) The hearer wants to know whether A/not-A is true.

Consider (8). Given definition (7), if A is an assertion counter to S, then not-A is sufficient for S. Hence, were (8) the case, uttering S would be otiose and thus out of order. Consider (9). Given definition (7), S is compatible with either A or not-A: Since A is not necessary for S, S does not entail A; since A is not sufficient for not-S, S does not entail not-A either. Hence, were (9) the case, S would again be out of order, not answering to the hearer's immediate interest in A. Thus, given that S is conversationally in order, and A an assertion counter to S, either (10) or (11) will be the case.

(10) The hearer believes A.

(11) The hearer has no special belief about or interest in A/not-A.

Consider now the speaker's explicit assertion of A. In the case of (10), this is simply otiose. In the case of (11), the speaker runs the risk of an argument beside the point, if the assertion of A turns out to be controversial. Given (11), the speaker's point is simply to assert S. Given that not-A is sufficient for S, it is quite beside this point to be put in the position of having to defend A. Where S is conversationally in order, then, an assertion counter to S is either otiose or risks controversy beside the point, and, hence, in either case, is conversationally *out* of order.

It is worth adding here that although (10) is the more usual case for uttering S, (11) is not extraordinary. For example, consider an utterance of the form of (12), where the F—but not NN—is antecedently known, and of interest, to the audience.

(12) NN (a proper name) does not have relation R to the F.

Consider, in particular, (13), where we are all interested in finding Smith's murderer (still at large!), but where the audience has never heard of Bill Jones.

(13) *Bill Jones is not the murderer of Smith.*

In this case, we really are reluctant to say (14).

(14) *Bill Jones exists but is not the murderer of Smith.*

Even where I am quite sure of Jones's existence, asserting this is quite beside my point—and I would not want to bother definding this—since my point is simply to eliminate Jones as a suspect. If he turns out *not* to exist, in a sense, all the better. Of course, this is not to deny that (14) is still (part of) the preferred interpretation of (13). It *is* to suggest, however, that in such a case, where (13) turns out to be true on the basis of Jones's nonexistence, (13)'s conversational unacceptability is not so marked. The explanation, I think, has to do with **relevance:** Since (13) concerns Smith's murder, a matter of interest to the audience, no matter whether Jones exists, Jones's nonexistence does not render (13) devoid of relation to the audience's interests. By contrast, were *Jones* the focus of audience antecedent interest—as he might be were this a case of (10)—the utterance of (13) on the basis of Jones's nonexistence would be *very* unhappy. Correspondingly, then, we may say in general that the strength of preference for the preferred interpretation of S will vary, roughly in proportion to the audience's antecedent interest in the fact conveyed by the assertion counter to S. Where the preference is very marked, then, this is because assertion of S on the basis of at least some of the alternative interpretations violates *both* the Maxim of Quantity *and* the Maxim of Relation.[8,9]

[8] A case in which it is conversationally permissible to say (13) on the basis of Jones's nonexistence is also a case in which it is permissible to say (i) on this basis.

(i) *It is false that Bill Jones is the murderer of Smith.*

Hence, the case can be perceived as a counterexample to the semantic presuppositional treatment of (13) and its contradictory. Of course, such counterexamples have been acknowledged in the literature at least since Strawson conceded that (ii) and (iii) are permissible, said on the basis of there being no lodger next door, or no king of France (cf. Strawson, 1954, pp. 225–226).

(ii) *It is false that the lodger next door offered you twice that sum.*
(iii) *It is false that your friend went for a drive with the king of France.*

Sentences (ii) and (iii) on this basis are permissible when *that sum* and *your friend* refer to items

Finally, I should discuss the cases in which we **would** say something like (1e) (or (14)). These will be cases in which we would not say (1) itself, since this will already be taken for granted, but in which we want to make clear the circumstances which make (1) true. Typical here would be the following exchange, taking for granted (1).

A: *What's John so happy about? Doesn't he know Bill is ill?*
B: *John knows that Bill is ill, but he's not sorry about it.*

Applying all this now to the example of (1), given the argument based on the Maxim of Quantity, we cannot interpret an utterance of (1) with (1a)–(1d). However, we can interpret it with (1e). A speaker uttering (1) on the basis of (1e) **could** have been more relevantly informative, but being more relevantly informative here requires an explicit assertion counter to (1), namely, (3), and such an assertion—being either otiose, or risking controversy beside the point—violates other principles of conversation. Of course, by the same token, we can interpret the utterance of (1) with something like (1f) or (1g).[10] It remains to develop the special position of (1e) with respect to these other allowed interpretations.

5. Consider, then, these other interpretations which embed an assertion counter to (1). I would claim that these are at least the interpretations obtained by conjoining an existential generalization on some term in (2) with (an elliptical representation of) (1) itself.[11] In (1f), for example, the existential generalization on Bill—that is, (15)—is conjoined with (1).

(15) *Someone regrets that Bill is ill.*

Other examples of interpretations thus produced are:

(1) h. *John regrets some condition of Bill's, but not his illness.*
i. *John has some attitude about Bill's being ill, but not regret.*
j. *John regrets that someone is ill, but not Bill.*
k. *Someone regrets something, but not John that Bill is ill.*

---

of antecedent interest to the hearer, and I think this fact is explained by my account in terms of relevance. It is worth adding that at least with respect to (iii), Strawson anticipates this account, characterizing (iii) as being "about" the friend in question.

[9] As Grice 1974 states the Maxim of Relation, it is simply the directive, "Be Relevant." Grice does not explicate relevance, nor have I. I have simply drawn on our intuitive understanding of this notion.

[10] It may be verified that (1f) embeds *Someone regrets that Bill is ill*, and that this is an assertion counter to (1)—and *mutatis mutandis* for (1g) with regard to *John regrets something*.

[11] In what follows, I will speak as though these interpretations thus characterized provide the only alternatives to (1e) embedding assertions counter to (1). This may be overly simple. However, I suspect—and would hope, at any rate—that whatever additional alternatives are discovered could be shown to exhibit the same epistemological asymmetry with (1e) that I develop for (1f)–(1k).

It may be verified that these interpretations all embed assertions counter to (1)—namely, these existential generalizations on terms in (2). We need to explain, then, why (1e) is preferred as an interpretation of (1), over (1f)–(1k).

To do this, we may appeal to an epistemological asymmetry which a hearer interpreting (1) will perceive between (1e) on the one hand, and (1f)–(1k) on the other. The asymmetry is this: Clearly, there is no difficulty in supposing the speaker to know (1e) without knowing any of (1f)–(1k). However, the hearer will find it difficult to suppose the speaker to know one of (1f)–(1k) without knowing (1e). The reason is that the hearer will have already eliminated (1a)–(1d) as interpretations of (1), given the Maxim of Quantity. So, in particular, the hearer will suppose the speaker not to know (1d′).

(1d′) *John does not know that Bill is ill.*

Had the speaker known (1d′), he would have said it. Consider, then, that if the hearer supposes the speaker to know (1)—on the basis of (1f)–(1k)—without knowing (1e), then he must suppose the speaker not to know (3) either.

(3) *John knows that Bill is ill.*

One cannot know (1) together with (3) and not know (1e). Hence, the hearer here will suppose (16); the speaker not knowing that John **does** know of Bill's illness and not knowing that John **does not.**

(16) The speaker does not know whether John knows that Bill is ill.

Now, given (16), the speaker is simply not privy to John's thoughts and feelings—such as they are—concerning Bill's illness.[12] Hence, supposing (16), the hearer must suppose that the speaker, knowing one of (1f)–(1k), has excluded John's regret of Bill's illness by:

[12] One might quibble here that (16) does not exclude the speaker's knowing whether John **believes** that Bill is ill. However, this worry—such as it is—can be taken care of by refining the account slightly, as follows: First, given the Maxim of Quantity, we eliminate (1d″) as an interpretation of (1), along with (1a)–(1d)/(1d′).

(1d″) *John doesn't believe that Bill is ill.*

Second, we substitute for (1e) the slightly weaker (1e′).

(1e′) *John knows, or* ***at least believes,*** *that Bill is ill, but is not sorry about it.*

Consider, then, that the hearer can conclude (16′) if he supposes the speaker uttering (1) to know one of (1f)–(1k) without knowing (1e′).

(16′) *The speaker doesn't know whether John* ***believes*** *that Bill is ill.*

We may note finally that the tendency to understand (1e′) as the stronger (1e) is natural, given that (1a)–(1d) are eliminated as interpretations. That is, it would be unusual for someone not knowing any of (1a)–(1d) to know (1e′) without knowing (1e).

In case (1f), considering all the people who regret that Bill is ill, knowing that this is all, and not finding John among them.

In case (1g), considering all the things that John regrets, knowing that this is all, and not finding Bill's being ill among them.

In case (1h), considering all the conditions of Bill's that John regrets, knowing that this is all, and not finding Bill's illness among them. And so on.

Such considerations, even where possible, are truly extraordinary. Hence, the hearer, having eliminated (1a)–(1d) as interpretations, will find it very difficult to suppose a speaker to know (1f)–(1k) without knowing (1e). In view of this, then, if we hear an utterance of (1), having narrowed the interpretations to (1e)–(1k), we may reasonably suppose that if the speaker is basing his utterance on any one of these, he is basing it on (1e).

Thus, we have our account of (1e) (carrying the presuppositions) as the preferred interpretation. We explain as well why (1f)–(1k), while **allowed** as interpretations, are not **preferred,** and, hence, why, for example, (15) is not perceived as a presupposition. Of course, had the speaker said (1) on the basis of (15)'s contradictory, that is, (1n), he would have been more cooperative in saying (1n) rather than (1).

(1n) *No one regrets that Bill is ill.*

Thus, (1) excludes the contradictory of (15) as an interpretation. However, since (1) does not prefer, but merely allows, (1f) as an interpretation, it merely allows that the speaker believes (15) itself.

6. Summing up, then, (1e) is determined to be the preferred interpretation in two stages. First, interpretations (1a)–(1d), as well as those like (1n), are eliminated by the argument from Grice's Maxim of Quantity. This leaves interpretations (1e)–(1k), the interpretations embedding an assertion counter to (1), as allowed. Second, the hearer notes the epistemological asymmetry among the allowed interpretations—given that (1a)–(1d) are eliminated—yielding the conclusion that if the speaker is saying (1) on the basis of one of (1e)–(1k), then he is surely saying it on the basis of (1e).

We may observe that the account, thus summarized, is applicable to other cases. For any sentence S, admitting alternative interpretations, if we find a class of interpretations which embed assertions counter to S, and if among these (where there is more than one) we find one interpretation exhibiting the epistemological asymmetry with regard to the others, then this one is explained as the preferred interpretation. Before discussing other cases, however, it is worth sharing some limitations of the account—at least at this stage of its development—which dictate that it must be applied piecemeal. First, it should be clear that the feature of embedding an assertion counter to S does not fit comfortably within any systematic schema of representation now available. Indeed, it seems to straddle all three levels of deep/semantic structure, surface structure, and pragmatics; being a function not only of entailment relations, but also of what we can say in surface English, and of what is relevant. Hence, I offer no *mechanical* way of recognizing this feature. Nor do I have a mechanical formula for recognizing epistemological asymmetry. From the case of (1), we might suppose it to have something to do with the general vis-à-vis the specific, but I have no great confidence that this would be true in all the relevant cases.

Be that as it may, if we confine our attention to the more familiar cases of presupposition, the application of the account is obvious enough. Consider, then, the following list of sentence-pairs:

(i) a: *John doesn't reget (isn't sorry, isn't happy, isn't surprised, isn't ashamed, etc.) that p.*
b: *John knows that p, but doesn't regret (isn't sorry, etc.) that p.*
(ii) a: *It isn't too bad (nice, strange, etc.) that p.*
b: *P, but it's not too bad (nice, strange, etc.) that p.*
(iii) a: *John didn't x quickly (slowly, carefully, etc.).*
b: *John x-ed, but not quickly (slowly, etc.).*
(iv) a: *John didn't x.*
b: *John exists, but he didn't x.*
(v) a: *John isn't a good soldier.*
b: *John is a soldier, but not a good one.*

In each pair, the second sentence expresses the presupposition-carrying interpretation of the first. It can easily be verified that this interpretation also embeds an assertion counter to the sentence it interprets—this assertion being just the presupposition—and that this interpretation exhibits the epistemological asymmetry relative to other interpretations embedding other such assertions. Hence, for each of these cases, my account explains the presupposition-carrying interpretation as the preferred one.

9. As a final application of this account, I would like to consider the more complicated case of sentences of the form of (17)—the great-granddaddy of the presupposition problem.

(17) *The F is G.*

I will suppose that (17) has the standard Russellian analysis, namely, (18).

(18)
a. *There is at least one F.*
b. *There is at most one F.*
c. *Any F is G.*

The task, then, is to explain why (19), with interpretations (19a)–(19e), has (19e) as its preferred interpretation.

(19) *The F is not G.*
a. *There is no F.*
b. *There is no G.*

c. *There is more than one F.*
d. *No G is F.*
e. *There is an F which is not G.*

To begin, we may note at once that (19a)–(19d) are eliminated by the Maxim of Quantity. This leaves (19e), and one might suppose that this interpretation embeds *There is an F*. This is not *quite* right, however, since, saying (19) on the basis of (19e), the speaker can also say (20), (21), or (22), and these are all relevant to (19) and appear to be logically stronger than (19) as well.

(20) *No F is G.*

(21) *All F's are not G.*

(22) *Any F is not G.*

However, as utterances here, (20)–(22) do not provide happy alternatives to (19e). All of (20)–(22) at least suggest that there is more than one F. Given that (19) is not being asserted on the basis of (19c), this is a misleading suggestion. *For practical purposes,* then, where (19) is said on the basis of (19e), the speaker cannot say something relevantly stronger than (19) unless he says (19e) itself. In this sense, (19e) *does* embed *There is an F,* and, of course, *There is an F* is an assertion counter to (19). Hence, (19e) is explained as the preferred interpretation.

I should say, finally, that given that the speaker is not taken to be uttering (19) on the basis of (19c), there is a general understanding that the existence of an *F* is thought to entail its uniqueness. Normally, the understanding is that *F* is filled in from context to attain this uniqueness (for example, for *the book* we read *the book* ***in question***), although sometimes *F*, as it stands, is taken to satisfy this condition (for example, where *F* is *world's tallest man*). Hence (19e) is naturally understood as the slightly stronger (19e′).

(19e′) *There is a unique F which is not G.*

I should note that where *F* is understood as filled out from context, the account is slightly more complicated. Letting "*F*-c-d" abbreviate "*F* as contextually determined," we give the account for (23) instead of (19).

(23) *The F-c-d is not G.*

This has as interpretations (23a)–(23h), (23h) being the preferred one.

(23)
a. *There is no F-c-d.*
b. *There is no F.*
c. *There is no G.*
d. *There is more than one F-c-d.*
e. *No F is G.*
f. *No G is F.*

g. *No G is F-c-d.*
h. *There is an F-c-d, but it is not G.*

In explanation of this, (23a)–(23g) are eliminated by the Maxim of Quantity. By an argument parallel to that for (19e), (23h) is determined for **practical** purposes to entail an assertion counter to (23), and hence to remain as the preferred interpretation.

Another variant is the case of (24), with interpretations (24a)–(24e).

(24) *The F is not the G.*
a. *There is no F.*
b. *There is no G.*
c. *There is more than one F.*
d. *There is more than one G.*
e. *There is an F and there is a G, but they are nonidentical.*

Clearly, the Maxim of Quantity eliminates (24a)–(24d). This leaves (24e), which can be seen to embed *There is an F and there is a G*—an assertion counter to (24)—at least for practical purposes. Of course, where (24) is said on the basis of (24e), the speaker *can* say one of (25)–(31).

(25) *There is an F, but it is not G.*

(26) *There is a G, but it is not F.*

(27) *No F is G.*

(28) *Any F is not G.*

(29) *Any G is not F.*

(30) *All F's are not G.*

(31) *All G's are not F.*

However, again by considerations parallel to those for (20)–(22), each of these may be seen to mislead as to the number of *F*'s or *G*'s. Hence, as desired, (24e) is explained as the preferred interpretation of (24).

REFERENCES

Grice, H. P. (1974). Logic and conversation. In D. Davidson and G. Harman (Eds.), *The logic of grammar*. London: Dickenson.

Kempson, R. M. (1975). *Presupposition and the Delimitation of Semantics*. Cambridge: Cambridge University Press.

Strawson, P. F. (1954). A reply to Mr. Sellars, *The Philosophical Review*, **63**, 216–231.

Wilson, D. (1975). *Presuppositions and Non-Truth-Conditional Semantics*. London: Academic Press.

PART III

METATHEORETICAL ISSUES

LOGICAL RATIONALISM AND FORMAL SEMANTICS OF NATURAL LANGUAGES: ON CONDITIONS OF ADEQUACY[1]

ASA KASHER

1. INTRODUCTORY: GOALS OF SEMANTIC THEORIES

1.1. Delimiting the Field of Semantics

It is a truism that semantic theories should specify and explain semantic properties of semantic objects and semantic relations between such objects. Since there is no broad agreement about what should be taken to be the na-

[1] This is a completely revised version of a paper read to the Third Groningen Round Table under the title "Original Sins and Possible Redemptions: Possible World Semantics and Possible Context Pragmatics."

Noam Chomsky's remarks on my paper "The Proper Treatment of Montague Grammars in Natural Logic and Linguistics" (Kasher, 1975) motivated the present study; I admit turning more skeptical with respect to some versions of formal semantics.

I am also indebted to Renate Bartch, M. Cresswell, Henry Hiż, Hans Kamp, Shalom Lappin, Hans-Heinrich Lieb, Helmut Schnelle, and Dieter Wunderlich for their comments on earlier presentations of this paper. I am particularly indebted to Frank Heny for suggesting many improvements.

The present work was supported by the Z. W. O. (Zuiver Wetenschappelijk Onderwijs) (The Netherlands).

SYNTAX AND SEMANTICS
VOLUME 10

ISBN: 0-12-613510-X

ture of semantic objects, properties, and relations, the question arises whether there are any nontrivial conditions which linguists and philosophers would agree must be imposed on semantic theories. The present paper is devoted to a discussion of two, seemingly simple conditions of this type. Some ways of satisfying these conditions are pointed out, which are of linguistic and philosophical interest.[2]

Any suggestion of adequacy conditions for semantic theories seems out of place as long as no general delimitation of semantics has been put forward, at least in broad outline. Indeed, a list of accurately formulated adequacy conditions often serves as a delimitation of a theoretical topic, but what gives such a list its significance is a pretheoretical conception of the intended subject matter. Yet it is pointless to try to delimit semantics precisely from the outset of any research program, and it is equally useless to make any attempt to find a significant theoretical similarity between different studies that have been labelled "semantic" during the last decades. The same holds for any attempt at a premeditated, exact delimitation of pragmatics. We are content to remain on the intuitive level of discussing semantic and pragmatic theories, which suffices for a broad outline of intricate theoretical interests.

Both semantics and pragmatics are centered upon specification and explanation of certain human competences or parts thereof, possibly abstracted parts. The competence related to semantics concerns the literal connections between sentences and their well-formed parts, on the one hand, and situations, courses of events, or whatever in a world, whether actual or possible, on the other hand. The pragmatic competence concerns the systematic use of linguistic means for effecting literal purposes.

A few clarifications would not be out of place. Notice, first, that we are not committed to the view that the subject matter of semantics is an independent competence.[3] First of all, our characterization of it is in terms of both sentences and worlds. Consequently, if semantics does not carry the burden of specifying what is a sentence and what is a world, then the system of semantic rules **either** depends on a grammar, which determines the class of sentences and their forms, and on a conceptual framework, which determines the class of possible worlds and their relations, **or** is part of a complex of interrelated rules which fixes sentences, worlds, and connections between them, perhaps inseparably. Moreover, it is a plausible assumption about the human network of systematic sentence–world connections that they are based, partly but intrinsically and significantly, on nonlinguistic knowledge and belief.[4] Still, the literal component of the sentence–world connections—

[2] Kempson (1975, Ch. 1.1 and p. 60, fn. 2) maintains that "there are four conditions which linguists working within the framework of a formal model of language would agree must be satisfied by a semantic theory," but philosophers will not consider the conditions she mentions as either theoretically neutral or philosophically innocent.

[3] A similar question arises with respect to pragmatics. See Kasher, 1976c and 1976d.

[4] See Davidson, 1973.

the one that does not depend on nonlinguistic knowledge and belief—might be determined by means of appropriate abstractions. The existence of a **clear-cut** distinction between literal and nonliteral sentence–world connections is an empirical, open problem, of course, to be settled by semantic theories; the latter are not a priori, but are judged, inter alia, by their adherence to hard facts.[5] The same methodological position will be adopted here toward the problem of whether our subject matter is a competence—an independently defined system of rules—a separable part of such a system of rules, or a definable abstraction from such a system.

Second, we are not committed at the outset, to any extreme position with respect to the role played in semantics by truth conditions. The importance of the latter, expressing as they do a central sentence–world relation, is undeniable, but we will claim neither that everything in semantics can be done in terms of truth conditions nor that everything in semantics can be done without resort to truth conditions, lacking conclusive evidence for either position.[6]

Among the sentence–world relations that semantics will probably be required to explain are institutional relations which grant many imperative and interrogative sentences, for example, their meanings. (By "institutional" we mean, here and throughout this paper, what belongs to a system of rules that define their own sphere, as distinct from a system of rules that regulate a given sphere of activity. Examples of the former type are games, both of nonlinguistic and linguistic nature—chess, football, assertion, and advice; examples of the latter type are traffic regulations, flight instructions, and many laws. (See Kasher, 1976d for further discussion). Since these linguistic institutions define the standard use of such nonindicative sentences, there seems no way of avoiding pragmatic considerations within basic semantic theories. In a sense, therefore, some parts of pragmatics are completely included within semantics.[7]

1.2. Empiricist Versus Rationalist Goals in Semantic Theory

What, then, is the subject matter of semantics of natural languages? What are the goals of semantic theories? Since it is impossible to extract from the current theoretical approaches to semantics any common answer to these questions, it is difficult to avoid taking sides with one party or another. We

[5] See Chomsky, 1976.

[6] See Davidson, 1976 for an articulated defense of this view.

[7] The treatment of indices (time, place, speaker, etc.) and of the institutional characterization of the meanings of nonindicative sentences, for example, should not be excluded from semantics, whereas the study of implicatures or the role of rationality principles in linguistic activity should be.

will proceed to point out a major dispute over goals of semantic theories and to present our own view of these goals.

To get some idea of the disputed goals, recall two dicta: Chomsky's characterization of linguistics as a branch of cognitive psychology versus Thomason's characterization of Montague's formal linguistics as a branch of mathematics.[8] Obviously, a branch of mathematics is not incompatible, in any significant sense, with a branch of a science, be it physics or psychology; the rivalry between the schools is only apparent, because **thus characterized** the two approaches do not share their theoretical goals.

The nature of the implicit dispute becomes much clearer when we move from a contrast between dicta to a comparison of explicit pronouncements. Chomsky argued that

> Linguistics is concerned with the nature, function and origin of a particular "mental organ" . . . the language faculty, a common human attribute, genetically determined, one component of the human mind. Through interaction with the environment, this faculty of mind becomes articulated and refined. We will therefore be particularly interested in properties of attained linguistic competence that are vastly underdetermined by available experience in general, but that nevertheless hold of the linguistic competence attained by any normal speaker of a given language, and in fact by all speakers of all natural languages . . . on the natural assumption of uniformity across the species. The commitment to formulate a restrictive theory of Universal Grammar is thus nothing other than the commitment to discover the biological endowment that makes language acquisition possible and to determine its particular manifestations [1975].

Three points should now be stressed. First, the major goal of linguistics is, according to this view, the characterization of the concept of a psychologically possible human language.[9] However, since human beings are psychologically restricted in various ways which reflect not only on language but on other human faculties of mind as well, the psychological specification of our cognitive abilities—language included—will incorporate both an account of linguistic competence and a theory of independently specified factors of linguistic performance.

Second, we should not disregard the obvious point that human beings seem to be unique in their linguistic capacity. Thus, another way of describing the major task of linguistics, according to Chomsky's view, is requiring linguistic theories to provide a framework for describing and explaining the unique properties of the psychologically possible human languages, in terms of structure, meaning, or use.

Third, notice that Chomsky's program is in the rationalistic tradition, being engaged in the pursuit of the *innate ideas* which characterize human languages, that is, the internal equipment which makes language acquisition

[8] Chomsky, 1972; Thomason, in the introduction to Montague, 1974.

[9] Ones interest in what is *psychologically possible* does not commit one to any negative attitude toward biological explanations of psychological systems.

possible in the manner shared by all normal infants. Such an endowment may take the form of a specification of the conditions, formal and substantive, which a system of rules must satisfy in order to count as an internal description of a natural language.

Montague's attitude stands in sharp contrast with Chomsky's view on each of these points, as shown by the very first sentence of each of his papers (1970a) and (1970b) according to which there is no important theoretical difference between natural and formal languages.[10] Now, importance of difference is relative to purposes of comparison, and what is of utmost significance when attempts are made to answer the question "What is a psychologically possible human language?" might be ignored when efforts are required to answer another question, namely, "What is a language, a symbolic system of no restriction?" Whereas Chomsky is interested in drawing an interesting distinction between what is linguistic and human and what is linguistic but not human, so to speak, Montague disregards it on purpose.

There is no genuine argument concerning empirical issues between Chomsky and Montague. It is not the case that Montague claimed that there is no significant difference between English and intensional logic, while Chomsky has maintained that such differences exist and are extremely important. From the generative point of view, the natural language is unlike the logical system in some respects which bear the marks of the distinction between what can serve as a human, natural language and what cannot. From the formal point of view, the same considerations are of no theoretical significance so long as both English and intensional logic can be described within the same framework.[11]

Notice also that Montague's approach carries some of the traits of empiricism, in contrast with the clear rationalistic nature of Chomsky's program. First of all, Montague was engaged in a project of reduction. Empiricists have always thought that reducing one theoretical system to a seemingly less problematic one is of major philosophical import. Thus, attempts to reduce psychology to biology, chemistry to physics, and mathematics to logic used to be freely recommended by empiricists. But whereas some argument for relating psychology to neurophysiology might be advanced, the reduction of an empirical domain to a mathematical field is doomed to failure, on a par with the futility of any attempt to reduce physics to mathematics. Correspondence with facts cannot be reduced to formal coherence.

Thomason has suggested that a less stringent relation holds between a Montague grammar and a natural language, namely, the relation mathematical geometry holds to physical geometry. However, there is a major point of difference between the two cases. Formulas of mathematical geometry are uninterpreted in themselves and it is physical geometry, which associates

[10] Scott 1970 expresses similar views.

[11] Notice the peculiar ambiguity of the term "universal grammar" which is given completely different meaning in Chomsky's and in Montague's frameworks.

"line" with "light beam," for example, which furnishes them with a standard interpretation. Under such an interpretation, physical reality is a **model** of the mathematical system; the interpretations of the axioms of the latter hold in the former. However, because of the intricate nature of various performance factors, natural languages cannot be regarded as models of Montague grammars, even under the assumption that such grammars are adequate theories of linguistic competence. Embodiment is not interpretation. Whereas a model calls for a structural similarity between two spheres—the model and the modeled—an embodiment does not require such a similarity. We do not consider a computer to be a model of the rules of arithmetic, though it embodies them by being able to come out with the right results under certain restrictions. (For a fuller discussion, see Kasher and Lappin, 1977).

Another dogma of empiricism is that of the respectability of mathematical appeal. Natural language is accordingly a legitimate object of empiricistic study, because it is identical with or reducible to a mathematical object. Though mathematical studies provide exemplary standards of clarity, preciseness, and expressive power, however, mathematics itself is not innocent of philosophical mines and pitfalls. Mathematics is not a philosophical resting point.

The divergence of Montague's approach from generative linguistics is reflected in different theoretical decisions that will be reached by the two induced research programs, under the same conditions.[12] There is no reason to assume that the fundamental divergence will turn into a practical convergence or a genuine contest, without a major shift in the theoretical goals of the Montaguean enterprise. Our concern, then, is not with any supposed incompatibility between the systems of Montague and Chomsky, but rather with the possibility of developing an approach to semantic theory which combines important aspects of both. This, we will call *logical rationalism in the realm of language,* where "logical" is taken to mean formal to the extent shown in current philosophical logic, Montague's intensional frameworks included, and where "rationalism in the realm of language" stands for the philosophical foundations, theoretical goals, and articulated methodology of Chomsky's linguistics. Is logical rationalism in the realm of language possible? The following discussion will be devoted to this question, special reference being made to semantics.

2. ELEMENTARY ADEQUACY

2.1. Elementary Adequacy as a Constraint on Theories

General as the preceding outline of goals might seem, it suggests an obvious adequacy condition imposed on every theory of human mental capacity,

[12] See Kasher, 1975 for the details.

namely, finite representability (in the general sense that applies to systems of rules). Every theory which is not a finite representation of a system of rules and which purports to specify a human competence, we shall describe as "elementarily inadequate."

The elementary adequacy of most syntactic theories is clear. A transformational grammar includes a finite class of syntactic rules, each of which is a computable function from finite labeled trees to finite labeled trees.[13] Some of the systems current in philosophical semantics are also elementarily adequate, when contrued as partial specifications of a "semantic competence."[14] Hintikka's model sets and model systems are cases in point.[15] Our own theory of pragmemes, which represent implicatures of sentences characterizing the sentences' literal mood (or use) in terms of standard preferences of some possible courses of events over others, is an example of an elementarily adequate theory in pragmatics.[16]

Taking steps to assess logical rationalism in semantics, the question arises of the elementary adequacy of Montaguean frameworks. As far as one can tell from Montague's own papers, he was not interested in imposing such restrictions on systems definable in the framework of his "universal grammar" and intensional logic. Nothing resembling elementary adequacy is suggested by Montague, even when fragments of English are under consideration; functions, sets, and operators of all kinds are used without restraint.

Rather than issuing an explicit, global limitation of formal armament, thus gaining elementary adequacy without illuminating any semantic matter, we suggest, in the following section, some particular restrictions which may enrich formal semantics with elementary adequacy.

2.2. Adverbs, Operators, and Elementary Adequacy

Consider an adverb, such as *slowly*. If one understands this word, then one has mastered a finite system of rules which governs the literal properties of the adverb, in terms of structure, meaning, and use. Moreover, such a system of rules should furnish one with an understanding of *walk slowly* as a function of one's knowledge of the literal properties of *walk,* and with an understanding of *talk slowly* as a function of one's knowledge of the literal properties of *talk*. Now, these functional expressions should differ from each other just with respect to their arguments, that is, properties of *walk* and of *talk*. Hence it is natural to construe an adverb as an operator from predicates to predicates, provided it satisfies some requirements which are not our business here. Mastering the literal properties of a certain adverb is, then, mastering a function of a certain kind.

[13] For a precise demonstration of computability some arithmetization of labeled trees is required, but this is carried out easily.

[14] See Cresswell, 1973 and 1976.

[15] See Hintikka, 1962, for example.

[16] See Kasher, 1974.

Now, there are two ways of treating functions within formal frameworks. According to the extensional approach, a function is a set of ordered pairs and its identity and properties depend on nothing but the identities of the pairs. According to the intensional approach, a function is identified not by a set of ordered pairs solely, but by a given specification of this set. It comes as no surprise that different intensionally given functions share their extensionally given set of ordered pairs.

If understanding an adverb involves mastering a function, then the mere compilation of ordered pairs, such as ⟨*%walk, %walk slowly*⟩,[17] is extremely implausible for a variety of reasons, not the least of which is that the list is most probably infinite. If such a function is represented within an elementarily adequate semantic theory, it is finitely representable, which means that there is a **uniform** method of deriving the properties of the modified predicate from properties of the predicate itself.[18] The following conjecture seems plausible:

(C1) The semantic values of adverbs in an elementarily adequate, formal framework, which are functions from appropriately represented verbal phrases to appropriately representable modified verbal phrases, are computable operators.[19]

(If an elementarily adequate, formal framework represents adverbs differently, a modification of this condition would be required.) It is important to notice here the difference between a computable operator and a computable predicate. If a predicate (of one argument) is semantically represented as a function from individuals to truth values, relative to possible worlds, then it is computable, in a sense, if there is a uniform procedure for computing the appropriate truth value, given a standard representation of an individual and a standard indication of the possible world relative to which the truth value is computed. The truth value depends on whether the predicate holds for the given individual or not. When an operator is under consideration, however, it does not involve a computation of a truth value for a given individual and a possible world, but a computation of an appropriate representation of a function, given a representation of another one. Evidently, predicates which are not computable are available in natural languages, for example, *is the last seven in the decimal representation of a fraction;* it is our conjecture that operators which are not computable are not available. If

[17] The mark "%" stands for "a representation of ______within the theory under consideration" or some similar expression.

[18] Such a derivation may involve uses of, say, common knowledge which is clearly not of a linguistic nature, but the way such information is used is uniform, in a sense.

[19] By "the semantic value" of any expression, we mean the formal entity which is ascribed to that expression by the theory.

Again, computability presupposes arithmetization, but this is of no particular importance here.

Notice that functions may generally be partial.

a predicate is represented as an operator from individual terms to sentences (or from individuals to propositions), a revised form of our conjecture (**C1**) will apply to it.

This conjecture seems to be generalized to every case in which the semantic properties of an expression are uniformly determined by the corresponding properties of its parts and by its structure. It has been commonly assumed since Frege, that this is always the case, but some examples have started to erode that assumption, and even if it has not been worn away, some significant modification seems to be required.[20] Hence, the general form of the computability hypothesis does not involve presupposing Frege's principle:

(**C**) In an elementarily adequate formal framework for semantics, all operators are computable.

A proof of such a conjecture or a counterexample to it requires a certain extent of formalization of the concept of computability for common linguistic entities, such as predicates, propositions, quantifiers, etc. Inputs and outputs of operators will be required to be finitely and uniformly represented, that is, built from an appropriate, finite collection of atoms through applications of operations drawn from a finite list, all explicitly given.

2.3. Two Apparent Problems for Logical Rationalism

Among arguments and parameters of semantic functions one finds expressions of different logical types, two of which deserve particular attention at this juncture, namely, proper names, however construed, and the theoretical terms which stand for possible worlds. At first glance, both seem to resist the suggested atomistic approach, for different reasons, and any attempt to develop logical rationalism in linguistics is bound to take up the challenge.

2.3.1. Proper Names

Many formal semanticists, including Montague, have adopted the view that the semantic value of a proper name is a set of properties or a cluster of sets of properties. According to one version of this approach, the sets define uniquely identifying descriptions of the object referred to by using the given proper name under certain circumstances. The definitions of these special descriptions are given in terms of finitely many given properties that the object has under those circumstances, and in terms of the relationships in which it then stands to other given objects. However, according to a different version of the same approach, the set of properties which the semantic theory assigns to a proper name is the set of all the properties of the corresponding object, in a certain context.[21] Bearing in mind the requirement of

[20] See Hintikka, 1976 and also Kasher, 1976a.

[21] This is Montague's approach, in a sense. The former position has been defended by Strawson, 1950 and Searle, 1958, and criticized by Kripke, 1972.

elementary adequacy, it seems that the first version should be preferred to the second one. However, the following apparent counter-argument should be considered: The set of **all** properties a certain object has under certain circumstances is independent of the state of belief and knowledge of anyone who uses the proper name which refers to the object, whereas sets of uniquely identifying descriptions vary with users and should, therefore, be excluded from **linguistic** theory, at least from semantics. The charge against logical rationalism at this point is that the requirement of finite representability blurs the limits of semantics.

In response to this charge one does not have to indulge in a defense of any delimitation of semantics or of any distinction between linguistic knowledge and any other kind of knowledge. Granting that uniquely identifying sets of properties may vary with speaker and context, and that the correspondence between proper names and such sets is dependent on beliefs and cognitions which could not be assumed to be shared by all speakers of the same natural language, one still imports a nonvacuous restriction by requiring elementary adequacy. Clearly, even if semantics does not provide a conceptual analysis of, say, some relation, it is only natural for the theory to impose some formal or material restrictions on the possible semantic values of the given term. When proper names are under consideration, the specification of characterizing sets of objects is relegated to whatever part of the mental equipment which suits such a task, but since semantics processes proper names, so to speak, it is obvious that theoretical restrictions are imposed on semantic values[22] of proper names *qua* inputs for linguistic operations. Such values may be produced elsewhere, but since they play a semantic role they are required to satisfy general semantic requirements, such as finite, uniform representability.[23]

2.3.2. Possible Worlds

It would be difficult to imagine logical rationalism completely deprived of possible worlds and everything else which is often defined by means of possible worlds, such as propositions.[24] For the present purposes, it is assumed that possible worlds will play an indispensable role in semantics and that it is our task to find out whether their use is reconcilable with elementary ade-

[22] Perhaps the term "cognitive value" would fit better, since parts of ascribed values are not purely linguistic in any way.

[23] It seems that a similar approach may provide useful theoretical devices for explaining inherent vagueness. The semantic value of a single common noun is related to several procedures for deciding whether an individual, given under a description or through an observation, belongs to a certain class. Usually, those different procedures will yield the same results when applied to the same individual, if appropriately given. Vagueness involves borderline cases where the procedures are not always in accord with each other.

[24] See Stalnaker, 1976 and Hintikka, 1975 for some arguments.

quacy. (Of course, if formal semantics could dispense with possible worlds, no reconciliation problem would arise.)

To get an idea of what we take to be a natural approach for logical rationalism, consider the proposal for identifying possible worlds with distributions of matter in space–time.[25] Assuming that matter is homogeneous, in the sense that for any single space–time point there is nothing to specify beyond mere existence or mere absence of matter at that point, a possible world is defined by the set of space–time points which include matter. When logical necessity or possibility are considered, it seems natural to impose no restriction whatsoever on those sets, but otherwise such generality is not obviously tolerable. When propositional attitudes, for example, are analyzed, only possible distributions of matter which show a certain extent of continuity, required by the existence of persons or similar individuals, are admissible. In such cases, all classes of possible worlds over which variables range include only possible worlds for which that continuity condition obtains. Moreover, the universe of discourse for that purpose, which will consist of all possible worlds satisfying the continuity condition (and perhaps some other conditions, similarly required), and all its proper subsets over which variables will range, should be finitely representable. Otherwise, it would be implausible to assume that the theory can be incorporated into an elementarily adequate theory of natural language.

In general, if any class of possible worlds plays any theoretical role in an elementarily adequate theory, it should be finitely representable. To see that this is not an obvious requirement, consider two extreme cases, namely, a class of just one possible world and the class of all possible worlds.

To characterize any single "object," whether a person, a state, or a possible world, is to provide theoretical means for its identification, by a uniquely identifying description of by an appropriate use of special institutional devices which do not produce such descriptions.[26] Notice here the difference between a complete description of an object and an identification of it. Clearly, a president of a state can be uniquely identified by a description from which nothing follows about most of his virtues and vices. Now, each case of such an identification involves a finite description or a finitely represented institution, but an infinity of independent cases of identification might involve infinite resources of description or institutional operation, which are not finitely representable.[27] Such a possibility is banned by the requirement of elementary adequacy; if infinitely many objects are individually characterized by an elementarily adequate theory, all the infinitely many characterizations must be within the expressive confines of a finite supply of predi-

[25] Quine, 1968 and Cresswell, 1973. At the moment, no commitment to any form of materialism should be taken to be involved.

[26] See Kripke, 1972 and Putnam, 1975.

[27] This possibility should not look far-fetched, because usually space–time points are represented in terms of real numbers.

cates, relations, operators, logical particles, or what have you; the same finite supply being used for all characterizations. In a sense, elementary adequacy requires uniform representability.

Turning now to the class of all possible worlds, recall the previously mentioned stipulation that a possible world is a distribution of matter in space–time. Thus, the requirement that the class of all possible worlds will be finitely representable is tantamount to the requirement that the concept of possible world will be (finitely) definable in terms of concepts of matter, space, time, and distribution. Constructing a **logical** space[28] each of whose dimensions is the matter–scale for a certain space–time point, we can represent every distribution of matter in space–time as a "point" in the logical space, but there is no reason to assume that even if every conceivable situation or course of events, which is the case or might have been, is representable in the general terms of matter distribution; our own conceptual framework is thus general. It is only plausible to maintain that our innate logical space is much more structured: It has dimensions of various kinds, it includes interrelated dimensions, and it can probably be extended in useful ways.[29]

Now, the subject matter of semantic theory consists of the relations between sentences and worlds. Obviously, all the concepts involved in these relations are drawn from the concurrent logical space, and are, therefore, restricted in their nature. Revealing the structure of our logical space—its dimensions, interrelations, and the possibilities of extending it—is a major task of logical rationalism in semantics.[30] If classes of possible worlds are employed in a semantic theory, they should not be considered as classes of "points," because possible worlds are not atomic and classes of possible worlds are not extensionally understood collections of such atoms. The classes of possible worlds used in a semantic theory should be appropriately representable within the logical space suggested by the theory; this fact about classes of possible worlds bears, indeed, characteristic consequences with respect to the elements of these classes: They have properties which are representable in the logical space under consideration.

Intensional logic is of theoretical significance because it provides a framework for representations of interesting semantic properties and relations, but it should not serve as a linguistic resting point. For intensional logic to constitute an essential part of an elementarily adequate semantic theory, its framework should be finitely and uniformly representable. The pursuit of ap-

[28] See Thomason, 1972 for an application which presents all the essentials of the concept of logical space. See also van Fraassen, 1967.

[29] This does not mean that distributions of matter in space–time do not play any role in the innate logical space, but just that space, time, and matter are not the only fundamental concepts of our mental nonformal equipment.

[30] Some parts of Katz's semantic theories can, it seems, be interpreted along these lines, thus vindicating at least their general goals.

propriate logical space seems the advisable way of trying to give intensional logic linguistic spirit, following the basic train of thought of logical rationalism.

3. EXPLANATORY ADEQUACY

In the preceding section of the present paper, the notion of elementary adequacy has been introduced. Whether a particular semantic theory is elementarily adequate depends only on certain global properties of the theory and on that alone. Different properties of such a theory, which involve not just this theory but others as well, underlie another adequacy condition, to which we now turn.

Theories are meant to **explain** families of phenomena in addition to describing them; giving details of some aspects of the field, and taking no notice of other aspects of it. However, we are quite in the dark about the precise nature of explanation, all the attempts philosophers of science have made to explain explanation notwithstanding, and for the moment there is no choice but to rely upon some basic, shared intuitions.

Chomsky has defined as **explanatorily adequate** those linguistic theories that succeed "in selecting a descriptively adequate grammar on the basis of primary linguistic data."[31] Explanation involves, in this case as well as in any other case, some enhancing of understanding by a systematic broadening of the theoretical scope. An explanation of a variety of phenomena requires not only a theory which includes interesting generalizations pertaining to these phenomena, but also a move from one theory and its set of goals to another theory which has a deeper and more intricate set of goals.

Different explanatory moves suggest themselves with respect to any linguistic theory. The assumption of homogeneity of the linguistic community can be waived giving rise to theories which might explain general facts about syntactic systems in terms of a system of syntactic resources that copes with the requirement to understand sentences produced by grammars of different idiolects.[32] The ideas of moving from a competence theory to a performance theory, or from a synchronic theory to a diachronic one also spring to mind. However, some theoretical moves are more natural than others. To see that, consider the analogous pursuit of understanding social systems. Assume that a social order is determined, in abstraction from some practical factors, by a system of social rules. Having a thorough understanding of the general form of such a social system of rules is a prerequisite to understanding several other social aspects, such as interrelationships between subsystems of

[31] Chomsky, 1965 (p. 25). Notice that this characterization does not bear any direct relation to Hempel's (1965) models of explanation.

[32] See Bierwisch, 1976 for an attempt in this direction.

the same society or between different societies. Moreover, supplementing a theory of abstract social systems by an analysis of the reasons for preferring certain systems to others will probably provide an explanation of social facts related to general goals societies have and their institutionalized realization. The contribution of an analysis of historical changes in social order to the understanding of the concept of social system itself should not be denied, but the introduction of historical changes into the theoretical framework for specifying and explaining social systems involves not just comparisons of social systems to each other but also an account of the interaction of such systems with completely different ones. One theoretical move is, therefore, more natural than another, in a certain theoretical context. Such considerations seem to justify Chomsky's conception of explanatory adequacy for syntactic theories.

Similar considerations suggest a general notion of explanatory adequacy for pragmatic theories, the subject matter of which is the human competence to use linguistic means for achieving certain standard, literal purposes. Our ability to use linguistic means for obtaining desired ends is in part just a special case of our general ability to use appropriate means for attaining given aims. Thus, principles of rationality such as that of effective means govern aspects of our linguistic activity, and an **explanation** of several regularities concerning conversational implicatures is provided by applying general rationality principles to the case of linguistic means and ends.[33]

Furthermore, linguistic activity is always institutional; different kinds of speech acts involve different systems of constitutive rules which define these institutions, and the very appropriateness of an utterance of a sentence by some speaker in a certain context depends on the existence of a linguistic institution which grants that speaker in that context an institutional role which enables him to operate within the institution to attain the particular aims he is entertaining, by uttering that sentence.[34] Now, the existence of institutions, incorporating characteristic goals, roles, means and products, is not confined to language, and it is again the application of the general notion of institution to a particular case which enables us to explain a variety of facts about linguistic activity.

The competence to use linguistic means cannot, then, be formed in one's mind without an understanding of institutions and of how to use means in general. Since there are nonlinguistic uses of means, nonlinguistic institutions, and even nonlinguistic intentional activities within nonlinguistic institutions, a connection is thus established between theories of language and other branches of that field of inquiry which is engaged in understanding our mental capacities. Such connections are able to provide explanations for

[33] See Kasher, 1976b.

[34] See Kasher, 1976c or 1976d.

regularities in the realm of linguistic pragmatics in terms of broader conceptual frameworks.

However, explanation of such facts can be sought elsewhere as well. If some linguistic means are sentences, which have syntactic structures, then the source of some interesting restrictions on the family of linguistic institutions might be discovered to be in the syntactic system. Explanations of established uses of artifacts are often provided by their structural properties, and the institutional uses of sentences in contexts are similar in that respect.

Our approach to the problem of specifying explanatory adequacy for semantic theories should now be clear. Semantic theories invoke conceptual frameworks, logical spaces, possible worlds and classes thereof, but all of these are used within semantic theories in such a general way that any attempt to treat them as specifically semantic is futile. If one admits that a certain logical space underlies the abstract internal representation of what is linguistically understood and expressed, and also that conceptual activity is involved in our perception, then it seems plausible to assume that the logical spaces of language and of perception are not independent of each other. And vice versa, it is only natural to put forward as a starting point for discussion the hypothesis that the logical space underlying perception is mostly included within the logical space which underlies natural language.[35] Indeed, such connections between the conceptual frameworks of two different mental capacities should naturally be expected to provide some explanations of semantic generalizations. Obviously, the possibility of explaining certain features of some devices used in semantic theories, such as classes of possible worlds or clusters of truth conditions, in terms of syntactic properties of sentences or in terms of substantive or formal properties of syntactic systems, should not be overlooked.

4. CONCLUSION: LOGICAL RATIONALISM

The problem we have set out to explore is that of the possibility of logical rationalism in semantics. In the preceding sections of the present chapter, we have pointed out two conditions that a serious attempt at launching into logical rationalism in semantics should meet. Uniform finite representability, which is at the heart of elementary adequacy, is a formal property of theories, and relevancy to nonlinguistic mental activity, which is at the center of explanatory adequacy, is a substantive property of theories, but both stem from the rationalistic factor of logical rationalism. The logical factor has supplied central notions, formal frameworks, and powerful tools; these must now be incorporated into a rationalistic conceptual theory.

Tentative as many conclusions seems at the present stage, it remains an

[35] See Hintikka, 1975 for an intentional (and intensional) theory of perception.

open question whether logical rationalism is possible in semantics, but any thrust into a position which provides better understanding of the relations between full-fledged formal logic and well-worn philosophical rationalism is worthwhile, for the possible benefit of both foundational linguistics and rationalistic philosophy.

REFERENCES

Bierwisch, M. (1976). Social differentiation of language structure, In A. Kasher (Ed.), *Language in Focus: Foundations, Methods, and Systems*. Dordrecht: Reidel.

Chomsky, N. (1965). *Aspects of the Theory of Syntax*. Cambridge, Massachusetts: MIT Press.

Chomsky, N. (1972). *Language and Mind*. New York: Harcourt Brace Jovanovich.

Chomsky, N. (1975). Conditions on rules of grammar. Mimeograph.

Chomsky, N. (1976). Problems and mysteries in the study of language. In A. Kasher (Ed.), *Language in Focus: Foundations, Methods, and Systems*. Dordrecht: Reidel.

Cresswell, M. J. (1973). *Logics and Languages*. London: Methuen.

Cresswell, M. J. (1976). Semantic competence. Mimeograph.

Davidson, D. (1973). Radical interpretation. *Dialectica*, **27**.

Davidson, D. (1976). Performance and meaning. *Dialectica* forthcoming.

Hempel, C. G. (1965). *Aspects of Scientific Explanation*. New York: Free Press.

Hintikka, J. (1962). *Knowledge and Belief*. Ithaca, New York: Cornell University Press.

Hintikka, J. (1975). *The Intentions of Intentionality and other Models for Modalities*. Dordrecht: Reidel.

Hintikka, J. (1976). A counter-example to Tarski-type truth-definitions as applied to natural languages. In A. Kasher (Ed.), *Language in Focus: Foundations, Methods, and Systems*. Dordrecht: Reidel.

Kasher, A. (1974). Mood implicatures: A logical way of doing generative pragmatics. *Theoretical linguistics*, **1**, 1–38.

Kasher, A. (1975). The proper treatment of Montague grammars in natural logic and linguistics. *Theoretical Linguistics*, **2**, 133–146.

Kasher, A. (1976a). Pragmatic representations and language-games: beyond intensions and extensions. In J. Hintikka (Ed.), *Rudolf Carnap, Logical Empiricist*, Dordrecht: Reidel.

Kasher, A. (1976b). Conversational maxims and rationality. In A. Kasher (Ed.), *Language in Focus: Foundations, Methods, and Systems*. Dordrecht: Reidel.

Kasher, A. (1976c). "What is a pragmatic representation?" In C. Butts and J. Hintikka, (Eds.), *Proceedings of Vth International Congress for Logic, Philosophy and Methodology of Science, London, Ontario 1975*.

Kasher, A. (1976d). What is a theory of use? *Journal of Pragmatics* **1**.

Kasher, A. (Ed.) (1976). *Language in focus: Foundations, methods and systems*, Dordrecht: Reidel.

Kasher, A. and Lappin, S. (1977). *Philosophical Linguistics: An Introduction*, Scriptor.

Kempson, R. M. (1975). *Presupposition and the Delimitation of Semantics*, Cambridge: Cambridge University Press.

Kripke, S. (1972). Naming and necessity. In D. Davidson and G. Harman, (Eds.), *Semantics of Natural Language*, Dordrecht: Reidel.

Montague, R. (1970a). English as a formal language. In R. Thomason (Ed.), *Formal Philosophy: Selected Papers of Richard Montague*. New Haven: Yale University Press.

Montague, R. (1970b). Universal grammar. In R. Thomason (Ed.), *Formal Philosophy: Selected Papers of Richard Montague*. New Haven: Yale University Press.

Montague, R. (1974). *Formal Philosophy: Selected Papers of Richard Montague*, R. Thomason (Ed.). New Haven and London: Yale University Press.

Putnam, H. (1975). The meaning of "meaning." In *Mind, Language and Reality,* Cambridge: Cambridge University Press.

Quine, W. V. O. (1968). *Ontological relativity,* New York: Columbia University Press.

Scott, D. (1970). Advice on modal logic. In K. Lambert, (Ed.), *Philosophical Problems in Logic: Some Recent Developments,* Dordrecht: Reidel.

Searle, J. (1958). Proper names. *Mind,* **67** [Also in C. E. Caton (Ed.), *Philosophy and Ordinary Language.* Urbana: University of Illinois Press.]

Stalnaker, R. (1976). Propositions. In A. F. MacKay and D. D. Merrill, (Eds.), *Issues in the Philosophy of Language.* New Haven and London: Yale University Press.

Strawson, P. F. (1950). On referring. *Mind,* **59** [also in C. E. Caton, (Ed.), *Philosophy and Ordinary Language* (1963) Urbana: University of Illinois Press.]

Thomason, R. (1972). A semantic theory of sortal incorrectness. *Journal of Philosophical Logic,* **1,** 209–258.

van Fraassen, B. (1967). Meaning relations among predicates. *Nous,* **1,** 161–179.

MONTAGUE GRAMMAR AND THE WELL-FORMEDNESS CONSTRAINT[1]

BARBARA H. PARTEE

1. MOTIVATION

1.1. The Well-formedness Constraint

The syntactic component in Montague's theory of grammar consists of a simultaneous recursive definition of the sets of well-formed expressions of each of the syntactic categories of the language, starting from the basic lexical expressions of the various categories, and building up the more complex ones by syntactic rules. The categories always include a category of sen-

[1] I am first of all grateful to the past and present linguistic colleagues and students who refused to let me ignore the absence of syntactic constraints in Montague's theory of grammar, especially Robin Cooper and Frank Heny. Many of the proposals for constraints presented here are the result of group efforts by the participants in a seminar on Montague grammar which Terence Parsons and I conducted at the University of Massachusetts at Amherst, in the spring of 1976; those whose contributions I am most aware of drawing on are Emmon Bach, Greg Carlson, Terence Parsons, and Richard Saenz.

The paper has benefited from discussion of its oral presentation at the Third Groningen Round Table, particularly from comments by Helmut Schnelle, and I am grateful to Frank Heny for organizing such a highly interactive conference. I especially wish to thank Emmon Bach for

SYNTAX AND SEMANTICS
VOLUME 10

ISBN: 0-12-613510-X

tences or formulas; what the remaining categories should be is an empirical question of descriptive linguistics. As has often been noted, Montague put virtually no restrictions on the form of the syntactic rules, which was consistent with his aim of providing a theory of all logically possible languages. If one wishes to pursue the linguist's goal of defining as narrowly as possible the class of possible grammars of natural languages, and if one wishes for independent reasons to explore the possibility of constructing such a theory within a Montague-like framework, the task of imposing further constraints on the syntactic component is a central one.

The thesis I wish to develop here is that within Montague's framework there is a very natural constraint that can be imposed on the syntax, a constraint I shall call **the well-formedness constraint,** which in conjunction with some further constraints I shall suggest on the form of the rules, can greatly narrow down the class of syntactic analyses and hence greatly strengthen the claims made by Montague's version of compositionality. I shall not give any formal arguments about generative capacity, but rather talk in terms of various existing analyses of particular constructions that are allowed or ruled out by the constraints in question.

What I mean by the well-formedness constraint is given in (1).

(1) **The well-formedness constraint:** Each syntactic rule operates on well-formed expressions of specified categories to produce a well-formed expression of a specified category.

My use of the term **well-formed expression** corresponds to Montague's use of the term **meaningful expression** in Montague 1970b and 1973 (henceforth UG and PTQ, respectively).

Montague imposes this constraint on his syntax, but not in quite such a direct way as (1) suggests.[2] In the theory specified in UG, the syntactic specification of a language L consists of the specification of a disambiguated language L′ and a binary relation *R* (which we might call an "ambiguating relation") that maps expressions of L′ into expressions of L. The constraint (1) is imposed by Montague on the **disambiguated language** L′, and he imposes no comparable constraint on the final or "surface" language L, nor on the relation *R* which maps L′ into L. In the case of a natural language, say English, L is English and L′ is a theoretical construct that the linguist hypothe-

numerous helpful discussions during the final writing of this paper. I retain responsibility for inadequacies, errors, and loose constraints.

The earliest stages of this work were supported by the National Science Foundation Grant GS 39752: Formal Syntax and Semantics for Natural Languages (with Terence Parsons); the completion of this study was made possible by a fellowship at the Center for Advanced Study in the Behavioral Sciences, Stanford.

I am presupposing basic familiarity with Montague's theory of grammar. For background, see Montague, 1974, and Thomason's introduction to that volume, and also Partee, 1975.

[2] In the oral presentation of this work, I incorrectly attributed the well-formedness constraint directly to Montague; I am grateful to Helmut Schnelle for pointing out my error.

sizes as a level of representation in the derivation of expressions of L. Whether an expression is well-formed in English is a question with relatively high empirical content (although certainly not completely theory-independent), but since L′ is permitted to differ from L in arbitrary ways, for example, by inclusion of abstract elements not found in L, the question of whether an expression is well-formed in L′ may have arbitrarily little empirical connection with the kinds of data normally available from native speakers of English. The constraint (1) will therefore be strongest if it is stated directly in terms of rules for generating L, and departures therefrom must be made with care if the empirical content of (1) is to be maintained.

If *R* is the identity mapping, L′ will equal L, and the well-formedness constraint will necessarily hold for the surface expressions of the language. If *R* is totally unconstrained and may be specified ad hoc for each language, then the relation of L′ to the actual surface language could be highly abstract, and the well-formedness constraint would become vacuous as an empirical constraint. In what follows, some of my suggestions can be regarded as proposals for a narrow specification of *R*; the differences between the disambiguated language L′ and the surface language L that I propose to tolerate include: (i) L′ contains labeled brackets, and *R* deletes them; (ii) L′ contains "variables" like he_3, and *R* erases the subscripts; (iii) L′ contains morphological representations such as *Past see* or *she*[+Acc], which *R* maps into the surface forms *saw* and *her*. The same proposals could be viewed as eliminating *R* completely but as adding certain devices such as labeled bracketing and a morphological component to the grammar to take over some of the intended functions of *R*. Either way, my aim will be to make the well-formedness constraint as strong as possible, consistent with the demands of descriptive adequacy.

An illustration of the effect of the well-formedness constraint is given in (2):

(2) a. OK: F_i ($\underset{\text{CN/CN}}{\textit{recent}}$, $\underset{\text{CN}}{\textit{immigrant}}$) = $\underset{\text{CN}}{\textit{recent immigrant}}$

b. NO: F_i (**immigrant who is recent*) = $\underset{\text{CN}}{\textit{recent immigrant}}$

That is, if the well-formedness constraint is taken seriously as part of the syntax, the semantically motivated analysis of at least some adjectives as common noun modifiers[3] is also selected on purely syntactic grounds over the classic transformational derivation of all prenominal adjectives from relative clauses, since the classic analysis requires the generation of ill-formed expressions as part of the derivation of well-formed ones. As this example illustrates, one effect of the well-formedness constraint is to automatically disallow obligatory rules.

[3] See Montague, 1970a; Parsons, 1970; Siegel, 1976a,b.

1.1. Three Kinds of Constraints

The well-formedness constraint is of interest, I believe, mainly because of its interaction with other constraints in a system of constraints such as that outlined in (3):

(3) (i) constraints on the form of syntactic and semantic operations (central to transformational theory)
(ii) compositionality (from Frege, central to Montague's theory)
(iii) the well-formedness constraint

Transformational grammarians have provided many fruitful and interesting hypotheses about constraints on the form of syntactic rules, and have lately begun at least to acknowledge the need for equally explicit constraints on the form of semantic rules. In terms of Montague grammar, this kind of constraint is a constraint on the form of the syntactic and semantic **operations,** and I shall suggest some constraints of this sort below.

The main constraint in Montague's theory, and the one that makes it so interesting semantically, is the compositionality constraint, that is, the requirement that there be a semantic interpretation rule corresponding to each syntactic formation rule, so that the interpretation of each expression is determined by the interpretation of its syntactic subparts and the rule by which they were combined. There are, of course, other forms a compositionality constraint might take. Katz and Fodor's original theory of projection rules (Katz and Fodor, 1963) was similar to Montague's in that projection rules were associated with both phrase structure rules and transformations; the theory elaborated in Katz and Postal 1964 and Chomsky 1965 intended a compositional semantics acting on deep structures alone; and one might argue, although I do not believe anyone has, for a compositional semantics acting just on the surface structures generated in some version of transformational grammar. (There have been recent proposals that the semantics might operate just on "enriched" surface structures,[4] but Chomsky, at least, appears to believe that the semantic interpretation rules are not compositional; that is, that they do not recursively build up interpretations of larger constituents from interpretations of smaller ones.)

Now, I do not believe that critics who have argued against the compositionality principle have ever suggested that it will be impossible to construct a compositional system, but rather, that to do so will require otherwise unmotivated elements in the syntax, and that the **best** syntax for a natural language will in general not be amenable to a compositional semantics. I think the fruitfulness of Montague's kind of approach, with compositionality as its primary constraint, is becoming increasingly confirmed by the actual analyses of linguistic phenomena that are being produced. But I am also sensitive

[4] See Chomsky, 1977.

to the fact that the theory by itself is not **forcing** the choice of analyses enough; I suspect that the compositionality constraint by itself hardly limits the choice of analyses at all,[5] and that the reasonableness of so many of the analyses that have been proposed results from the linguist's or philosopher's implicit adherence to informal notions of what constitutes a reasonable set of rules, more than from any constraints imposed by the theory. (Some of my colleagues have dubbed this implicit working constraint the "seriousness constraint," that is, "no funny business.")

As I mentioned above, one important class of constraints to introduce are constraints on the form of the syntactic and semantic rules, and there are some suggestions for such constraints in what follows. But I believe that the well-formedness constraint, if taken seriously, can also provide an extremely strong limitation on the syntax, and one which I think is largely independent of constraints on the form of rules. I do not believe the well-formedness constraint puts any constraint at all on the total class of generable **languages,** or even on the total class of possible grammatical rules, since there is no a priori determination of what strings might be well-formed expressions of some category in some language. But from the point of view of the child trying to determine a grammar compatible with the data of the **particular** language he is being exposed to, that is, considering the class of available grammars compatible with primary linguistic data, it seems clear that that class of grammars is drastically reduced if every step of a derivation must produce a well-formed expression of the given language. Abstractness in the syntax is thereby limited to the syntactic apparatus itself: the rules, the derivation trees and/or labeled bracketings, and whatever other auxiliary syntactic devices are required.

So I believe that if we can defend the well-formedness constraint and some strong constraints on the form of rules as well, and still maintain compositionality, then we will have taken a big step toward showing that an empirically strong linguistic theory can be formulated as a subtheory of Montague's theory of universal grammar, Montague's own lack of interest in these particular goals notwithstanding.

1.2. The Initial Implausibility of the Well-formedness Constraint

I have stressed the potential importance of the well-formedness constraint because I believe it has not in general been taken very seriously, largely be-

[5] In earlier work (Partee, 1975, pp. 230–231), I suggested that compositionality could force the choice of analyzing restrictive relative clauses as combining with common noun phrases rather than with full term phrases; Bach and Cooper (1976) show that without additional constraints, the choice is still open. Since that was the clearest example I had of the power of the compositionality constraint to choose among competing analyses, Bach and Cooper's work underscores the need for additional constraints on the grammar.

cause it seems on the face of it to be much too strong. Two immediate counterexamples have often been noted:

(4) a. *He$_3$ will talk.*
b. *John Pres have en be ing walk.*

Example (4a) is from Montague's PTQ; surface English certainly does not contain variables like *he$_3$*, so I at least have tended to assume that Montague himself did not take the constraint seriously when it conflicted with an otherwise elegant analysis.[6] Example (4b) is from Chomsky 1957, and all the linguists I know consider it much better to allow such an analysis at some stage of a derivation than to be forced into a family of rules like Montague's **S17** in PTQ, with a separate rule for each possible combination of auxiliaries so as to produce surface forms directly.

1.3. Complicating Factors

Even supposing we can overcome the initial implausibility of the well-formedness constraint, there are still a lot of complicating factors that may make it impossible to make the constraint completely precise. For instance, the "interpretive grammar" variant of PTQ developed in Cooper and Parsons 1976 eliminates indexed variables in the syntax, but it adds a separate indexing component to the grammar. The number of distinct components and the divison of labor among them is bound to have great effects on the interpretation of the well-formedness constraint. If, for instance, surface filters are allowed, it is difficult to keep the well-formedness constraint from becoming vacuous unless the surface filters are severely limited in power. I shall suggest that (4b) can be tolerated as syntactically well-formed if we separate morphology from syntax; but that requires careful specification and limitation of the class of possible morphological rules. As another example, the use of syntactic features could also add excessive power if not severely constrained; (4a) could be gotten around if numerical indices were treated as features, and even (4b) could undoubtedly be handled by a moderately powerful feature system, just as the definite and indefinite article have sometimes been treated as features on nouns that later get spelled out as morphemes.[7] And last but not least among such difficulties is the fact that there is no pretheoretic classification of kinds of deviance into syntactic, semantic, pragmatic, etc. Thus, it cannot even be determined by looking at a syntax alone whether it does meet the well-formedness constraint; I believe this

[6] In the fragment of English in UG, Montague included expressions such as *he* v_3 in the disambiguated language, and the "ambiguating relation" R was in that case a function which, among other things, mapped *he* v_i into *he*, so that no logical variables or numerical subscripts occurred in the surface expressions. One can thus see the similarity between the role of the relation R and such devices as late obligatory transformations and surface filters.

[7] See Postal, 1966 for an example of a treatment of articles as features that become spelled out as morphemes by later rules.

can only be judged after an entire system of syntax, semantics, pragmatics, and whatever else is needed is constructed, when we can see whether all of our pretheoretic intuitions about deviance are satisfactorily explained somewhere in the system. But in the absence of any such complete theory, I shall assume that we can make enough good, educated guesses about which sorts of deviance are syntactic to at least seriously examine the effects of the well-formedness constraint to see if it has any chance of being tenable. I have already indicated some reasons for thinking it would make an extremely strong addition to the theory, and I shall devote most of the remainder of this paper to an examination of the consequences of adopting it. Some of the consequences look good, some bad, and about some I cannot judge; but it should become clear that there certainly **are** strong consequences. Before discussing the well-formedness constraint itself, however, I want to suggest some constraints on the form of the syntactic and semantic rules and on the correspondence between them, so as to have a more concrete framework established. The constraints I shall suggest are tentative and not fully formalized; I would hope to encourage other researchers who are interested in the potential of Montague grammar as a linguistic theory to join in the effort of finding and formalizing as strong constraints as possible.

2. CONSTRAINTS ON THE FORM OF RULES

2.1. The Basic Form of Rules

We shall start from the basic form of grammar exemplified in PTQ, in which each syntactic rule has an associated translation rule that translates the derived natural language expression into an expression of intensional logic. For simplicity, the form of the rules will be given for binary constructions, but the generalization to one argument or more than two arguments is straightforward. We shall, however, depart from PTQ and the general theory specified in UG by permitting the syntactic rules to specify **partial functions;** this follows standard linguistic practice and does not, as far as I am aware, have any undesirable consequences for the rest of the system.[8] The basic

[8] Montague (1970a) does not require that syntactic operations be defined for all expressions of a given category, nor does he even require that they have a unique output where they are defined; they are given instead as *relations*, and the form of a typical rule is:

$$\text{If } \alpha \in P_A \text{ and } \beta \in P_B \text{ and } (\alpha, \beta, \gamma) \in R_i \text{ then } \gamma \in P_C.$$

Thomason (1976) also follows this practice of specifying relations rather than total or partial functions, although of course some of the relations do turn out to be functions. I have not so far found any compelling reasons to choose between allowing a single rule to have more than one output (using relations) and having separate rules each with a unique output (using partial functions). I have found it quite artificial to try to restrict the operations to total functions, however (in the same sort of way that it is artificial to force every singular definite description to have a unique referent).

form of the syntactic rules and their associated translation rules will then be as in (5) and (6) below.

(5) Syntactic Rule n: If $\alpha \in P_A$ and $\beta \in P_B$ and α, β are of the form SA_n, then $\gamma \in P_C$, where $\gamma = F_i(\alpha, \beta)$.

(6) Translation Rule n: If α, β translate into α', β', respectively, then $F_i(\alpha, \beta)$ translates into $G_k(\alpha', \beta')$.

The parameters to be specified for the particular rules are the following: (i) the categories A and B of the input expressions and the category C of the resulting expression; "P_A" is to be read as "the set of phrases (expressions) of category A"; (ii) the "structural analysis" expression SA_n, which may put further requirements on the input expressions beyond their category membership; (iii) the syntactic operation F_i, which may be as simple as concatenation or as complex as a transformational operation; (iv) the "semantic" operation G_k, which is actually a syntactic operation mapping two expressions of the intensional logic onto an expression of the intensional logic; but I shall continue to refer to it as a semantic operation to distinguish it from the object language syntactic operation F_i.

I shall have nothing more to say about the specification of the categories. Most of my tentative proposals for constraints on the form of the rules can be grouped according to which part of the rule they constrain: the structural analysis part, SA_n (Section 2.2), the syntactic operation F_i (Section 2.3), the semantic operation G_k (Section 2.4). In addition, I shall suggest in Section 2.5 some possible constraints on the pairings of translation rules of particular forms with syntactic rules of particular forms.

2.2. Constraints on Structural Analyses

The structural analysis, SA_n, plays a role similar to that of a "domain statement" or "structural description" part of a transformational rule; it specifies further properties that the input expressions must have for the rule to be applicable to them, usually in terms of internal structure. The constraints I shall suggest here are partly borrowed from transformational theory, within which there is, of course, much controversy over the analogous questions. I shall be somewhat sketchy here; I intend to provide an expanded and more rigorous treatment in a projected further paper.

(i) In Partee 1975, I proposed the addition of labeled bracketing to the generated expressions, and gave examples of rules in which labeled bracketing was used in structural analyses. The labeled bracketing gives the surface immediate-constituent structure of the generated expressions of each category, assigning to each subconstituent an appropriate category label. The structural analysis can then include such statements as "α is of the form ${}_{IV}[{}_{TV}[\delta]_T[\epsilon]\pi]$," to indicate that α must be an intransitive verb phrase consisting of a transitive verb (phrase) followed by a term phrase followed by an arbitrary expression (possibly null).

I should hope by including labeled bracketing to be able to rule out any reference to derivational history or any other "global" properties of the derivation tree. (This contrasts with Thomason 1976, who permits reference to derivational history but does *not* add labeled bracketing.) I also rule out, by means of the well-formedness constraint, the use of abstract markers or "triggers" whose sole purpose is to mark certain expressions as allowable or disallowable inputs to particular rules. (Bennett 1976 and Montague in UG make use of such markers and do without either labeled bracketing or reference to derivational history.)

(ii) I also wish to permit structural analyses to make use of variables over numerical subscripts, so as to say, for example, "α is of the form he_i δ." The indexing of words which translate into variables is a central feature of Montague's system, and I shall suggest various constraints on the use of this mechanism but shall not try to eliminate it.[9]

(iii) the final kind of statement to be allowed in structural analyses is reference to **recursively defined properties of expressions;**[10] by this device I believe it is possible to capture both the linguist's use of syntactic features and Montague's use of "auxiliary notions" such as the **main common noun occurrence** of a common noun phase (see Montague 1970a (pp. 219–220).[11]

Montague's form of syntax, starting with basic (lexical) expressions and building up larger expressions by recursive rules, provides a very natural way to unify the linguist's use of features on lexical items and features on nonterminal nodes, if both kinds of features are viewed as properties of expressions. Consider gender, for example, in the kind of language where predicate adjectives agree in gender with the subject (for example, Romance languages). Each lexical common or proper noun has a gender specified in the lexicon; as larger common noun and term phrases are built up, the gender of the resulting phrase can be determined on the basis of the gender of the constituent phrase(s) and the syntactic rule applied. When a common noun phrase is combined with a relative clause, for instance, the gender of the resulting phrase is always the gender of the constituent common noun phrase. When a sentence is transformed into a *that*-clause type of term phrase, the gender of the resulting term phrase is always a certain fixed gender (neuter in those Romance languages that have neuter, otherwise masculine, I believe). When two term phrases are conjoined, each language has a particular rule for determining the gender of the resulting phrase (for example, the gender of the constituents if they are alike in gender, otherwise mas-

[9] See Cooper and Parsons, 1976, and Cooper, 1975, for some alternative treatments in which the use of numerical subscripts is eliminated from the object language syntax.

[10] The analysis of syntactic features as recursively defined properties of expressions was suggested by Terence Parsons in the seminar mentioned in the acknowledgements.

[11] Page references to Montague's papers refer to the collection, Montague, 1974.

culine.) Such a method of systematically defining the syntactic properties of expressions at the same time that they are built up should obviate the need for special devices such as "feature climbing" (which is straightforward only for the simplest endocentric constructions anyway).

Another attraction of the device of recursively defined properties is its potential for bringing these feature-like notions together with notions like "head noun of a common noun phrase," etc., which cannot be treated simply as binary features and which are at best very difficult and awkward to define simply in terms of labeled bracketing and related configurational properties of derived structures.[12] There are clearly some systematic connections between, for example, which noun is the head noun of a noun phrase and what the gender of the noun phrase is, although there are noun phrases without heads (that is, some nominalizations) to which some gender is nevertheless assigned, and there are conjoined noun phrases with more than one head but a single gender for the entire phrase. I believe that what is needed is a system of rules that run in parallel with the formation rules, recursively defining all of the properties to which other rules may need to refer, starting with lexically specified properties (= lexical features) and specifying properties of derived expressions on the basis of the properties of the constituents and the rule that is combining them.

One place to look for generalizations and restrictions, so as to prevent arbitrary and completely ad hoc, recursively defined properties, is in the correlation between the form of a syntactic formation rule and the form of the associated recursive property specifications. Research on Chomsky's X-bar notation is directed in part toward this goal, and depends on having nonarbitrary names for the various categories. It remains to be seen whether there is any simple way to fit such notions into a categorial-based grammar like Montague's PTQ.[13]

Putting together the three kinds of specifications to be included in structural analyses, a (fictitious) example of the "if" part of a syntactic rule is given in (7):

[12] See for example Stockwell, Schachter, and Partee, 1973 (p. 212), where the head noun of a noun phrase is identified by adding a condition "4 is immediately dominated by NOM, has no sister NOM, and no right sister N (i.e. is head N of NP)." This problem is often ignored, probably in large part because of the absence of any systematic way of handling it. Montague makes use of recursively defined auxiliary predicates of this sort in UG (pp. 238–239), although the particular rules he gives there violate the well-formedness constraint.

[13] As an example of a potential difficulty, consider the fact that in PTQ, *slowly* is of category IV/IV, *try* of category IV//IV: Both combine with a verb phrase to make a new verb phrase, but *try* becomes head of its verb phrase, while *slowly* does not. This difficulty is related to the fact that there is no categorial unity in PTQ among all of the items one would want to call "verbs." Of course *verb* could itself be a recursively defined property, but to make it so without more systematic considerations would be begging some important questions.

(7) If $\alpha \in P_T$, $\varphi \in P_t$, and α, φ are of the form:
(i) α has x gender
(ii) $\varphi = {}_t[{}_T[he_i]_{IV}[\delta]]$
then . . . .

2.3. Constraints on Syntactic Operations

In Partee 1976b there are some suggestions for constraints on the power of the syntactic operations F_i and the semantic operations G_k, which I shall largely repeat in this and the following section. What I propose for the syntax is to require each syntactic operation F_i to be a composition of subfunctions, where each subfunction is itself a composition of certain primitive operations (to be specified later) defined over the structural analyses of the input expressions. The reason for bringing in the intermediate-level subfunctions is to provide a way to capture generalizations involving recurrent operations specific to a particular language that are smaller than a whole rule but more complex than a primitive operation.[14] An example for English in the PTQ framework would be to define a subfunction PRO(α, i, δ) which substitutes for every occurrence of he_i or him_i in δ a pronoun whose gender and number are the gender and number of α and whose case is the case of the given occurrence of he_i or him_i. This subfunction appears in PTQ as part of the relative clause rule, **S3**, and as part of the quantification rules, **S14–S16,** and would probably show up in other rules as well in extensions of PTQ.

For the stock of primitive operations, I would suggest the operations listed in (8) (these are quite tentative and not fully precise). Note that since we are generating labeled bracketed expressions, or equivalently, trees, all of the operations are to be understood as operations on labeled bracketings or trees, and not simply on linear concatenations of words.[15] I shall say more about each operation below.

(8) Primitive Syntactic Operations
(i) concatenation
(ii) node building
(iii) simple substitution
(iv) "everywhere" substitution
(v) property specification

(i) Since **concatenation** here means concatenation of trees, it is essen-

[14] The introduction of these "subfunctions" was suggested by Emmon Bach (personal communication); it is related to some of the ideas in Bach 1965, and also bears strong resemblance to the introduction of subroutines in computer programming.

[15] In making these operations more precise, I would expect to draw heavily on work in the formalization of transformational grammar, including Peters and Ritchie, 1973a; Ginsburg and Partee, 1969; Friedman, 1971; and Bresnan, 1976.

tially the same as sister adjunction. In a bottom-up system like Montague's, there is no place for a distinction between a phrase structure operation of concatenation and a transformational operation of sister adjunction. However, concatenating two trees will not by itself provide a "parent node" to connect them; for that reason, a bottom-up system necessarily requires a "node-building" operation as well.

(ii) **Node building** should be limited, if possible, to the addition of a new labeled node or bracket dominating the entire structure formed by a given rule. For example, for a rule like **S6** of PTQ which concatenates a transitive verb with its object term phrase to make an IV-phrase, node building would be the operation providing the IV-node dominating the result. In Partee 1975 (p. 262), the formulation of *Tough* Movement makes use of "internal" node building as well, but I would hope that the creation of new internal nodes could be either eliminated or restricted to cases that would follow from general principles of structure preservation.

(iii) **Simple substitution** is to be defined as the substitution of one single-rooted tree or subtree for another single-rooted subtree having the same category label; for example, a term phrase for a term phrase as in the substitution of ${}_{T}[$*every man*$]$ for a specified occurrence of ${}_{T}[he_0]$. If we introduce a convention that the empty tree $\varnothing$ counts as belonging to every category, then deletion of any single-rooted subtree would be subsumed under simple substitution. Further constraints such as some form of recoverability constraint (see, for example, Chomsky, 1965; Peters and Ritchie, 1973a) should, of course, be added here.

(iv) **"Everywhere" substitution** is to be defined as the substitution of a specified constituent (again, this could be $\varnothing$) for all occurrences of a given indexed element such as the free variable he_i of PTQ. Such an operation is familiar to logicians and is employed by Montague, but linguists will rightly demand justification for its inclusion as a syntactic primitive. I would strongly suggest that it either be added as a primitive operation or disallowed altogether; in either case, the other primitives should be kept sufficiently restricted so that an operation as powerful as "everywhere" substitution will not be definable in terms of the other primitives.[16] I shall not attempt a full defense of this operation here; in its favor I shall note only that (*a*) an explicit formulation of the generative semanticists' rule of quantifier lowering requires such an operation, and (*b*) an analogous operation will be needed in the interpretive rules if it is not included in the syntactic rules. Such rules seem to be part of our linguistic competence in one form or another; they seem to be closely tied up with variable binding, which is probably expressible in some form in all languages.

[16] In particular, I want to disallow the use of the Kleene star (representing iteration) in structural analyses or syntactic operations, as exemplified in one of the transformational reformulations of PTQ in Cooper and Parsons 1976 (pp. 318, 319).

(v) **Property specification** is to be an operation roughly similar to what linguists have called "feature adding" or "feature copying." It will be used, for instance, for Romance language predicate adjective agreement, to assign to a predicate adjective the gender and number properties that will have been recursively defined for the subject term phrase. It can also be used to assign accusative case to objects of transitive verbs and prepositions.

Given this set of primitive syntactic operations, it can be seen that the subfunction PRO mentioned earlier is definable in terms of simple substitution, property specification, and "everywhere" substitution. It might be simplest to include in this definition a smaller subfunction, which we could call PROFORM(α), which is a simple substitution for α of the pronominal form of α, where the pronominal form of α can be represented by an "archipronoun," say SHE, plus gender and number property specifications determined by α. PRO(α, i, δ) could then be defined simply as the "everywhere" substitution of PROFORM(α) for he_i in δ.[17]

This system of syntactic operations obviously requires much more formal specification and further constraints, and perhaps further primitives will be necessary. But I hope this brief sketch will be enough to provide some substance to the discussion of the well-formedness constraint in Section 3.

2.4. Constraints on the Semantic Operations

The semantic operations G_k map α' and β' (the translations into intensional logic of the syntactic constituents α and β) onto a new expression γ' of intensional logic which is the translation of the expression γ which is formed from α and β. The intensional logic is itself an interpreted, disambiguated language with its own syntax and model-theoretic semantics, and we can use Montague's specification of its syntax to give a set of constraints on the form of the operations G_k. We shall hypothesize that the logic gives, in effect, a universal inventory of primitive semantic operations analogous to the primitive syntactic operations, and we can then require that each semantic operation G_k be definable as a composition of those primitive operations. Further research on semantic constraints would include centrally the question of whether Montague's list, given in (9), can be further narrowed down. [There is no point-by-point correspondence between the primitive syntactic operations given in (8) and the primitive semantic operations of (9); some constraints on their connection at the level of whole rules are suggested in Section 2.5.]

[17] There needs to be some way to insure that if he_i or its dominating T node has been assigned "accusative" by property specification, that specification will be retained. I do not know the best way to formalize this, and I have simply left it out of the final description of "PRO."

(9) Primitive Semantic Operations
(i) connecting formulas with propositional connectives (*and, or, not,* material implication)
(ii) applying modal and tense operators to formulas
(iii) connecting expressions of like type by "="
(iv) quantification (over all types)
(v) λ abstraction
(vi) from an expression α, denoting a function, and an expression β, denoting an argument of the appropriate type, forming the expression $\alpha(\beta)$, denoting the application of the function to the argument
(vii) from an arbitrary expression α, forming the expression $^\wedge\alpha$ whose extension is the intension of α
(viii) from an expression α of "intensional" type, forming the expression $^\vee\alpha$ whose extension at a given world is the extension picked out by α at that world

Some discussion of these operations (which come from PTQ, pp. 256–259) can be found in Partee 1975. It seems reasonable to look for subfunctions in the semantics as well as in the syntax, to capture generalizations about recurring semantic operations larger than a primitive but smaller than a whole rule. One recurring subfunction in PTQ is $\alpha'(^\wedge\beta')$, a composition of the intension operator applied to β' and function–argument application applied to α' and $^\wedge\beta'$; the resulting expression denotes the application of the function α' to the intension of β'.

Little has been done by linguists to develop arguments for choosing a best set of primitive semantic operations. The general tendency has been to assume some form of logic, often simply first-order predicate logic, with little or no argument. The great variety of formal logics available, however, requires the development of new tools for investigating the "psychological reality" of competing logics. Montague's intensional logic is relatively parsimonious in primitive operations while allowing a great deal of expressive power; for example, it permits the treatment of *seek* as an intensional transitive verb, without requiring decomposition into a complex expression such as *try to find*. But in the absence of any criteria beyond observational adequacy of the overall system, and because the main topic of this paper is syntax, I will not try to argue for this particular choice of primitives here.

2.5. Constraints on the Correspondence between Syntactic Rules and Translation Rules

In the preceding two sections, I suggested various possible constraints on the syntactic operations F_i that occur in the syntactic rules and on the semantic operations G_k that occur in the translation rules. Montague's system includes the requirement that there be a translation rule corresponding to each syntactic rule, and imposes constraints on the correspondence between the syntactic categories of the input and output expressions and the semantic types of their translations, but there are so far no constraints on the relation between the syntactic operation F_i of a given rule and the semantic operation G_k of the corresponding translation rule. The theory would be further enriched if we could find some substantive universals of this sort. In this section, I will suggest a few such constraints; these are quite tentative, and are intended primarily as an indication of a direction in which substantive universals could profitably be sought.

2.5.1. Rules of Functional Application

In PTQ, Montague lists a subset of the syntactic rules and of the corresponding translation rules under the heading "Rules of Functional Application"; these are the rules that most closely resemble phrase structure rules, combining expressions of major categories basically by concatenation, sometimes with additional operations such as accusative case marking or subject–verb agreement. The translation rules for all of these use the same semantic operation G_k: $\alpha'({}^\wedge\beta')$. This regularity can be taken to represent the hypothesis that all of the basic grammatical relations are interpreted as intensional function–argument application. I discussed this hypothesis in Partee 1974, and argued that it was an interesting and reasonable alternative to the view that intensionality is always a result of sentence-embedding operations, and that it appears to hold for all of the major grammatical relations except the subject–predicate relation. (There too the semantic interpretation is function–argument application, but it appears to be purely extensional; see Partee, 1974.)

Ignoring for the present the apparent exceptionality of the subject–predicate relation, there remains the problem of making the hypothesis precise. We cannot simply say that any syntactic operation F_i that includes concatenation as one of its component primitives will be interpreted by a semantic operation G_k of the form $\alpha'({}^\wedge\beta')$, since nearly all of the syntactic operations that take two arguments include concatenation (the rules of quantification, which are basically substitution rules, are the only exceptions in PTQ). There are various ways we might try to define the relevant class of syntactic rules, and I will mention two without trying to make them precise. The first involves looking at the whole syntactic rule and not just at the operation F_i, and this probably comes closest to what Montague had in mind:

(10) If syntactic rule n involves two expressions α and β whose categories are of the form A/B and B, respectively, and the syntactic operation F_i can be decomposed into

the concatenation of some operation on α with some operation on β, then G_k in translation rule n will be $\alpha'(\hat{}\beta')$.

The second approach involves a narrower specification of the syntactic operation F_i, and is independent of Montague's way of representing categories. We could seek to define a notion of "morphological operation," which would undoubtedly be useful elsewhere in the grammar as well; morphological operations might be limited to those operations which involve just property specification, but it may well be necessary to introduce a notion of "affix" and/or a definition of "word." I cannot undertake such a far-reaching task here, so I will just assume we have a notion of morphological operation, and represent morphological operations by M_i, M_j, etc. (I assume, for example, that putting a verb phrase into the appropriately agreeing present tense form is a morphological operation; but adding a *will* or a *have* or any other whole word is not.) Then we could formulate the second version of the constraint as shown in (11).

(11) If syntactic rule n involves a binary operation $F_i(\alpha, \beta)$, and there are morphological operations M_1 and M_2 such that $F_i(\alpha, \beta)$ is the concatenation of $M_1(\alpha)$ with $M_2(\beta)$, then G_k in translation rule n will be $\alpha'(\hat{}\beta')$.

These two versions of the functional application constraint are not equivalent, although a suitable definition of morphological operations would make them coincide with respect to Montague's PTQ. The question to explore is whether either one can be refined into a genuine substantive constraint that holds across all languages.

2.5.2. Unbounded Rules and Variable Binding

Ross (1967) introduced the notions of unbounded rules and essential variables; the existence of unbounded transformational rules has been challenged by Chomsky (1978) and defended by Bresnan (1977). Part of Bresnan's argument consists in showing that there is a general constraint on proper analyses, the "Complementizer Constraint on Variables," which can be formulated only in terms of essential variables. To the extent that the constraint has explanatory force, it provides strong indirect evidence for the existence of unbounded rules with essential variables. Similar indirect evidence might be provided if we could show that all and only the unbounded rules (assuming we can characterize this notion in a Montague framework in terms of the structural analyses SA_n and the syntactic functions F_i) have semantic interpretations of a certain form (in terms of the semantic operations G_k).

Although there are a number of unclear or dubious cases in both the "all" and "only" directions, I believe there is a substantial amount of evidence to support such a constraint, which I would state very roughtly and tentatively as in (12):

(12) All and only unbounded syntactic rules are interpreted semantically by rules that bind variables which were free in one or more of the input expressions.

Variable binding may be either by quantification or by lambda abstraction; unbounded rules will not be easy to characterize precisely in this framework until we have a more exact notation for stating structural analyses and syntactic operations. At the least, all instances of "everywhere" substitution should qualify, and perhaps in addition all operations that (delete or) substitute something for a constituent that is preceded by an essential variable not surrounded directly by a pair of brackets.[18] This characterization is undoubtedly deficient in many ways, perhaps in part because I have no primitive operation that corresponds directly to "movement" in the transformational sense; in future work I hope to try to adapt many of the technical notions of Bresnan 1976 to this framework, which I hope will make a precise characterization of the relevant class of syntactic rules possible.

Among existing analyses that support the intended interpretation of the suggested constraint are Montague's original treatment of relative clause formation (PTQ), and later extensions of it by Rodman 1976 and Thomason 1976; the quantification rules of PTQ; Quantifier Lowering as in G. Lakoff 1971; the treatment of *wh*-question formation by Karttunen (1975); the treatment of Comparative Deletion and Subdeletion by Davis and Hellan (1975), which follows quite closely the syntax of Bresnan 1973, and which uses a semantic framework sufficiently close to Montague's that I think it reasonable to include it as evidence for the proposed constraint; and the Derived Verb Phrase rule of Partee (1975). Among the problematical cases is *Tough* Movement as formulated in Partee 1975, which appears to be syntactically unbounded but meaning preserving (that is, G_k is just the identity operation). If *Tough* Movement were reformulated to map a sentence into a predicate, turning *to please him*$_3$ *is hard* into *be hard to please,* rather than into *he*$_3$ *is hard to please,* then the semantic interpretation would involve lambda abstraction and the rule would fit the proposed constraint. But then I see no reason not to give a similar reformulation to Passive, Subject Raising, and other rules that move a free variable into subject position, and these are not, in general, unbounded.

I will not try to resolve these issues here; I cannot yet offer a strong de-

[18] By an **essential variable** is meant one which could not be replaced by a finite disjunction of nonvariables. And since a transformational structural description such as (i) would be represented in this framework as (ii), I have to speak of variables not surrounded directly by brackets rather than simply of variables.

(i) $_S[X \text{ NP } Y]$

(ii) $_S[\alpha\ _{NP}[\beta]\ \gamma]$

fense of this particular constraint, but I suspect that behind the wide range of examples which appear to support it may lie the fact that variable binding itself is a very powerful, special, and in some sense, unbounded semantic device.[19] Tying two such powerful devices together by a universal constraint on the connection between them would be a fruitful step toward limiting the places where such power is to be expected.

2.5.3. Conjunction and Other Particular Universals

Some universals that have been proposed for syntax are implicitly equally semantic, as in the case of the suggestion that all languages have sentence conjunction. (I am not concerned here with whether the claim of universality is in fact correct in this case, but rather with the nature of the proposed universal.) On a purely syntactic level, such a universal claim would be simply that every language has sentence connectives; the more interesting and usually intended claim is rather that every language has a syntactic rule for combining sentences φ and ψ whose corresponding semantic interpretation is $[\varphi' \wedge \psi']$.[20]

Other particular universals of this sort could be proposed, varying in levels of detail and strength. For example, the conjunction universal just stated could be further strengthened by putting more constraints on the form of the syntactic rule that combines the sentences, for example, by saying that it always involves the concatenation of the two sentences with the addition of a fixed morpheme either between them or inside one or both of them at a specifiable position (again I do not know whether this would in fact be correct and am only making proposals about the form of the universals.)

Among the areas where further particular universals involving the relation between syntactic and semantic operations could be profitably sought are negation,[21] relative clause formation,[22] proper names, and various kinds of nominalizations of sentences or verb phrases to form names of propositions, events, actions, etc.

[19] If it is true, as I have often heard reported but never seen proved, that the set of closed sentences of first-order predicate calculus is not context free, it is clearly the "unboundedness" of variable binding that is responsible, since the full set of well-formed formulas is clearly context free.

[20] In PTQ, conjunctions are not treated categorially, so the correspondence is between a syntactic operation on pairs of sentences which concatenates them with *and* in between, and a semantic operation which concatenates their translations with "$\wedge$" in between; but an equivalent correspondence could be defined in a straightforward fashion between a conjunction and its translation.

[21] The treatment of negation has so far received relatively little attention in the Montague framework; see Lee 1974 for a discussion of some inadequacies of the PTQ treatment and some proposals.

[22] The semantic differentiation of restrictive and nonrestrictive relative clauses and the fact that not all languages appear to have both types (or to make them syntactically distinct) complicates this area considerably.

2.6. Summary

Much of this section has been programmatic. My primary aims here have been twofold: (i) to sketch the beginnings of an answer to the well-founded criticism that the Montague framework, with or without transformational extensions, has so far been too unconstrained to be a reasonable candidate for a theory of natural language whose aim is to characterize all and only the possible human languages; and (ii) to suggest by examples that within a theory containing an explicit and compositional semantics, it becomes possible to formulate and investigate constraints of a new sort, namely, constraints on the relation between syntactic rules and the corresponding semantic interpretation rules. The establishment of such constraints as these is important in its own right as part of the quest for a universal theory of natural languages; from the point of view of this paper, it is crucial if the well-formedness constraint is to escape vacuity.

3. EFFECTS OF THE WELL-FORMEDNESS CONSTRAINT

The well-formedness constraint is very different in kind from the constraints on the form of rules discussed in the previous section. It is a constraint on the output of the rules at successive stages of a derivation, and it presupposes a notion of well-formedness for expressions of all of the syntactic categories of a language and not just for sentences. We can be sure in advance that there will be unclear cases, and Sections 3.1 and 3.2 will be concerned with two sorts of unclear cases, but I will endeavor to show in the succeeding sections that there are enough clear cases with clear consequences to make the constraint a substantive one.

In Section 3.1, I will propose weakening the constraint sufficiently to allow some abstractness on a "morphological" level; in Section 3.2, I advocate the tolerance of "indexed" variables such as he_3 in well-formed expressions under certain constraints. In Section 3.3, I discuss the effect of the well-formedness constraint in ruling out "abstract markers" and forcing a more careful consideration of the domain and power of syntactic operations. The remaining sections are devoted to particular syntactic constructions: Section 3.4 to three cases that go against the well-formedness constraint in that I know of no well-supported analysis that conforms to the constraint; Section 3.5 to three cases that support the constraint, in that it rules out certain proposed analyses that I think should be ruled out; and Section 3.6 to a handful of cases where proposed analyses are ruled out by the constraint, but where I do not have enough evidence to decide whether the effect is a good one or a bad one.

In most cases in Sections 3.4–3.6, the proposed analyses in question are in transformational terms; I will be implicitly assuming a recasting of these in the framework outlined in Section 2. Note that the well-formedness constraint has no plausibility at all in a standard transformational grammar where phrase structure rules are all applied before any transformations, since a deep structure is almost **never** a well-formed surface structure. It is only in the Montague framework, with transformation-like operations included at all stages in the bottom-up derivation, that such a constraint makes sense, so if the constraint can be maintained, it can indirectly provide a potentially strong kind of syntactic support for the Montague framework.

3.1. Low-Level Abstractness

The syntax is much simplified if we count as well-formed such strings as (13):[23]

(13) *John see Pres she*
$\begin{bmatrix} \text{3 pers} \\ \text{-pl} \end{bmatrix}$ [+ acc]

These can be characterized as strings which are not directly pronounceable but which can be turned into pronounceable strings by the application of "spelling" rules which operate on terminal symbols and their "features" or "recursively defined predicates." (The rules may be sensitive to labeled bracketing, as argued by Chomsky and Halle for stress and by Selkirk for liaison.)

The advantages to allowing such representations, rather than following Montague's practice of, for example, "replacing the first verb in δ by its negative third person singular present perfect [PTQ, p. 253]," are numerous. To cite some examples, (a) the familiar Subject–Aux inversion process for questions requires the isolation of the Tense morpheme; (b) if there is a rule of Subject-to-Object Raising (*believe that he* · · · → *believe him to* · · ·), it requires deletion of a Tense morpheme and a change from nominative to accusative if the moved NP is a pronoun; (c) if Passive is a sentence-to-sentence rule, it changes the case marking on both NPs, inserts *be* and *en* into the sequence of stems and affixes, and may change the number agreement of the verb phrase with the subject;[24] (d) the Derived Verb Phrase rule argued for in Partee 1975 changes a present tense sentence into a tenseless verb

[23] I have used standard feature notation for what I described in Section 2.2 as recursively defined predicates, for want of a better standard notation. It is conceivable that this set of problems would disappear under an adequate formalism for recursively defined predicates, but I have not explored the issue far enough to tell.

[24] If Passive is a sentence-to-sentence rule, then either Affix Hopping has to be undone and redone in going from an active sentence to its passive, or else sentences in which Affix Hopping has not been applied at all should be counted as well-formed. This may be taken as an argument for direct generation of passive sentences (as in Thomason, 1976, or Bennett, 1976), but it may also be taken as an argument for including Affix Hopping among the "morphological rules."

phrase, which is most simply stated as deletion of a *Pres* morpheme (a similar operation is required in Rule 11 of Thomason, 1976, p. 93).

But if we are to allow representations like (13) without seriously weakening the well-formedness constraint, we must find ways to severely limit the sort of abstractness to be tolerated. Basically, I would view this as a search for a general characterization of a class of rules and representations to be labeled "morphological" and distinguished from the purely syntactic. As possible constraints, stated very roughly, the following might be suggested:

(14) (i) Disallow morphological features that never affect the phonological representation.
(ii) Disallow terminal symbols all of whose allomorphs are phonologically null.
(iii) Allow obligatory morphological rules, but constrain them so they are all "local" and "late" rules.

3.2. Indexing Notations

Another class of apparent counterexamples that must be dealt with immediately if the well-formedness constraint is to have any plausibility are those involving the use of indices, as in (15).

(15) a. *he*$_3$ *believes that he*$_3$ *will find a unicorn* (Montague)
b. *that*$_2$ [*he*$_2$ *shaves him*$_6$] (Thomason)

Indices have also been used in various forms of transformational grammar, on terminal symbols (*John*$_i$), nonterminals (NP$_i$), and sometimes with logical variables (x_i, or just x, y, . . .). I will not discuss the use of indices on nonterminals here, since that has no direct analog in the Montague framework, and it is only the presence of indexed terminal symbols that affects the well-formedness constraint.

One response to the problem would be to argue that *he*$_i$ or x_i are perfectly well-formed terminal symbols because such elements occur normally in mathematical or logical (quasi-) English. But this leaves unexplained the fact that all spoken natural languages manage with just a finite number of pronouns.

Supposing that we acknowledge that indexed elements are not ordinary well-formed terminal symbols, and therefore cannot be tolerated by the well-formedness constraint without some specific weakening of the constraint, there are still a number of alternatives that could be explored. I will list four, and suggest reasons for preferring the fourth, which involves relaxing the constraint in as limited a way as possible.

(i) In EFL, Montague included a category of **formulas** in the recursive syntactic rules, and at the end of all the rules characterized the set of **sentences** of the fragment as the set of formulas containing no variables.[25] But the syntactic notion "sentence" plays no role in the recursive syntactic rules on this approach; the distinction between sentences and formulas is drawn only at the final surface level. Montague draws no corresponding distinction among the "name phrases" or the common noun phrases, although these too may contain variables or not, and in general such a distinction is completely ad hoc from a purely syntactic point of view. However, it is not ad hoc semantically: It was one of Frege's basis insights that a compositional semantics requires that open sentences be included among the constituents of complex syntactic constructions; Frege even cited relative clauses among the examples requiring such treatment.[26] Montague's use of the distinction between formulas and sentences could be construed as suggesting that a violation of the well-formedness constraint is necessary here: In the construction of well-formed sentences, we must include as intermediate stages expressions that are ill-formed by virtue of containing (free) variables.

(ii) A linguist might naturally turn to "surface filters" as a device to achieve the same effect: Simply "star" at the final surface level all sentences containing indexed expressions. But the introduction of surface filters would be likely to render the well-formedness constraint vacuous, or nearly so. For example, the well-formedness constraint rules out the use of abstract markers as "triggers" for transformations, but if surface filters were added and ill-formed expressions tolerated, whenever there was a surface filter to eliminate them later, then any abstract markers could be treated in the same way as indexed expressions: Allow them in intermediate stages, and "star" at the surface any expression which still contains one or more.

I will indeed advocate [in (iv)] allowing indexed expressions to count as exceptions to the well-formedness constraint, but by a stipulation much narrower than the introduction of surface filters.

(iii) Another alternative is to take indexed variables out of the syntax altogether, as in Jackendoff's interpretive semantics or the related I-grammar approach detailed in Cooper and Parsons 1976. Cooper and Parsons' I-grammar generates English pronouns directly in the syntax, and includes a separate indexing component which operates on the output of the syntax and produces an input to the (compositional) semantics. (Whether the indexing component should be viewed as syntactic or semantic seems to be a purely terminological question.) They show that their I-grammar is equivalent to

[25] Montague 1970a includes variables v_0, v_1, etc. in the English syntax; the syntactic rules of quantification and relative clause formation, which have the semantic effect of binding variables, all replace the variables by English phrases (*every man, that man*). The formulas of English which contain variables are all interpreted as the logician's "open sentences" and those without as "closed sentences." Closed sentences have truth conditions which are independent of the choice of assignment of values to variables; open sentences in general do not, although they may happen to in special cases.

[26] See Frege, 1892 (p. 96) in the translation reprinted in Feigl and Sellars, 1949.

the grammar of PTQ, but in an indirect way that bears directly on the treatment of variables: The syntactic component generates some sentences to which the indexing component cannot assign a "viable indexing," and which are therefore in a sense filtered out by the indexing component. The sentences thus filtered out are just those which contain pronouns in positions in which they could not have been generated by the rules of PTQ (which generates pronouns only through the "variable binding" rules of quantification and relative clause formation). The sentences filtered out are not syntactically ill-formed; rather, they contain pronouns which would be understood as demonstrative or anaphoric to preceding discourse. If it is correct, as argued by Wasow (1975 and this volume), that all uses of pronouns should be treated uniformly in the syntax, then the I-grammar kind of approach is to be preferred and no exception need be made to the well-formedness constraint for indexed expressions. But the strengthening of the well-formedness constraint gained by such an approach must be balanced against the need for an additional indexing component and the concomitant weakening of the compositionality principle.

(iv) I do not have any substantive arguments against the interpretive solution just discussed; I believe that a comparative evaluation must await further work in alternative frameworks on the various uses of pronouns, of the sort illustrated by a number of papers in this volume (Wasow, Cooper, Bartsch, Hausser, Potts). What I will suggest here is a way to weaken the well-formedness constraint enough to allow the use of certain indexed expressions without weakening it to the extent that would be done if, say, surface filters were admitted to the system. All uses of he_i and similar devices are included to account for variable binding, a phenomenon that must be accounted for somewhere. Attempts to keep it out of the syntax require an extra indexing component; specifying narrowly how such devices are permitted to work in the syntax can be as big a restriction on the total class of grammars as adding a restricted additional component.

My proposal is to limit the use of indexed expressions to those expressions that have indexical (that is, deictic, or demonstrative) counterparts, which would allow for such English elements as the personal pronouns he_i, she_i, it_i, and $they_i$, as well as $then_i$, $there_i$, perhaps $Past_i$ (see Partee, 1973), and perhaps the determiner/pronoun $that_i$. (If it turns out that there are expressions that act as bound variables that have no demonstrative homonyms —perhaps *do so* is such a case—they would provide counterexamples and a different criterion would have to be sought.)

On the semantic side, I would propose the constraint that any indexed expression in the syntax must be translated into an expression of the logic containing a corresponding free variable.[27] (I would be tempted to strengthen this to an "all and only" provision if that could be done in a way that was

[27] "Containing" rather than simply consisting of, because of such translations as $\hat{P}P\{x_i\}$ for he_i.

compatible with Cooper's (this volume) suggestions for the treatment of certain pronouns as disguised definite descriptions containing a free property variable.) The syntactic and semantic restrictions are alike in admitting the indexed pronouns of PTQ and ruling out the indexed to_i and $that_i$ of Thomason 1976. If they coincide everywhere, so much the better, but we cannot have much basis for confidence without studying more phenomena of variable binding in categories other than term phrases.

3.3. Abstract Markers

One of the main effects of the well-formedness constraint is that it rules out the use of "abstract markers" in the syntax, elements which are posited simply to make the rules work but which never appear at the surface. The use of such markers is fairly widespread in the Montague framework, starting with Montague (UG).[28] Although abstract markers do not contribute directly to excessive generative capacity in the way that, say, unconstrained deletion can, their use in the Montague framework in particular has seemed entirely ad hoc so far, and I believe that behind all of them lie important issues about the domain and power of syntactic operations that should be dealt with more systematically.

I will begin by listing a sample of the abstract markers that have appeared in Montague's work and extensions of it, all of which would be ruled out by the well-formedness constraint. Their uses will be discussed below.

(16) UG: *, (̵)̵, (̵̵)̵̵, (̵̵̵)̵̵̵, (̵̵̵̵)̵̵̵̵
Bennett (1976): $, #, *, (#), (* . . .)
Partee (1975): *
Thomason (1976): [,]
Rodman (1976): the R of he_i^R

Unlike the indexing devices discussed in the previous section, none of these elements are meaningful, so there is no semantic justification for them. If any of them are syntactically essential, then the well-formedness constraint cannot be maintained.

The abstract elements used by Montague in UG can all be eliminated in the framework of Section 2 by virtue of the addition of labeled brackets and of recursively defined predicates ("features"). Montague's "*" was used in defining the auxiliary relation Y_{10}, where $\delta' Y_{10} \delta$ is interpreted as saying that δ' is the main verb location in the verb phrase δ (UG, p. 239); "*" in fact never appeared in any generated expressions of any category, but only among the relata of the auxiliary relation Y_{10}. The discussion of recursively

[28] The use of abstract morphemes has a long history in transformational grammar as well, but to discuss their use in both frameworks together would bring in too many extraneous issues to be fruitful, so I will limit my attention in this section to their use in the Montague framework. In Section 3.6, I will mention some of the transformational analyses that could not be directly "imported" into the kind of Montague transformational theory sketched in Section 2.

defined properties in Section 2.2 above was concerned in part with just such cases as this; if syntactic rules in general need to make use of such notions as main verb of a verb phrase, there should be a systematic way of defining such notions so that ad hoc markers need not be introduced for such cases. Montague's special parentheses in UG, like Thomason's brackets, are used to mark constituent boundaries of certain sorts, and are needed only if the syntactic operations are defined over strings of terminal symbols rather than over labeled bracketed strings. Although the introduction of labeled bracketing is an added source of power that needs defense, I believe that work in transformational syntax and phonology has provided strong evidence for its inclusion, and there has been some evidence from psychological experiments in support of the "psychological reality" of at least the bracketing, if not the labels.

Thomason (1976) allows syntactic rules to refer to earlier stages of the derivation tree, although not to labeled bracketing; Bennett (1976) allows neither, operating entirely on strings (as did Montague). Bennett uses "#" to mark finite verb positions so that the application of subject–verb number agreement will work properly. Montague's rules in PTQ did not handle conjoined verbs correctly, since he applied the tense morpheme only to the first verb of the main verb phrase, giving (17a) instead of (17b).

(17) a. *John walks and talk.*
b. *John walks and talks.*

I argued in Partee 1975 (p. 259), that the most natural correction to the rule would require reference to labeled bracketing in order to distinguish (18a) from (18b), both of which are built up from verb phrases whose terminal string form is simply *try to walk and talk.*

(18) a. *John tries to walk and talk.*
b. *John tries to walk and talks.*

Bennett accomplishes the distinction without labeled bracketing by using the marker "#"; the verb phrases entering into (18a) and (18b) are represented distinctly in his system, as in (19a) and (19b), respectively.

(19) a. *#try to walk and talk*
b. *#try to walk and #talk*

Note that the addition of labeled bracketing does not in itself eliminate the need for Bennett's device: The rule that puts the tense and number agreement morpheme onto the main verb(s) when the subject and verb phrase are combined needs to know not only that the verb phrases of (18a) and (18b) are distinct, but also which verb or verbs are the main verbs; hence, the recursively defined predicate mentioned above is needed as well.

Bennett's other two main devices, "$" and "*," also relate to the potential need for recursively defined predicates. The marker * is used to designate certain indexed pronouns (variables) as being in "reflexivizable position,'' so that those pronouns will be changed to reflexive form by certain rules if a like indexed pronoun is introduced at the appropriate point. Thomason (1976) accomplishes the same thing by appeal to derivational history (that is, to properties of the analysis tree). Partee (1975) treats reflexives by a transformational rule appealing to the notion of ''simplex sentence,'' which is not made explicit; moreover, the rule is obligatory and thus in direct violation of the well-formedness constraint. Thus, none of the authors mentioned (including the present author's former self) have a treatment of reflexives that conforms to the constraints advocated here. Such a treatment would have to (a) introduce reflexives in the formation of sentences wherever they are required, not by a later, obligatory rule; (b) do so without reference to derivational history; (c) do so without abstract markers representing ''reflexivizable positions.'' Two possibilities seem to be open: (i) to introduce a recursively defined property of term phrases which is equivalent to Bennett's ''*''; (ii) to find a way to explicitly define the reflexivizable positions in terms of labeled bracketing (for example, he_i is in a reflexivizable position in α if it does not occur within a pair of sentence brackets ($_t[\ .\ .\ .\ ]$) in α).

Bennett's marker ''\$'' is prefixed to plural common noun phrases and plural terms, and affects both the choice of possible quantifiers and number agreement on the verb. It is not equivalent to the plural morpheme that shows up morphologically on nouns, since, for instance, a pair of conjoined singular terms gets the marker ''\$'' but no plural morpheme, and collective nouns like *class* and *herd* are marked with ''\$'' when they are used with plural verbs. This is a case that could be handled quite naturally by a recursively defined predicate of the sort discussed in Section 2.2.

The use of ''*'' in Partee 1975 is a different sort of case, but certainly one that counts as ad hoc and that should be ruled out by the well-formedness constraint. It arises in the Derived Verb Phrase rule, which deletes a free variable subject he_i, changes all other occurrences of he_i in the remaining verb phrase to he^*, and interprets the verb phrase by lambda abstraction with respect to x_i. There is then a ''metarule'' stating that whenever a rule combines a term phrase with an expression containing any starred variables, those variables are to be replaced by pronouns whose gender is that of the given term phrase. The peculiarity in this situation is that the starred variables are semantically bound by the lambda operator, not by any quantifier, so semantically they have no ''antecedent''; their syntactic gender is determined at a later stage by a term phrase that does not have any expressed semantic relation to them at all. This complication would not arise if English reflexives were not always marked for gender; the rules proposed in Partee 1975 would work much more neatly if English reserved its gender-marked reflexives for cases where the variable was bound by a quantifier phrase as in (20a), and left the reflexive unmarked in cases where the corresponding variable was bound by lambda abstraction, as in the pseudo-English (20b) and (20c).

(20) a. *Every man saw himself.*
b. (*) *Every man tried to see self.*
c. (*) *Mary* $\left\{\begin{array}{l}\textit{got from}\\ \textit{gave}\end{array}\right\}$ *John permission to absent self.*

The work by linguists on the "control problem" [see Jackendoff, 1972 (Ch. 5)] and on "sloppy identity" bear on closely related issues; the sorts of difficulties that arise in this area have led to a number of proposals for extensions to transformational theories (absolute exceptions, networks of obligatory coreference, sloppy identity conditions, thematic hierarchy, etc.), and it may well turn out that there is no acceptable replacement for the "Starred Variable Convention" of Partee 1975 that meets the constraints proposed here. But if that is the case, it would be preferable to seek principled extensions of the theory (in the manner of Jackendoff 1972, for instance) rather than to tolerate such an unprincipled abstract marker as the "*" of Partee 1975.

Some of the other abstract markers just discussed, as well as the "*R*" marker of Rodman 1976, relate to constructions for which transformational linguists have suggested extensions to the standard theories. Rodman's "*R*" marker plays a role in capturing the Ross constraints on movement rules and relating them to restrictions on scope possibilities; if the restrictions are universal, which has been argued, then Rodman, like Ross, could state them as part of the theory and omit any special marker from the syntax, but if not, then they are a problem. Just to mention briefly some other parallels, consider the relation of Bennett's "#" to "across-the-board rules"; of Bennett's "$" for plural terms to devices of "feature climbing" and "feature spreading"; of Bennett's "*" for reflexivizable positions to the notion of "clause-bounded rules."

The one virtue in the ad hoc markers whose use I have been arguing against is that they are so visibly ad hoc; they make it easy to pick out the unsolved problems in a given analysis in a way that is not so easy when rules are written in an elegant form but require appeal to inexplicit provisions to interpret them in the intended way (for example, the Reflexive rule of Partee 1975, which is more elegant looking than the complex but explicit treatment of reflexives using the "*" marker in Bennett 1976, but uses the undefined and possibly undefinable notion of "simplex sentence"). I am therefore not criticizing the cited authors for using such markers; I hope they would agree that the most "descriptively adequate" theory would not include them.

3.4. Particular Constructions that Seem to Have Ill-Formed Sources

In this section, I will discuss three constructions which offer particularly difficult challenges to the well-formedness constraint: Sluicing, Comparative

Deletion and Subdeletion, and polarity items. These have all been discussed in the transformational literature and have not so far been treated in the Montague framework.[29] Whether they can be accounted for without weakening the well-formedness constraint is an important open question.

3.4.1. Sluicing

Sluicing is the rule proposed by Ross 1969 that derives (21b) from (21a).

(21) a. **Irv and some woman were dancing together, but I don't know which woman Irv and _____ were dancing together.*

b. *?Irv and some woman were dancing together, but I don't know which woman.*

Although the usual case is for both the input and the output of Sluicing to be grammatical, (21) illustrates a case where the (questionably) well-formed sentence (21b) has a source which is ill-formed because of a violation of the Coordinate Structure Constraint (Ross, 1967). This leads Ross to propose that the movement constraints of Ross 1967 must be interpreted as derivational constraints, since a bad structure can be improved by the subsequent application of rules (particularly where the subsequent rules remove the source of ungrammaticality from the surface structure).

The well-formedness constraint prohibits deriving a well-formed sentence from an ill-formed source. I can imagine two lines of defense for the constraint against the challenge posed by Sluicing: (i) As Ross acknowledges, the sentences formed by Sluicing from ungrammatical sources are generally judged to be ungrammatical to some extent, though often less than their sources. It might be plausible to argue that the grammar should not generate them at all, and the lesser strength of the ungrammaticality judgments might be explainable in terms of the greater amount of processing required to expose their deviance. (There might be a theory of why they sound all right to our "surface ear" but not to our "deep ear," figuratively speaking.) (ii) Sluicing is a rule that can operate across discourse; Williams (1977) has suggested that such rules should be separated out from the rules of "sentence grammar," and that different principles may apply to them. (But this line of defense may just push the same problems into another domain.)

I have no general solution to propose, but the seriousness of the problem is at least somewhat lessened by the fact that it is extremely hard to find any clearly good Sluiced sentences with clearly bad sources.

[29] Comparatives have been treated in frameworks similar to Montague's in many respects, by Davis and Hellan (1975), Cresswell (1976), but (a) neither discusses the problem cases at issue here, and (b) more importantly, the frameworks of these authors involve something like "underlying" representations which are not intended to be well-formed surface structures.

3.4.2. Comparative Delection and Subdeletion

Comparative Deletion as formulated by Bresnan (1975) is the rule that derives (22b) from (22a); Comparative Subdeletion, which may or may not be part of the same rule, derives (23b) from (23a).

(22) a. (*)*John saw more women than Bill saw x many women.*
b. *John saw more women than Bill saw.*

(23) a. (*)*John saw more women than Bill saw x many men.*
b. *John saw more women than Bill saw men.*

I am not concerned here with the fact that the source sentences (22a) and (23a) are ill-formed; I assume that by making use of a source sentence with an indexed variable in the "degree" modifier slot, for example, (24), a Montague-type derivation could be constructed without violating the well-formedness constraint.

(24) *Bill saw that*$_i$ *many women.*

The problem arises in cases where the rules of Comparative Deletion and Subdeletion themselves **produce** ill-formed sentences which can be transformed into well-formed sentences by subsequent Comparative Ellipsis. Bresnan (1975), citing Higgins 1973a as the source of this observation, gives examples such as the following:

(25) a. **Fewer of the women came to the party than* ______ *of the men came to the party.*
b. *Fewer of the women came to the party than* ______ *of the men.*

(26) a. **These constraints should be revised more radically than that they should be (revised)* ______ *has been suggested.*
b. *These constraints should be revised more radically than* ______ *has been suggested.*

Sentence (25a) is produced by Comparative Subdeletion and (26a) by Comparative Deletion; both violate Kuno's (1973) "Incomplete Subject Constraint." In (25b), Comparative Ellipsis has "ameliorated" the sentence by removing the incomplete subject. Bresnan notes that such "amelioration" effects suggest that Kuno's constraint should perhaps be reformulated as a surface structure constraint rather than as a constraint on the applicability of rules. Unless further analysis of these constructions leads to a different kind of solution, they would seem to require the admissibility of ungrammatical intermediate stages. (Note that the derivations in question give semantically reasonable sources, so any reanalysis has a strong semantic as well as syntactic challenge to meet.)

3.4.3. Polarity Items

There is at least *prima facie* evidence that the generation of the (b) sentences in the pairs (27)–(29) requires the generation of the (a) sentences as component parts, and the (a) sentences are ill-formed.

(27) a. **Anyone saw us there.*
b. *John doubts that anyone saw us there.*

(28) a. **Anyone will ever see that many blue whales again.*
b. *We saw more blue whales in the ocean last year than anyone will ever see again.*

(29) a. **I took anything just now.*
b. *The first person who claims that I took anything just now will be shot.*

In this case the problem is wide open, since there is still no satisfactory analysis of polarity items that I know of. One plausible suggestion that would be compatible with the well-formedness constraint is that the (a) sentences above are not **syntactically** ill-formed;[30] their deviance is certainly different from, for example, violations of Ross constraints, where the meaning of the ill-formed sentences remains perfectly clear. But I do not know of any way to decide whether a given case of deviance is syntactic or semantic short of seeing what the "best theory" says about it, and so far we do not seem to be very close to a "best theory" of polarity items.

A further problem can be raised in connection with polarity items, for these also exhibit the "amelioration" effect, where a bad sentence is made good by subsequent ellipsis.

(30) a. **John didn't implicate anyone yet, but he will implicate anyone.*
b. *John didn't implicate anyone yet, but he will.*

Here, the semantics plus the well-formedness constraint would seem to

[30] This idea has been urged by Richard Saenz (personal communication). Hintikka (1975) has argued that English syntax cannot be either autonomous or recursive, assuming that the (a) sentences above are syntactically ill-formed, arguing that it requires transderivational and sometimes undecidable semantic computation to tell *which* sentences with *any* are ill-formed. Hintikka's own analysis does not handle all cases of *any*, however, and even if it did, one might still reasonably challenge the hypothesis that the ill-formedness of the (a) sentences should be counted as syntactic deviance—it may well be that the "best theory" counts them as only semantically deviant.

favor restoration of the *some–any* rule, but strong arguments against it were given by R. Lakoff 1969. It is furthermore unclear how the ellipsis rule that yields (30b) could be made to regard *someone* and *anyone* as identical, so I do not consider that a very hopeful alternative.

Having posed a number of problems that the well-formedness constraint faces, I will turn to a discussion of some examples that appear to support the constraint.

3.5. Examples That Support the Well-formedness Constraint

3.5.1. Adjectives from Relative Clauses

The derivation of (31b) from (31a) has a long-standing acceptance in transformational grammar.

(31) a. **a convert who is recent*
b. *a recent convert*

Such examples were noted by Winter (1965), but were generally dismissed by transformationalists on the grounds that "rule features" requiring certain adjectives to obligatorily undergo Relative Clause Reduction and Adjective Preposing were preferable to the postulation of two separate sources for prenominal adjectives, the relative clause source for at least some of them being independently supportable by the broad applicability of the two steps of the transformational derivation (since participial phrases can also be preposed). The well-formedness constraint requires direct prenominal generation of adjectives like *recent,* however, so if there is independent support for prenominal generation of such adjectives, the constraint can be seen as helping to force the right choice.

Siegel (1976a) shows that, for Russian, there is striking syntactic, morphological, and semantic evidence for a dual categorization of adjectives into the categories CN/CN of common noun modifiers and *t*/*e* of simple predicates. In addition, Siegel (1976b) gives syntactic and semantic arguments for dual categorization of adjectives in English (although the fact that there are many "doublets," that is, apparently identical words in both categories, stands in need of further explanation).

In this case, the compositionality constraint and the well-formedness constraint converge in rejecting (31a) as a source for (31b); the fact that in Russian even the morphological form of predicate and adnominal adjective differ, along with Siegel's additional evidence from Russian and English, gives strong support to both constraints.

3.5.2. Tough Deletion Versus Tough Movement

The classical derivation of *tough*-sentences like (32a) involves a movement rule operating on a source like (32b), which in turn is derived from a source like (32c).

(32) a. *John is easy to please.*
b. *It is easy to please John.*
c. *To please John is easy.*

Alternative proposals that have appeared recently include the deletion analysis of Lasnik and Fiengo 1974, which postulates (33) as the source of (32a), and the *wh*-movement analysis of Chomsky (1977) which includes (34) as an intermediate stage.

(33) **John is easy to please John.*

(34) **John is easy whom to please.*

I have argued elsewhere that the phenomena used by Lasnik and Fiengo to support their analysis can all be accounted for in a Montague-type analysis that makes use of "derived adjective phrases" (Partee, forthcoming) without such ill-formed and semantically incoherent source structures as (33), and Bresnan (1977) has strong syntactic arguments against Chomsky's attempt to replace all unbounded movement rules by successive–cyclic applications of *wh*-movement. I therefore consider it an advantage that both of these analyses are ruled out by the well-formedness constraint.

3.5.3. Pseudo-Clefts

There have been a great many proposals in the literature for the analysis of pseudo-cleft sentences such as (35).

(35) *What John ate was the meat.*

Among the proposed sources for such sentences are at least the following:[31]

(i) $_{NP}$[*what John ate*] *was* $_{NP}$[*the meat*]
(ii) $_{NP}$[(*it*) $_{S}$[*John ate the meat*]] *was* $_{NP}$[*the meat*]
(iii) $_{NP}$[*the thing* $_{S}$[*John ate wh-something*]] *was* $_{NP}$[(*it*) $_{S}$[*John ate the meat*]]
(iv) $_{NP}$[(*it*) $_{S}$[(*wh*) *John ate the meat*]] *was* $_{NP}$[Δ]

Of these, only (i) meets the well-formedness constraint; probably not coin-

[31] There has been a longer oral than published tradition surrounding pseudo-clefts, so attributions of proposals are somewhat uncertain. Peters and Bach (1968) argue against (iv), which they attribute in general terms to Chomsky (1970), and against (i) which they suggested as a straw person, and in favor of (iii), which they credit to Ross. Akmajian (1970) posits essentially (i), and Higgins (1973b) gives many arguments in favor of (i) over all the alternatives.

cidentally, it is also the only source that satisfies the compositionality principle.[32] Peters and Bach (1968) argue against (i) and in favor of (iii) on the basis of the behavior of reflexivization in examples like (36).

(36) *What the missile damaged was itself.*

However, the force of such examples is vitiated by the occurrence of similar reflexives in sentences not of the pseudo-cleft form, such as (37).[33]

(37) *The only weapon that the missile damaged was itself.*

The reflexives of (36) and (37) appear to be emphatic reflexives, hence not subject to the usual clausemate restrictions. The available evidence strongly favors hypothesis (i), argued for by Higgins (1973b), and if this is the correct choice, it provides another case where the well-formedness constraint leads to a syntactic preference for the analysis that is also semantically selected by the compositionality principle.

In all of the examples just discussed, the advantage that the well-formedness constraint gives the child in her acquisition task is surely an additional argument in its favor. The complexity of the rules in a bottom-up derivation system of the sort proposed in Section 2 does not seem any greater, in a formal sense, than in a transformational grammar, and the choice among possible sets of rules is greatly narrowed down if the constituents produced at each step are required to be ones that the child could actually encounter. The system is still highly "abstract" in that the rules themselves are strongly structure dependent, but the locus of abstractness is more restricted than in a system which permits arbitrary departures from surface structure in its underlying and intermediate representations.

It also seems clear, although I cannot offer a formal proof, that the compositionality principle and the well-formedness constraint together place much stronger restrictions on the class of possible grammars for a given language than either could provide alone (given some fixed specification of the possible forms of rules). In the examples just discussed, the two principles reinforce each other; it is very likely that ways could be found to "get around" either constraint in isolation, even with the limitations on the form of rules suggested in Section 2, but I believe it would be much harder to find alternative analyses within that system of rules that satisfy both constraints together.

[32] The apparent well-formedness of (35′) would suggest that (iii) also meets the well-formedness constraint in spite of its apparent assertion that John ate a proposition:

(35′) *What John ate was* (pause) *he ate the meat.*

However, the special intonation required for (35′) and the fact that it is not acceptable in all dialects suggest that different rules, perhaps including some sort of redundant reduplication, are involved in its formation.

[33] I am indebted to Josh Ard for making me aware of this observation, which can also be found in Higgins, 1973b.

3.6. Other Analyses That Are Ruled Out by the Well-formedness Constraint

In this section, I will mention briefly a number of proposed analyses of constructions in various languages that are ruled out by the well-formedness constraint. I will not take a stand on whether these cases argue for or against the constraint; in some cases, the proposals are specific to alternative frameworks that are not directly comparable to the framework proposed here, and in other cases, I simply do not have enough independent evidence in favor of or against the suggested analysis. I am including them here as further evidence that the constraint is a powerful one, although examples such as these may lead to arguments that it is in fact too powerful and cannot be maintained.

3.6.1. Quantifiers as Higher Verbs

The proposal that quantifiers are a species of predicates is no longer advocated by most generative semanticists; it would be eliminable on purely syntactic grounds by the well-formedness constraint because of the impossibility of sentences like (38).

(38) $$\text{*(}\textit{The}\text{)} \begin{Bmatrix} \textit{men} \\ \textit{man} \end{Bmatrix} \textit{ who } \begin{Bmatrix} \textit{love} \\ \textit{loves} \end{Bmatrix} \textit{ Mary } \begin{Bmatrix} \textit{are} \\ \textit{is} \end{Bmatrix} \textit{ every.}$$

3.6.2. Abstract Lexical Items

The postulation of abstract lexical items such as CAUSE, not identifiable with any actual lexical item such as *cause,* is ruled out by the well-formedness constraint. This requires that any lexical decomposition that is semantically justifiable must take place at the interpretive level, for example, through the translation of single lexical elements into complex expressions in the intensional logic. (See Dowty, 1976 for a discussion of lexical decomposition and Montague grammar.)

3.6.3. Lexically Governed Obligatory Extraposition

It is often maintained that (39) is the source for (40), and that the verb *appear* must have a "rule feature" that makes Extraposition obligatory for it.

(39) **That John is asleep appears.*

(40) *It appears that John is asleep.*

Such an analysis is impossible under the well-formedness constraint, as is any lexically governed obligatory rule. To defend the constraint, we would need independent evidence either in favor of giving up Extraposition altogether or else evidence in favor of multiple derivations of the same kind of "extraposed" surface structure, both by Extraposition and by direct generation in place.

3.6.4. German Verb-Final Word Order

The well-formedness condition would seem to require that "verb second" be the basic structure of German clauses, unless arguments could be given that the notions of "well-formed clause" and "well-formed sentence" should be kept distinct. There is a danger that admitting such a distinction might greatly weaken the effects of the well-formedness constraint. Such a weakening may nevertheless be necessary to account for phenomena that occur only in the highest clause in a sentence.

3.6.5. Romance Language Clitic Pronouns

In the Romance languages, there are special clitic pronouns that may occur only in clitic position, attached to a verb. These pronouns serve ordinary noun phrase "functions," typically direct and indirect object functions, so it would certainly seem simplest to generate them in ordinary noun phrase positions and move them into clitic position by an obligatory rule. Such a derivation would violate the well-formedness constraint, however. Whether one can justify a rule that moves a clitic pronoun into clitic position at the same time that it is formed, or in some other way preserve the well-formedness constraint in this case, remains an open question.

3.6.6. Trace Theory

Although the framework which includes trace theory is so radically different from the sort of framework proposed here that direct is comparison impossible, it may be worth noting that the well-formedness constraint immediately rules out elements like the trace "*t*" from inclusion in the terminal vocabulary, along with "*," empty nodes, "Δ," and other purely abstract elements that have been posited within some recent interpretivist theories (for example, Chomsky, 1977)

3.6.7. Obligatory Reflexive Transformation

Partee (1975) includes an obligatory transformational rule for deriving reflexive pronouns from nonreflexive ones. As mentioned earlier, such a rule seems simpler than the complex apparatus used by Bennett (1976) to generate reflexives directly as the sentence is built up. But obligatory rules are disallowed by the well-formedness constraint, so it is crucial to find additional evidence for or against the generation of reflexives by obligatory rule.

This list could doubtless be expanded in many ways; a number of transformational analyses involving ungrammatical intermediate stages have become so well entrenched that we may cease perceiving the intermediate stages as ill-formed. Further research on the tenability of the well-formedness constraint will require a careful formulation of a wide variety of extensions to the fragments of natural language that have been treated in the Montague framework so far.

4. SUMMARY

It would be nice to be able at this point to reach a definite conclusion about the validity of the well-formedness constraint. I hope to have shown that it does indeed rule out a great many otherwise possible syntactic analyses, but whether the analyses it permits include all the "right" ones must remain an open question for now. The constraint interacts with a number of very large-scale problems relating to the organization of the grammar and the permitted forms of rules; the most I can hope to have accomplished here is to clarify the issues involved, identify some of the potentially problematical cases and some of the most supportive cases, and provide arguments that the constraint is worth serious attention.

It has been argued here that even if the well-formedness constraint does not restrict the total class of languages or the total class of grammatical rules it severely restricts the class of grammars compatible with a given set of "primary data" and therefore could be as important as constraints on the form of the rules in its impact on the acquisition problem. The examples discussed above should make it clear that if the constraint is correct, it eliminates a great many proposed analyses for English (and other languages), merely by inspection of the terminal strings of intermediate stages of derivations. These are analyses that a child operating under the constraint could reject almost immediately (consider the sophistication of the arguments of Bresnan (1977) against Chomsky's reanalysis of Comparative Deletion as a case of successive–cyclic *wh*-movement; it would certainly be nice to end up with a theory that lets the child choose easily among the hypotheses that linguists need all their ingenuity to evaluate.)

The fact that the constraint interacts crucially with the rest of the framework, coupled with the multiplicity of current alternative frameworks, makes it difficult to examine the consequences of the constraint directly. But it would probably be fruitful to compare the effects of analogous constraints in other frameworks. Note in this connection that *Syntactic Structures* contains a similar restriction in the requirement that structures underlying well-formed kernel sentences be the sole building blocks of complex sentences. The elimination of abstract markers from the category of terminal symbols could be proposed in any framework, and various other effects of the well-formedness constraint could similarly be isolated for scrutiny across varying theories.[34]

[34] For a relevant parallel which has influenced my own thinking, see the discussion of the proposed "local-grammaticality constraint" in Bach, 1974. Bach relates his proposal to the "local-filtering constraint" proposed by Peters and Ritchie, which can be found in Peters and Ritchie, 1973b.

Faced with a proposed constraint which has great potential power but which may simply be incorrect, how should one proceed? It will be difficult if not impossible to find absolute counterexamples, since these would have to take the form "There is **no** analysis of such-and-such a construction within the given theory that is compatible with the well-formedness constraint," and even then a modification of the theory might be preferable to abandoning the constraint. The approach I would advocate would be to try to operate within the limits of the well-formedness constraint; and if it does not need to be weakened, to try to keep the permissible kinds of violations to it under as tight control as possible.

REFERENCES

Akmajian, A. (1970). On deriving cleft sentences from pseudo-cleft sentences. *Linguistic Inquiry,* **1,** 149–68.

Bach, E. (1965). On some recurrent types of transformations. *Georgetown University Monograph* **18,** 3–18.

Bach, E. (1974). Explanatory inadequacy. In D. Cohen (Ed.), *Explaining Linguistic Phenomena,* Washington, D.C.: Hemisphere Publishing Corporation. Pp. 153–171.

Bach, E. and Cooper, R. (1976). *The NP-S analysis of relative clauses and compositional semantics.* Manuscript, University of Massachusetts, Amherst.

Bennett, M. (1976). A variation and extension of a Montague fragment of English. In B. Partee (Ed.), *Montague Grammar.* New York: Academic Press. Pp. 119–63.

Bresnan, J. (1973). Syntax of the comparative clause construction in English. *Linguistic Inquiry,* **4,** 275–343.

Bresnan, J. (1975). Comparative deletion and constraints on transformations. *Linguistic Analysis,* **1**(1), 25–74.

Bresnan, J. (1976). On the form and functioning of transformations. *Linguistic Inquiry,* **7,** 3–40.

Bresnan, J. (1977). Variables in the theory of transformations. In P. Culicover, A. Akmajian, and T. Wasow (Eds.), *Formal Syntax.* New York: Academic Press.

Chomsky, N. (1957). *Syntactic Structures,* The Hague: Mouton.

Chomsky, N. (1965). *Aspects of the Theory of Syntax.* Cambridge, Mass: MIT Press.

Chomsky, N. (1970). Remarks on nominalization. In R. Jacobs and P. Rosenbaum (Eds.), *Readings in English Transformational Grammar,* Waltham, Mass: Ginn and Company.

Chomsky, N. (1977). On Wh-movement. In P. Culicover, A. Akmajian, and T. Wasow (Eds.), *Formal Syntax.* New York: Academic Press.

Cooper, R. (1975). Montague's Semantic Theory and Transformational Syntax. Unpublished doctoral disseration, University of Massachusetts, Amherst.

Cooper, R., and Parsons, T. (1976). Montague grammar, generative semantics, and interpretive semantics. In B. Partee (Ed.), *Montague Grammar.* New York: Academic Press.

Cresswell, M. J. (1976). The semantics of degree. In B. Partee (Ed.), *Montague Grammar.* New York: Academic Press. Pp. 261–292.

Davis, C. and Hellan, L. (1975). *The syntax and semantics of comparative constructions.* Unpublished manuscript, Notre Dame University, Notre Dame, Indiana.

Dowty, D. (1976). Montague grammar and the lexical decomposition of causative verbs. In B. Partee (Ed.), *Montague Grammar.* New York: Academic Press.

Frege, G. (1892). Über Sinn and Bedeutung, *Zeitschrift für Philos. und Philos. Kritik* **100.** [Trans. as "On sense and nominatum." In H. Feigl and W. Sellars (Eds.) (1949), *Readings in Philosophical Analysis,* New York: Appleton. Pp. 85–102.]

Friedman, J. (1971). *A Computer Model for Transformational Grammar*. New York: American Elsevier.

Ginsburg, S., and Partee, B. (1969). A mathematical model of transformational grammars. *Information and Control,* **15,** 297–334.

Higgins, F. R. (1973a). On J. Emonds's analysis of extraposition. In J. Kimball (Ed.), *Syntax and semantics* 2, New York: Academic Press.

Higgins, F. R. (1973b). *The Pseudo-cleft Construction in English*. Unpublished doctoral dissertation, MIT, Cambridge, Mass.

Hintikka, J. (1975). *On the limitations of generative grammars*. Unpublished manuscript, Helsinki.

Jackendoff, R. (1972). *Semantic Interpretation in Generative Grammar,* Cambridge, Mass: MIT Press.

Karttunen, L. (1975). *Syntax and semantics of questions*. Unpublished paper, University of Texas, Austin, Texas.

Katz, J., and Fodor, J. (1963). The structure of a semantic theory. *Language,* **39,** 170–210.

Katz, J., and Postal, P. (1964). *An Integrated Theory of Linguistic Descriptions*. Cambridge, Mass: MIT Press.

Kuno, S. (1973). Constraints on internal clauses and sentential subjects. *Linguistic Inquiry,* **4,** 363–385.

Lakoff, G. (1971). On generative semantics. In D. Steinberg and L. Jakobovits (Eds.), *Semantics: An Interdisciplinary Reader in Philosophy, Linguistics, and Psychology,* Cambridge, England: Cambridge University Press.

Lakoff, R. (1969). Some reasons why there can't be any *some–any* rule. *Language,* **45,** 608–15.

Lasnik, H., and Fiengo, R. (1974). Complement object deletion. *Linguistic Inquiry,* **5,** 535–71.

Lee, K. (1974). The Treatment of Some English Constructions in Montague Grammar. Unpublished doctoral dissertation, University of Texas, Austin, Texas.

Montague, R. (1970a). English as a formal language. In B. Visentini *et al., Linguaggi Nella Società e Nella Tecnica,* Milan: Edizioni de Comunità, [Reprinted in Montague (1974).]

Montague, R. (1970b). Universal grammar. *Theoria,* **36,** 373–398. [Reprinted in Montague (1974).]

Montague, R. (1973). The proper treatment of quantification in ordinary English. In J. Hintikka, J. Moravcsik, and P. Suppes (Eds.), *Approaches to natural language,* Dordrecht: D. Reidel. [Reprinted in Montague (1974).]

Montague, R. (1974). *Formal philosophy: Selected papers of Richard Montague,* edited and with an introduction by Richmond Thomason, New Haven: Yale University Press.

Parsons, T. (1970). Some problems concerning the logic of grammatical modifiers. *Synthèse,* **21,** 320–334.

Partee, B. (1973). Some structural analogies between tenses and pronouns in English. *The Journal of Philosophy,* **70,** 601–09.

Partee, B. (1974). Opacity and scope. In M. Munitz and P. Unger (Eds.), *Semantics and Philosophy,* New York: New York University Press. Pp. 81–101.

Partee, B. (1975). Montague grammar and transformational grammar, *Linguistic Inquiry,* **6,** 203–300.

Partee, B. (Ed.), (1976a). *Montague grammar,* New York: Academic Press.

Partee, B. (1976b). Semantics and syntax: The search for constraints. In C. Rameh (Ed.), *Georgetown University round table on languages and linguistics 1976,* Washington, D.C.: Georgetown University Press. Pp. 99–110.

Partee, B. (forthcoming). John is easy to please. To appear in a collection edited by A. Zampolli, North-Holland Press (title unknown).

Peters, P. S., Jr., and Bach, E. (1968). Pseudo-cleft sentences. In Report to NSF, *On the Theory of Transformational Grammar.* Department of Linguistics, University of Texas, Austin, Texas (1971). Pp. 171–190.

Peters, P. S., Jr., and Ritchie, R. W. (1973a). On the generative power of transformational grammars. *Information Sciences,* **6,** 44–83.

Peters, P. S., Jr., and Ritchie, R. W. (1973b). Nonfiltering and local-filtering transformational grammars. In J. Hintikka, J. Moravcsik, and P. Suppes (Eds.), *Approaches to Natural Language,* Dordrecht: D. Reidel. Pp. 180–194.

Postal, P. (1966). On so-called "pronouns" in English, *Georgetown University Monograph Series on Language and Linguistics,* No. 19, 177–206. [Reprinted in Reibel and Schane, (Eds.) (1969), *Modern Studies in English: Readings in Transformational Grammar,* Englewood Cliffs, New Jersey: Prentice-Hall, Inc.]

Rodman, R. (1976). Scope phenomena, "Movement Transformations," and relative clauses. In B. Partee (Ed.), *Montague Grammar.* New York: Academic Press. Pp. 165–176.

Ross, J. R. (1967). Constraints on Variables in Syntax. Unpublished doctoral dissertation, MIT, Cambridge, Massachusetts.

Ross, J. R. (1969). Guess who?. In Robert Binnick *et al.* (Eds.), *Papers from the Fifth Regional Meeting of the Chicago Linguistic Society.* Chicago: Department of Linguistics, University of Chicago. Pp. 252–86.

Siegel, M. E. A. (1976a). Capturing the Russian adjective. In B. Partee (Ed.), *Montague Grammar.* New York: Academic Press. Pp. 293–309.

Siegel, M. E. A. (1976b). Capturing the Adjective. Unpublished doctoral dissertation, University of Massachusetts, Amherst.

Stockwell, R., Schachter, P., and Partee, B. (1973). *The Major Syntactic Structures of English.* New York: Holt.

Thomason, R. (1976). Some extensions of Montague grammar. In B. Partee (Ed.), *Montague Grammar.* New York: Academic Press. Pp. 77–117.

Wasow, T. (1975). Anaphoric pronouns and bound variables. *Language,* **51,** 368–383.

Williams, E. (1977). Discourse and logical form. *Linguistic Inquiry,* **8,** 101–139.

Winter, W. (1965). Transforms without kernels. *Language,* **41,** 484–489.

A PROJECTION PROBLEM FOR PRAGMATICS[1]

ROBERT M. HARNISH

> It is astonishing what language can do. With a few syllables it can express an incalculable number of thoughts, so that even a thought grasped by a human being for the very first time can be put into a form of words which will be understood by someone to whom the thought is entirely new. This would not be possible were we not to distinguish parts in the thought corresponding to the parts of the sentence, so that the structure of the sentence serves as an image of the structure of the thought . . . . If, then, we look upon thoughts as composed of simple parts, and take these, in turn, to correspond to the simple parts of the sentences, we can understand how a few parts of sentences can go to make up a great multitude of sentences, to which, in turn, there correspond a great multitude of thoughts [Frege, 1923, p. 2].

1. INTRODUCTION

At least since Frege, philosophers of language and logicians have adhered to the principle that the semantics of a complex expression is a function of the semantics of its constituents and their grammatical relations. Even phi-

[1] For their helpful remarks on an earlier draft of this material, I would like to thank Adrian Akmajian and Frank Heny.

SYNTAX AND SEMANTICS
VOLUME 10

ISBN: 0-12-613510-X

losophers of language generally unsympathetic to the details of logical semantics since Frege can be seen to subscribe to some such **compositionality** principle. In what follows I will suggest that compositionality is a constraint in the theory of speech acts. Also, I will briefly survey some ways in which syntax and speech acts might be adjusted to conform to the compositionality constraint.

2. PROJECTION AND PRAGMATICS

Following Stalnaker (1970, p. 275), we will assume that pragmatics is the study of speech acts. What, then, is the projection problem for pragmatics? There are two distinct ways of arriving at it, each with its own virtues and liabilities, though the final formulation is the same.[2] It is instructive to look at each in turn.

One way of arriving at the projection problem for pragmatics is by construing semantics as, in part, the theory of meaning or sense, and then subscribing to a **use** theory of meaning. The argument would go like this:

(i) The Compositional Idea: The meaning of a syntactically complex expression is determined by two things: (a) the meaning of its (immediate?) constituents, and (b) their grammatical relations.

(ii) The projection problem for semantics is the problem of precisely formulating the principles that determine the meaning of a complex expression on the basis of (a) and (b).

(iii) But the meaning of an expression is (or is determined by the rules for) its **use** in the linguistic community.

(iv) Thus, the compositional idea for speech acts: the **use** of a syntacti-

[2] Curiously, Fodor and Katz's (1964) original formulation of the projection problem for *semantics* follows both routes, though they officially commit themselves to only a portion of the first. They say,

> A synchronic description of a natural language seeks to determine what a fluent speaker knows about the structure of his language that *enables him to use and understand its sentences*. Since a fluent speaker is able to *use and understand* any sentence drawn from the *infinite* set of sentences of his language, and since, at any time, he has only encountered a *finite* set of sentences, it follows that the speaker's knowledge of his language takes the form of rules which project the finite set of sentences he has fortuitiously encountered to the infinite set of sentences of the language. A description of the language which adequately represents the speaker's linguistic knowledge must, accordingly, state these rules. The problem of formulating these rules we shall refer to as the *projection problem* [p. 482].

Notice that although "projection" here is from *encountered* sentence of L to *sentences of L*, the projection rules as formulated by Katz and Fodor conform to the description given in (i), and project from *constituents* of a complex expression to the complex expression *itself*—in the spirit of the quotation from Frege which heads this paper.

cally complex expression is determined by two things: (a) the use of its (immediate?) constituents, and (b) their grammatical relations.

(v) And thus the projection problem for pragmatics: to precisely formulate the principles that determine the use of a syntactically complex expression on the basis of (a) the use of its (immediate?) constituents, and (b) their grammatical relations.

In summary, if one is committed to the projection problem in semantics, and also subscribes to a **use** theory of meaning, then it looks like one is committed to (this formulation of) the projection problem for pragmatics.

The second way of arriving at the projection problem for pragmatics does not depend on a use theory of meaning; it arises from applying the same sort of arguments that motivate the projection problem for semantics to speech acts. These arguments are of two sorts: considerations concerning **repeatable components,** and considerations concerning **projection to new cases.**[3]

First, our ability to perform speech acts is compositional in the sense that complex speech acts can be seen to be analyzable into simpler, distinct, and repeatable sub-acts. One test for this is the possibility of changing the complex act by changing one of the constituent acts. Suppose, for example, that I **assert** that *Carter is happy being president;* I would have performed a different speech act had I done any of the following instead:

(1) **Asked** about his happiness.
(2) Referred to **Vance.**
(3) **Not** referred at all, but just said that somebody was happy.
(4) Attributed **boredom** to him.

The result of following the recipes implicit in (1)–(4) would be to:

(1′) Ask whether Carter is happy being president.
(2′) Assert that Vance is happy being president.
(3′) Assert that someone is happy being president.
(4′) Assert that Carter is bored being president.

These results are automatically accounted for on the hypothesis that one performs the complex act of **asserting** that *Carter is happy being president* in virtue of performing the relevant constituent acts in their relevant relationships. Put in the linguistic mode, the use of the sentence *Carter is happy being president* (or *I claim that Carter is happy being president*) to perform the complex act just mentioned is determined by: (a) the uses of its (immediate?) constituents to perform the subacts just mentioned, and (b) the grammatical relations between them. The projection problem for speech acts is to identify the elements, and specify the principles of composition.

[3] See, for example, Davidson, 1974 (p. 242), Wallace, 1970 (p. 145) for the first consideration, and Davidson, 1964 (pp. 387–388) for the second consideration.

Second, our ability to perform and identify speech acts seems to be projective in the sense that we can perform and identify speech acts we have never performed or identified before. For instance, to my knowledge I have never performed **any** of the previously mentioned complex speech acts, though I have performed the constituent acts. Analogously, the normal hearer is able to identify these complex acts even though he has only encountered their constituent acts before.

So far, we have considered only one mode of compounding acts of reference, predication and assertion—embedding—and it will remain our primary type of example. It should be noted, though, that there are simpler ways of compounding acts that illustrate both of the previously mentioned symptoms of compositionality. For instance, it is possible to perform various acts *conjointly*. One might assert that Mary is in the next room and request that you get her and promise to be nice to her and . . . . Similarly, one might refer to John and refer to Mary and refer to . . . and say of them that they met last week.[4] Or one might refer to John and proceed to predicate being tired of him and being hungry of him and so on.[5]

These remarks can probably be extended to other modes of compounding acts,[6] but this appears sufficient to make plausible the existence of a projection problem for speech acts—even if meaning is not use.

3. SEARLE'S THEORY

Although there are a variety of important differences between speech act theories,[7] there also seems to be general recognition of the fact that such a theory must be compositional.[8] Searle is fairly representative on this point when he writes,

(5) The meaning of a sentence is determined by the meanings of all its meaningful components. . . . The meaningful

[4] The predicate *met* precludes analyzing such cases as elliptical assertions. See Harnish, 1976a for further discussion and references.

[5] So far, I have given no argument that conjoining predications is not an elliptical form of conjoining assertions, requests, etc. Though I think one can be given, it involves the details of Searle's rule of predication and is best left to another occasion.

[6] Disjunction offers the rule-based speech act theorist some curious problems. For instance, in cases of disjoined references and of disjoined predications it is not clear how to distinguish their speech act interpretation from the *conjunctive* cases. This is because if one says that either John or Mary is tired, one has referred both to John and Mary, which is true for conjoined reference as well. It might be that differences will emerge at the level of truth (for what is asserted) fulfillment (for what is requested) etc., but speech act theorists have not yet offered us an account of these matters.

[7] For Grice, Strawson, Schiffer, Bach, and Harnish, the notions of intention, belief, desire, etc. are central. For Austin, Alston, and Searle the notions of rules and conventions are central.

[8] See, for example, Grice, 1968 (p. 65), Alston, 1964 (p. 60) and 1974 (p. 43ff).

> components of a sentence include also its deep syntactic structure . . . [1968, pp. 415–416].

What I intend to do in the main body of this paper is investigate the notions of composition and projection in Searle's theory. I pick Searle's theory because it is, I think, the most adequate overall theory of speech acts that we have at present. The basic idea behind Searle's theory of speech acts is that,

(6) The semantics of a language can be regarded as a series of systems of constitutive rules and . . . illocutionary acts are acts performed in accordance with these sets of constitutive rules [1965, p. 225].

There are three ideas in this excerpt that I would like to draw attention to:

(i) Illocutionary acts (like promising, ordering, stating, advising, etc.) are performed in accordance with **rules.**
(ii) These rules are **constitutive** rules.
(iii) These rules are the **semantic–pragmatic** rules of a language.

Let us accept for the moment the first and last claim, and say a few words about the second one.

There really is not space to review Searle's discussion of constitutive and regulative rules in detail. Let me just repeat what Searle says at one point. Regulative rules—like parking regulations—govern **antecedently existing** forms of behavior. One can park a car without there being parking regulations. But systems of constitutive rules **define** a new form of behavior. One cannot, so the story goes, make a touchdown antecedent to the existence of rules for football. In brief, constitutive rules constitute or define the activity they govern. Consequently, constitutive rules for performing illocutionary acts will **define** the act that they are rules for performing, in the sense that to conform to the rules guarantees the successful performance of the act.

Supposing, then, that these rules for performing illocutionary acts are somehow construed as semantic–pragmatic rules; rules of meaning–use for some (possible) language. How are these rules related to the language? Roughly, Searle's proposal seems to be this: From the point of view of semantic/pragmatic analysis, each sentence can be represented as divided into two (not necessarily syntactically distinct) parts, the illocutionary force indicator $F[\ ]$ and a content indicator C. The content indicator can be of various kinds: It can be just a **name** (for example, *Manchester, hurrah!*); just a **predicate** (for example, *How many are there?, Leave!*); a **proposition** (for example, *I promise that I will go*); or nothing (for example, *Ouch, Hello*).

The force indicator shows "what illocutionary force the utterance is to have [1969, p. 30]," and where the content is a proposition, the content indicator is called the "propositional content indicator" and it indicates what proposition is being expressed by the speaker in the utterance of that sentence. Although Searle does distinguish quantified from subject–predicate propositions, his discussion is almost exclusively devoted to the latter; I will follow him on this.[9]

One begins to see, then, that a semantic–pragmatic theory for a language must represent its sentences as having one or more of the above forms, and that one gives the meaning or use of the sentence by giving the semantic or pragmatic rules which attach to the elements of the sentence corresponding to these forms. Thus, every sentence that has any content at all will be of the general form $F[C]$, and will have at least two batches of semantic rules attached to it; one batch for F and one batch for C—keeping in mind the possibility that the same syntactic item (word or morpheme, for instance) might be doing both jobs.

Summarizing, sentences will be pragmatically analyzed as being of the following forms, where F stands for the device for performing some illocutionary act, R the device for referring, P the device for predicating:

$F(R,P)$:	*I promise that I will be there on time.*
$F(R)$:	*Manchester, hurrah!*
$F(P)$:	*Leave!*
F:	*Hello.*

So far we have not said what these constitutive rules look like. According to Searle,

(7) The meaning of a sentence is determined by rules, and those rules specify both conditions of utterance and also what the utterance counts as [1969, p. 48].

For convenience we will call rules that specify the conditions of utterance "U-rules," and rules that specify what the utterance counts as "C-rules." Thus, to specify the meaning (or meaning-determining aspects) of a sentence is to specify its U-rules and C-rules. As a format for such meaning–use specifications we will write:[10]

(R-SPEC) (U) x is to be uttered only if ______.
(C) The utterance of x counts as ______.

[9] It is a good exercise to try to figure out how a Searle type speech act semantics could be given for quantifiers. The least revisionary line to take would be the Fregean one of treating quantifiers as second-level predicates, then adjust the rule of predication accordingly. In any case, the problem is to formulate rules of the form demanded by (R-SPEC), governing expressions like:

(i) Some x such that . . . x . . .
(ii) Every x such that . . . x . . .

It is *very* unclear how, or even that, such rules can be stated.

[10] I will sometimes refer to what fills in the blanks as the **content** of the U-rules, and the **content** of the C-rules.

If we put together this notion of meaning as determined by rules for use with the ideas underlying (5), we get the following result embracing the compositionality of meaning/use:

(8) The meaning of a sentence is entirely determined by the meaning of its elements both lexical and syntactical. And that is just another way of saying that the rules governing its utterance are determined by the rules governing (the utterance of) its elements [1969, p. 61].

From this remark, as well as (5) and (7), we can, I think, extract the following semantic–pragmatic principles (governing nonidioms) which Searle thinks any plausible semantic–pragmatic theory must conform to.

SEMANTIC–PRAGMATIC PRINCIPLES

For Sentences

(i) MEANING–USE COMPOSITION: The meaning–use of a sentence is determined by:
- (a) The meaning–use of its (meaningful) constituents, and
- (b) their syntactic structure and relations.

(ii) SEMANTIC–PRAGMATIC RULES: The meaning–use of a sentence is determined by rules which specify:
- (U) Its conditions of utterance (U-rules).
- (C) What (speech act) the utterance of that sentence counts as (C-rules).

(iii) RULE COMPOSITION: The rules governing the use of a sentence are determined by the rules governing its (rule-governed) constituents.

For Constituents[11]

(iv) MEANING–USE COMPOSITION: The meaning–use of a constituent (of a sentence) is determined by:
- (a) The meaning–use of its (meaningful) constituents.
- (b) Its syntactic structure and relations.

(v) SEMANTIC–PRAGMATIC RULES: The meaning–use of a constituent (of a sentence) is determined by rules which specify:
- (U) Its conditions of utterance (U-rules).
- (C) What (speech act) the utterance of that constituent counts as (C-rules).

(vi) RULE COMPOSITION: The rules governing a constituent (of a sentence) are determined by the rules governing its constituents.

[11] Searle only explicitly mentions **sentences,** but the considerations that motivate compositionality at the sentential level apply as well to complex phrases: Relative clauses and verb–complement constructions can each involve (apparently unbounded) subordinate references and predications.

Principles (i)–(vi) thus constitute a set of **compositionality constraints** that any rule-based theory of speech acts must meet. It is natural to ask if Searle's theory can meet them, and if in meeting them the theory is projective—if it solves the projection problem for speech acts. In particular, does it answer the question "How are the speech act rules for a whole sentence determined by the speech act rules for its constituents and their syntactic relations?" To make progress on this question, I want to break it down into three subquestions:

(i) How does Searle specify the syntactic structure and relations of a sentence?

(ii) What are the primitive constituents in the linguistic description of a sentence to which the semantic–pragmatic rules attach, and what is the manner of their attachment?

(iii) How do the syntactic structure and relations of a sentence interact with its semantically–pragmatically interpreted constituents to determine the rules for the use of the whole sentence?

3.1. Linguistic Structure

Searle, of course, wants his theory to apply to natural languages like English. Since a linguistic description of a natural language is (in part) a theory of the syntactic structure of its elements, it would be natural for Searle to suggest that speech act semantics be integrated into a linguistic description, and this he does. However, there is a problem in deciding what form of linguistic theory we are to take these rules as being related to. Given the time of Searle's writing and other remarks (1969, p. 64), I will temporarily assume that the grammar with which we try to integrate these rules should be what is now called the *Aspects* model of the "Standard Theory" (cf. Chomsky, 1965).

Since Searle does not discuss the problem of how speech act rules relate to a grammar, it will be helpful if we have some simple target examples to work with. Suppose we concentrate on the so-called explicit performative sentences such as (9), (11), and (12), comparing them with (10).

(9) *I (hereby) promise (you) (that) I will leave soon.*

(10) *I promised (you) (that) I will leave soon.*

(11) *I (hereby) state that I promised (you) (that) I will leave soon.*

(12) *I (hereby) state that Liberace is a bachelor.*

Certainly Searle would want to be able to handle cases of this kind. As a condition of adequacy on such a speech act theory, I suggest that the theory must yield the result that an utterance of (9) can literally and directly count as a promise, but not so with (10) or (11); and that an utterance of (10), (11), or (12) can literally and directly count as a performance of the speech act of stating.

3.2. Interpretation of the Primitive Constituents

As was observed earlier in the summary of Searle's basic suggestions, only three kinds of rules are really discussed at any length by him: rules for referring, predicating, and for performing various particular illocutionary acts. For convenience and future reference, I will give Searle's (1969) rules for referring, predicating, stating and promising:

REFERENCE

The semantical rules for the use of an expression *R* to make singular definite reference are:

U-Rules

RULE 1: *R* is to be uttered only in the context of a sentence (or some similar stretch of discourse) the utterance of which could be the performance of some illocutionary act.

RULE 2: *R* is to be uttered only if there exists an object *x* such that either *R* contains an identifying description of *x* or Speaker is able to supplement *R* with an identifying description of *x*, and such that, in the utterance of *R*, Speaker intends to pick out or identify *x* to Hearer.

C-rule

RULE 3: The utterance of *R* counts as the identification or picking out of *x* to (or for) Hearer.

Notice that, like other systems containing essential rules, these rules are ordered: Rule 2 only applies if the previous rule 1 is satisfied, and rule 3 only applies if rules 1 and 2 are satisfied.

PREDICATION

Rules for the use of any predicating device *P* (to predicate *P* of an object *x*):

U-Rules

RULE 1: *P* is to be uttered only in the context of a sentence or other stretch of discourse *T*, the utterance of which could be the performance of some illocutionary act.

RULE 2: *P* is to be uttered in *T* only if the utterance of *T* involves a successful reference to *x*.

RULE 3: *P* is to be uttered only if *x* is of a type or category such that it is logically possible for *P* to be true or false of *x*.

C-Rule

RULE 4: The utterance of *P* counts as raising the question of the truth or falsity of *P* of *x* (in a certain illocutionary mode determined by the illocutionary force-indicating device of the sentence).

PROMISING

U-Rules

RULE 1: *Pr* is to be uttered only in the context of a sentence (or larger stretch of discourse) *T*, the utterance of which predicates some future act *A* of Speaker.

RULE 2: *Pr* is to be uttered only if Hearer would prefer Speaker's doing *A* to his not doing *A*, and Speaker believes Hearer would prefer Speaker's doing *A* to his not doing *A*.

RULE 3: *Pr* is to be uttered only if it is not obvious to both Speaker and Hearer that Speaker will do *A* in the normal course of events.

RULE 4: *Pr* is to be uttered only if Speaker intends to do *A*.

C-Rule

RULE 5: The utterance of *Pr* counts as the undertaking of an obligation to do *A*.

These rules are ordered: Rules 2–5 apply only if rule 1 is satisfied, and rule 5 applies only if rules 2 and 3 are satisfied as well.

ASSERTING

Rules for the use of any device ⊢ for asserting (that *p*):

U-Rules

RULE 1: ⊢ is to be uttered only in the context of a sentence or other stretch of discourse *T*, the utterance of which expresses the proposition that *p*.

RULE 2: Speaker has evidence (reasons, etc.) for the truth of *p*.

RULE 3: It is not obvious to both Speaker and Hearer that Hearer knows (does not need to be reminded of, etc.) *p*.

RULE 4: Speaker believes that *p*.

C-Rule

RULE 5: The utterance of ⊢ counts as an undertaking to the effect that *p* represents an actual state of affairs.

Our task at this stage boils down to determining what elements Searle thinks these three kinds of rules attach to, as well as the manner of their attachment. With one eye on Searle's speech act schema $F[R,P]$, we can turn now to the problem of formulating an adequate manner of attachment.

3.2.1. Referring Expressions

Although Searle has many interesting and provocative things to say about singular definite reference, he has little to say about the nature of the connection between these rules and devices for referring. He does, however, note that English has a (superficial?) variety of such devices.

> Paradigmatic referring expressions in English fall into three classes as far as the surface structure of English sentences is concerned: proper names, noun phrases beginning with the definite article or a possessive pronoun or noun and followed by a singular noun, and pronouns [1969, p. 28].

Now, if these are devices for referring, it seems natural that it should be to these items that the rules for referring attach. But where and how is this to be done? As far as I can tell there are only three places in a grammar where this pairing can be effected: in the lexicon, at some level of underlying structure, and at surface structure. It need not be assumed, by the way, that each kind of referential device will be associated with its rules at the same place. For instance, it seems possible for proper names to be associated with their rules in the lexicon in a way that is not possible for singular definite noun phrases. Furthermore, it seems that with respect to pronouns we must distinguish demonstrative as opposed to anaphoric **occurrences** in a sentence, and so we probably must go beyond the lexicon here, too.[12]

Assuming, then, that the pairing will take place somewhere in the derivational history of the sentence, where? Although there are some intuitions which suggest that the rules for referring should not be assigned prior to operations which delete (and perhaps move) referential noun phrases, we will proceed on the assumption that this assignment takes place on the underlying P-marker.[13]

[12] Actually, this poses an interesting problem for speech act theory. As many people have noticed, there seem to be linguistically determined relations of coreference, or at least of intended identity of reference:

(i) *Everyone there said that he likes beans.*

In (i) *he* can be used demonstratively or anaphorically. However, on the anaphoric reading, *he* cannot just replace *everyone* (*there*), because (i) does not mean the same as:

(ii) *Everyone there said that everyone* (*there*) *likes beans.*

How exactly is this to be reflected in a speech act theory of Searle's type?

[13] For instance, it is the judgment of some speakers that in uttering the sentence *I promise to leave early,* the speaker is only referring to himself once. The converse of this would presumably hold for sentences obtained by copying transformations. If the notion of being spoken in real time made sense when applied to underlying structures, then one might say that the assignment of rules of referring must come after movement transformations, since, in uttering the sentence *John was kicked by Mary,* I refer to John *before* I refer to Mary. Not so in underlying structure. This is only meant to be suggestive, since I cannot make any clear sense of *temporally* ordered constituents in underlying structures.

The next problem is to identify, in the underlying P-marker for a sentence, those constituents which are referential. Now, it is not clear that this can be done formally,[14] but then it is not clear that it cannot. I am going to leave that question aside for now, in order to get the theory going. I will simply assume for the sake of speech act semantics that there is some effective way of determining the referential occurrences of a constituent and further assume that they are all noun phrases which have been marked in some fashion as referential—say with a superscript "*r*." We can then define our convention for assigning the requisite rules of reference as follows:[15]

(Vr) If $m_1, \ldots, m_k$ is the maximal string of morphemes–formatives in the underlying P-marker dominated by a node NP^r then the result of interpreting this string is:

(U) $m_1, \ldots, m_k$ is to be uttered by Speaker to Hearer only if there exists an object x such that either $m_1, \ldots, m_k$ contains an identifying description of x or Speaker is able to supplement $m_1, \ldots, m_k$ with an identifying description of x, and such that in the utterance of $m_1, \ldots, m_k$, Speaker intends to pick out or identify x to Hearer.

(C) The utterance of $m_1, \ldots, m_k$ counts as the identification of, or the picking out of, x, to (or for) Hearer.

For simple cases this seems to give the results that Searle wants. For instance, let $m_1, \ldots, m_k$ be *Liberace* in sentence (12). We then get the following result:

(13) (U) *Liberace* is to be uttered by Speaker to Hearer only if there exists an object x such that either *Liberace* contains an identifying description of x or Speaker is able to supplement *Liberace* with an identifying description of x, and such that in the utterance of *Liberace,* Speaker intends to pick out or identify x to Hearer.

(C) The utterance of *Liberace* counts as the identification of, or the picking out of, x to (or for) Hearer.

This seems to accord with Searle's remarks, and it gives us a phrase marker for (12) which looks something like the following:[16]

[14] See for instance, K. Lehrer, 1972.

[15] Notice that we have dropped the first clause in Searle's rule for referring: that the referring expression occur in the context of a well-formed sentence. It is now redundant, being syntactic.

[16] In the case of *I*, it seems plausible to require of the rule not only that it say that there must be a referent, but that the referent be the speaker. Thus, the rule for referring for *I* would have the clause "x = the Speaker" inserted after the first *such that*. What about the rule as it applies to *Liberace?* This is more controversial, since we are not raising the question of what characteristics are **linguistically** related to the use of the referring expression. For a detailed analysis of

(14)

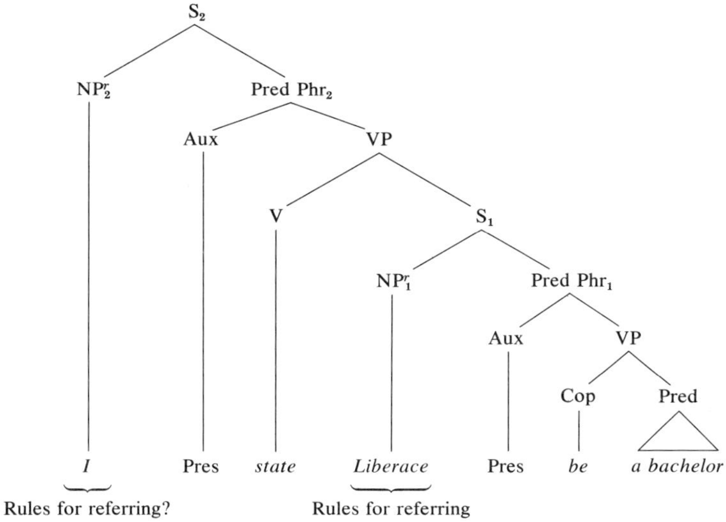

3.2.2. Predicate Expressions

Searle does not give any general method for identifying predicates in a sentence, and he seems in his discussions to use both logical and (traditional) grammatical criteria. However, if we limit ourselves to examples where these do not significantly diverge, we can still see how the theory might work (or not work).

As before, we will adopt a fairly mindless strategy, and simply identify the predicate of a sentence with the string of morphemes (in deep structure) dominated by the node *Pred-Phrase*. As far as I can tell there is no more attractive candidate in the Standard Theory.

Thus, we will simply assign Searle's rules of predication to each predicate phrase. How might we do this? As before, we can define a convention **(VPr)** as follows:

(VPr) If $m_1, \ldots, m_k$ is the maximal string of morphemes–formatives in the underlying P-marker of a sentence S dominated by the node Pred-Phrase, then the result of interpreting this string is:

(U) $m_1, \ldots, m_k$ is to be uttered (in S?) only if

---

this, however, we need to turn to Searle's theory of proper names, something beyond the scope of this discussion.

Notice that I am leaving open the question of whether *I* gets rules for referring at this stage for reasons that will become clear in a short while. But if we **are** to so assign them, notice something curious, namely, there seems to be no difference in the rules governing the use of *I* and *Liberace*. That is, so far our semantic–pragmatic rules do **not** distinguish the use of these two expressions. The natural question, then, is how we can modify our procedure to make the obvious distinction in use between *I* and *Liberace*.

(i) the utterance of S involves a successful reference to x.

(ii) x is of a type or category such that it is logically possible for $m_1, \ldots, m_k$ to be true or false of x.

(C) The utterance of $m_1, \ldots, m_k$ counts as raising the question of the truth or falsity of $m_1, \ldots, m_k$ of x (in a certain illocutionary mode determined by the illocutionary force indicating device of the sentence).

Again, for simple cases this seems to yield results similar to the examples Searle discusses. For example, letting $m_1, \ldots, m_k$ be *is a bachelor* in sentence (12), we get as a result the following:

(15) (U) *is a bachelor* is to be uttered (in S?) only if:

(i) The utterance of S involves a successful reference to x.

(ii) x is of a type or category such that it is logically possible for *is a bachelor* to be true or false of x.

(C) The utterance of *is a bachelor* counts as raising the question of the truth or falsity of *is a bachelor* of x (in a certain illocutionary mode determined by the illocutionary force-indicating device of the sentence).

For our sample sentence (12), we now have the phrase marker:

(16)

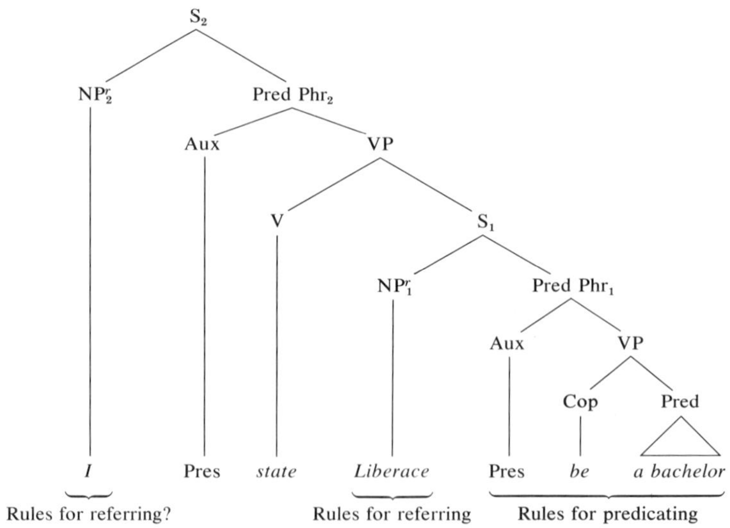

So far we seem to be assigning the right rules to the right constituents, at least at the propositional level. We turn now to the force indicators.

3.2.3. Force Indicators

In his book, Searle asks, "To which elements in an actual linguistic description of a natural language would rules such as 1–5 [for promising] attach? [1969, p. 64]." To this question Searle seems to have two (tentative) types of answer. The first, which has two variants, I will call the "syntactic" proposals; the second, I will call the "semantic" proposal.

SYNTACTIC PROPOSALS

After drawing the "semantic" distinction between force and content indicators, Searle remarks that,

> If this semantic distinction is of any real importance, it seems likely that it should have some syntactic analogue,[17] even though the syntactical representation of the semantic facts will not always lie on the surface of the sentence. For example, in the sentence, "I promise to come", the surface structure does not seem to allow us to make a distinction between the indicator of illocutionary force and the indicator of propositional content. In this respect, it differs from, "I promise that I will come", where the difference between the indicator of illocutionary force ("I promise") and the indicator of propositional content ("that I will come") lies right on the surface. But if we study the deep structure of the first sentence, we find that its underlying phrase marker, like the underlying phrase marker of the second, contains, "I promise + I will come". In the deep structure we can often identify those elements that correspond to the indicator of illocutionary force quite separately from those that correspond to the indicator of propositional content, even in cases where, e.g., deletion transformations of repeated elements conceal the distinction in the surface

[17] In accordance with the usual meaning of "analog" it would be natural that the claim would be:

> *If the meaning of a sentence S is determined by distinct sets of rules for force indicating, referring, and predicating, then there are syntactically distinct elements to which these rules attach.*

But this is a position Searle **seems** to deny when he says,

> From this semantical point of view we can distinguish two (not necessarily separate) elements in the syntactical structure of the sentence . . . illocutionary force indicating devices in English include at least: word order, stress, intonation contour, punctuation, the mood of the verb, and the so-called performative verbs [1969, p. 30].

It would seem from this passage that a single syntactic element can have different semantic functions, and the same semantic function can be fulfilled by different syntactic devices. It is a mystery to me why Searle would want to call such a many–many relation an "analog."

Remember that Searle gives no reasons for thinking that an important semantic distinction should (in general) be reflected in a syntactic distinction, or why it should obtain in this particular case. Now, it might in fact be true that in the ultimate grammar such a correspondence exists, but we are given no reason to believe it. In fact, there are some (weak) reasons for thinking that the general claim is false. Suppose that the distinction between analytically true (and false), and synthetically true (and false) statements is an important semantic distinction. Then, by the general claim, this would necessitate a syntactic distinction between them. But there is no reason at present for giving the sentences *Bachelors are married* and *Bachelors are miserable* any relevantly different syntactic analysis, let alone a special carrier for the distinction.

> structure. This is not to say, of course, that there is in general some single element in the underlying phrase marker of every sentence which marks its illocutionary force. On the contrary, it seems to me that in natural languages illocutionary force is indicated by a variety of devices, some of them fairly complicated syntactically [1969, pp. 30–31].

We now want to fashion some way of assigning, to the performative prefix in each performative sentence, the rules for performing the indicated illocutionary act. There seem to be a number of ways of proceeding, but I want to focus on just two. The first way, which I will name the "holistic" treatment, seems to be the one favored, or at least suggested, in the passage just cited from Searle. The second I will call the "compositional" treatment.

The Holistic Treatment. On this method, the performative prefix clause *I promise* (*you*) (*that*) is, as a whole, assigned the requisite rules for promising, it being the force indicator in the case of, for example, (9). Now, I must confess that I am not sure that such a treatment is possible in general without being extremely ad hoc and heavy-handed. To see this, consider the following way of pressing on.

First, we determine (formally if possible) those matrix prefix clauses that can be performatively used. At first blush this seems not too difficult. After all, there are only a finite number of performative verbs in the lexicon (marked, say [+perf]) and we simply take the performative prefix clauses to be all those which are first person, simple present declarative.[18] Thus, it would seem that the number of performative prefix clauses will be finite, and so, at least by brute force, effectively identifiable.[19] Continuing in this fashion, we simply mark those clauses meeting the requisite conditions in some way—say by subscripting the dominating S-node with f for *force* indicator. Given this, we now try the following convention for assigning the requisite speech act rules to the performative prefix clause, letting ϕ range over performative verbs and letting S range over sentences.

(Vf) If $m_1, \ldots, \phi, \ldots, m_k$ $(=F)$ is the maximal string of morphemes/formatives dominated by NP-(Adv)AUX-V in the underlying P-marker of S dominated by S_f, and S_f is the highest clause in S, then the result of interpreting the force indicator of S is:

(U) $m_1, \ldots, \phi, \ldots, m_k$ is to be uttered by Speaker to Hearer only if . . . [the U-rules for ϕ-ing] . . .

[18] Of course we want an analysis of the notion "is a performative verb," but for the purposes of description, lots of clear cases will do for a start, and this is just a start.

[19] Trouble emerges when we consider sentences beginning *I, who have been given the authority vested in me by . . . who . . .* ***order*** *you to leave this building*. In such cases, the verb is indefinitely far away from the "agent indicator," and this relieves us of an upper bound on performative prefix clauses. One way out would be to suppose that these nonrestrictive relatives originate somewhere else (Ross once suggested from conjunctions) and so do not get in the way.

(C) The utterance of $m_1, \ldots, \phi, \ldots, m_k$ by Speaker to Hearer counts as . . . [the C-rules for ϕ-ing] . . . .

By this convention, the rules for promising are assigned to the segment *I (hereby) promise (you) (that)* in (9) and the rules for stating are assigned to the segment *I (hereby) state (that)* in (11) and (12). For instance, with (12), we get the following results:

(17) (U) *I state (that)* is to be uttered by Speaker to Hearer only if
- (i) Speaker has evidence (reasons, etc.) for the truth of the proposition *p* expressed.
- (ii) It is not obvious to both Speaker and Hearer that Hearer knows (does not need to be reminded of, etc.) *p*.
- (iii) Speaker believes that *p*.

(C) The utterance of *I state (that)* by Speaker to Hearer counts as,
- (i) An undertaking to the effect that *p* represents an actual state of affairs.

This all seems in accord with Searle's remarks. Returning to example (12), we can now represent our sentence as follows:

(18)

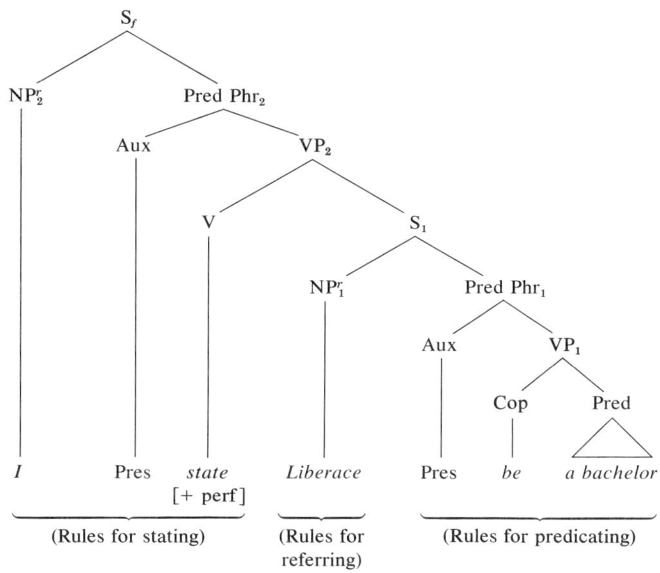

To me, the holistic treatment is unsatisfying in a number of respects. First, it fails to indicate or reflect the fact that what is **done** in the utterance of (9) is what is **reported** in the utterance of (10). It does not make a connection between performative and nonperformative uses of performative verbs (on one reading). Presumably these verbs do not change their meaning in these two occurrences, yet no such fact is reflected in the theory—or if it is, it is not clear **how** it is. Some way will have to be found to get this right, because the truth of the report *he promised* is obviously intimately related to the successful performance of the act of promising by uttering *I promise,* when *he* in the first sentence is being used to refer to the speaker of the second sentence.[20]

Second, this treatment gives us no explanation for why the confluence of syntactic facts (person, tense, mood, etc.) that make up performative prefixes have the "semantic" properties that they do. The holistic theory simply assigns these configurations the rule(s) for ϕ-ing. Yet one feels (perhaps wrongly) that this should be seen as a **result** or **product** of more elementary processes or operations. In short, the holistic treatment violates Searle's own compositionality constraints embedded in the proffered semantic–pragmatic principles.

The Compositional Treatment. One natural way of developing the compositional idea is to somehow get the rules for illocutionary force into the analysis of the performative prefix *via* the performative verb. Unfortunately this involves a theory of word meaning, and Searle does not make many suggestions in that regard.

To see what happens (and perhaps make headway), let us just assign the rules for promising to *promise* in the lexicon. Then, wherever one sees fit, assign the rules for referring to *I*, (*you*) and perhaps *Pres,* depending on one's view of temporal reference.[21] After this is completed, the problem of getting the requisite semantics for the whole prefix will be of a piece with the compositional problem for sentences in general—to which we will turn shortly. The advantage to this approach is that the performative prefix clause gets its rules as a function of the rules of its elements and so is compositional. Also, *promise* in both (9) and (10) will get rules that reflect their common meaning. But unless more is said, this strength will become a serious liability, because we do not want (10) to be given the rules for **promising;** rather, we want it to be given the rules for **reporting** a promise.

SEMANTIC PROPOSAL

About thirty pages after the first discussion of attachment for illocutionary rules, Searle returns to this question. What he says there is interestingly different.

[20] Notice also that if "explicit performative" sentences are without truth value, then we must reckon subject position in declaratives like *He promises to leave* to be *referentially opaque,* since substitution of coreferential singular terms can take one from truth to truth-valuelessness. But this conflicts with other generalizations about opacity. See Harnish 1976b.

[21] Searle (1969, p. 28) hedges on this.

> Let us assume for the sake of argument that the general outlines of the Chomsky–Fodor–Postal account of syntax and semantics are correct. Then it seems to me extremely unlikely that illocutionary act rules would attach directly to elements (formatives, morphemes) generated by the syntactic component, except in a few cases such as the imperative. In the case of promising, the rules would more likely attach to some output of the combinatorial operations of the semantic component [1969, p. 64].

According to this proposal then, the rules would be assigned in most cases to some string or constituent in semantic representation. As far as I can tell, this proposal is invariant between adequate interpretive and generative semantics. The problem here is understanding clearly what it means to assign **semantic** rules to **semantic** elements; after all, these rules are supposed to determine meaning. This two-tiered conception of semantics makes good sense only if one or the other tier is incomplete. Does Searle think that his theory is incomplete? On the one hand he says[22] that a Katz-type theory is a useful and interesting **adjunct** to a speech act semantics (though the above seems to reverse this). On the other hand, he says in his book that "A great deal can be said in the study of language without studying speech acts, but any such purely formal theory is necessarily incomplete [1969, p. 17]." Taken together, these remarks might add up to his claim that "the two approaches, construed not as theories but as approaches to investigation, are complementary, and not competing [1969, pp. 18–19]." As laudable as I find this ecumenical spirit to be, it is extremely difficult to put into practice. We cannot just take x in our schema (R-SPEC) as now ranging over parts of semantic representation, because this has the result that a **concept, property, attribute,** or whatever has a certain use—that **it** is to be uttered only if . . . and **its** utterance counts as . . . and this is unintelligible. More thought must be given to the nature of such a connection in a comprehensive description of a language by anyone who wants to work on the assumption that these two approaches will be complementary. For now I will assume some syntactic proposal to be operative.

SUMMARY

At this stage in the analysis of performative sentences we are assuming that we have the following four things represented in (18):

(i) A syntactic description of the sentence S.
(ii) The assignment of the semantic rules for performing the illocutionary act type, to the performative prefix of S.
(iii) The assignment of the semantic rules for referring, to the (relevant) NP(s) of S.
(iv) The assignment of the semantic rules for predicating, to the (relevant) Pred-Phrase(s) of S.

[22] Searle, (1972, Section V.)

This completes our sample interpretation of the syntactic elements. Notice that it does have the virtue of conforming to semantic principles (ii) and (v), but we must still meet the compositional constraints embodied in principles (i), (iii), (iv) and (vi) (p. 321). Keep in mind, however, the following defects of the account given so far. First, in the case of stating, we were not able to assign these rules to any sequence which is a **single constituent** of the sentence—there is no single node uniquely dominating just those elements to which the rules of stating are assigned. Second, we did not assign rules of predicating to Pred-Phrase$_2$ in (18) because it is not clear what this would have meant on Searle's account. Somehow this must be blocked in a nonarbitrary fashion.

3.3. Projection and Syntactic Structure

Since Searle makes almost no commitment regarding the details of the attachment of his semantic rules to linguistic elements, it is very difficult to imagine the details of their composition. Certainly, Searle himself says nothing in this regard. Nevertheless, it is instructive to make **some** attempt at an integration if only to see more clearly exactly what we do not yet know. Again, it should be emphasized that we are no longer **describing** Searle's theory, but **elaborating** it.

In accordance with the semantic–pragmatic principles, we want to determine how the rules governing the meaning of these interpreted elements are combined to give the rules governing the whole sentence. One thing that I am going to assume without much argument is that there is something to be done here; that the set of rules for the whole sentence is not just a **list** or conjunction of all the rules for its elements. One reason for this is that such a set of rules would violate the natural definition of synonym by making a sentence and its inverse synonymous, in virtue of the fact that the two sentences would have the **same set** of rules assigned. But clearly *I state that John kissed Mary* and *I state that Mary kissed John* are not synonymous.

A glance back at (R-SPEC)—our intended output—suggests that at least we want all of the U-rules together, and all of the C-rules together. Is this all? Recall the general form of sentences schematized by Searle as $F[R,P]$, where the material in the brackets represents the expression of a proposition, or better, the structure of a subject–predicate propositio**ning.** It is also one of Searle's claims that the same proposition(ing) can be a constituent in many different kinds of illocutionary acts: I might **promise** or **state** that I will go. These two observations go hand in hand to suggest that the rules for reference and predication first amalgamate together as a unit before the result of that is amalgamated with the force indicator. This seems intuitively plausible given that *Liberace is a bachelor* in (12) does seem to be a **semantic–pragmatic** unit in a way that, say, *promise that Liberace* is not. Thus, our principles of projection will apply upward through the underlying P-marker, first at the level of the propositional indicator—the second highest clause in our paradigms—then to the whole sentence. We turn now to these two stages.

3.3.1. Propositional Amalgamation

Since the requisite elements have been assigned their rules, we can work our way up the tree until we reach the first node dominating the first pair of interpreted nodes. At this point the rule composing reference and predication takes effect:

(R1) If $m_1, \ldots, m_k$ $(=R)$ is the maximal string of morphemes–formatives assigned rules of referring and dominated by NP^r; and if $m_m, \ldots, m_n$ $(=P)$ is the maximal string of morphemes–formatives assigned rules for predicating and dominated by Pred-Phrase, then

(U) $m_1, \ldots, m_k; m_m, \ldots, m_n$ $(=[R,P])$ is to be uttered by Speaker to Hearer only if:

(i) . . . (content of the U-rules for R) . . . .

(ii) . . . (content of the U-rules for P) . . . .

(C) The utterance of $m_1, \ldots, m_k; m_m, \ldots, m_n$ $(=[R,P])$ by Speaker to Hearer counts as:

(i) . . . (content of the C-rules for R) . . . .

(ii) . . . (content of the C-rules for P) . . . .

It is of course to be understood that the above order of contents be respected when applying this principle.[23] The result of applying (**R1**) to sentence (12) is:

(19) (U) *Liberace is a bachelor* is to be uttered by Speaker to Hearer only if:

(i) There exists an object x such that either *Liberace* contains an identifying description of x or Speaker is able to supplement *Liberace* with an identifying description of x, and such that in the utterance of *Liberace*, Speaker intends to pick out or identify x to Hearer.

(ii) (a) The utterance of *I state that Liberace is a bachelor* involves a successful reference to x.
(b) x is of a type of category such that it is logically possible for *is a bachelor* to be true or false of x.

(C) The utterance of *Liberace is a bachelor* counts as:

(i) The identification of, or the picking out of, x to (or for) Hearer.

[23] Searle (1969, p. 63) also claims, at least in some cases, this his rules are "ordered."

(ii) Raising the question of the truth or falsity of *is a bachelor* of x (in a certain illocutionary mode determined by the illocutionary force indicating device of the sentence).

As far as I can tell, this conforms with Searle's intentions.

3.3.2. Force Amalgamation

We next turn to the problem of composing the propositional indicator with the force indicator. We can formulate this principle in exactly the same say as (**R1**), but replacing references to NP^r and Pred-Phrase with propositional indicator and S_f (as suggested earlier):

(**R2**) If $m_1, \ldots, m_k$ $(=P)$ is the maximal string of morphemes–formatives assigned rules for expressing a proposition and dominated by S; and if $m_m, \ldots, m_n$ $(=F)$ is the maximal string of morphemes–formatives assigned rules for performing some illocutionary act and dominated by S_f, then

(U) $m_1, \ldots, m_k; m_m, \ldots, m_n$ $(=F[P])$ is to be uttered by Speaker to Hearer only if:

(i) . . . (content of the U-rules for P) . . . .

(ii) . . . (content of the U-rules for F) . . . .

(C) The utterance of $m_1, \ldots, m_k; m_m, \ldots, m_n$ $(=F[P])$ by Speaker to Hearer counts as:

(i) . . . (content of the C-rules for P) . . . .

(ii) . . . (content of the C-rules for F) . . . .

The result of composing our previous propositional indicator with the performative prefix clause in, for example, (12) would be:

(20) (U) *I state that Liberace is a bachelor* is to be uttered by Spearker to Hearer only if:

(i) There exists an object x such that either *Liberace* contains an identifying description of x or Speaker is able to supplement *Liberace* with an identifying description of x and such that in the utterance of *Liberace*, Speaker intends to pick out or identify x to Hearer.

(ii) (a) The utterance of *I state that Liberace is a bachelor* involves a successful reference to x.

(b) x is of a type or category such that it is logically possible for *is a bachelor* to be true or false of x.

(iii) (a) Speaker has evidence (reasons etc.) for

the truth of the proposition that Liberace is a bachelor.

(b) It is not obvious to both Speaker and Hearer that Hearer knows (does not need to be reminded etc.) that Liberace is a bachelor.

(c) Speaker believes that Liberace is a bachelor.

(C) The utterance of *I state that Liberace is a bachelor* counts as:

(i) The identification of, or picking out of, *x* to (or for) Hearer.

(ii) Raising the question of the truth or falsity of *is a bachelor* of *x* (in a certain illocutionary mode determined by the illocutionary force-indicating device of the sentence).

(iii) An undertaking to the effect that *Liberace is a bachelor* represents an actual state of affairs.

I think it is clear that the preceding does, at least in a crude way, produce an output of the sort required by (R-SPEC) for the whole sentence. Moreover, it seems to do this compositionally in a fashion satisfying the suggested semantic–pragmatic principles. Thus the process has a certain prima facie plausibility about it as a partial solution to the projection problem for speech acts.

One defect of the foregoing is that (**R1**) and (**R2**) state related facts twice and so fail to reflect a single principle at work in both of these cases. The generalization seems to be roughly this:

(**R**) If semantic–pragmatic rule R_i is being composed with semantic–pragmatic rule R_j, then the result of the composition is the sequence: the U-rules of R_i + U-rules of R_j + C-rules of R_i + C-rules of R_j.

This general convention, if correct, could then replace the two more specific and overlapping conventions (**R1**) and (**R2**).

However, there is a problem. In the case of composition at the propositional level, R_i must be the rule for referring and R_j, the rule for predicating: This cannot be reversed if we want to avoid stranding free variables throughout the rules for expressing a proposition. Unfortunately, (**R**) does not, as it stands, block this possibility. It is perfectly compatible with (**R**) to have the sequence U-rules for predicating + U-rules for reference. To block this, we might add the convention that (**R**) applies from left to right (R_i is to the left of R_j), as well as cyclically from bottom to top. Let us call this modified version of (**R**) the rule (**AR**). But at the level of force indicator-plus-propositional content (**AR**) applies "backward," and is defective for reasons comparable to those which made (**R**) unsuitable at the level of the proposition. At the moment I do not know how to restrict (**R**) suitably; hence, the retention of the special projection rules (**R1**) and (**R2**).

3.4. Summary and Conclusions

In this section, I have tried to sketch various ways in which Searle's semantic–pragmatic principles might be put into effect in the course of giving the semantics–pragmatics of a particular natural language (English). It is to be emphasized that defects of these proposals are not necessarily intrinsic to Searle's theory, since he has not yet discussed these problems. These proposals are my elaborations of his theory, not his. Recall that the projection problem for pragmatics resolved into two stages: first, to determine to which primitive elements of the language the speech act rules attached, as well as the manner of attachment; second, to state principles composing such rules on the basis of the syntactic relations between the elements to which these rules attach. Among the problems we encountered at the first stage (Section 3.2.) are the following:

(i) We need an account of referential occurrences of NPs in a sentence.
(ii) We need, and do not have, some way of blocking the assignment of rules for predicating to predicate phrases which should not get such rules; in our sample case, blocking the rules for predicating from being attributed to the phrase *state that Liberace is a bachelor*. Alternatively, some account must be given of the intelligibility of such predication within Searle's theory.
(iii) On the holistic treatment of the force-indicating device, we do not have a node marking off *I state* (*that*) as a single constituent to which the rules for stating can attach. There seems to be no motivated reason for taking any one of many ways out of this.
(iv) The holistic treatment does not connect performative with nonperformative uses of phrases with the same verb.
(v) The holistic treatment does not reflect the internal structure of the performative prefix.

Some of the problems encountered at the second stage (Section 3.3) are the following:

(vi) The compositional treatment of the force-indicating device foundered on the lack of a viable theory of word meaning, or at least of meaning below the level of noun phrase and predicate phrase.
(vii) We were unable to state our principles of projection maximally simply.
(viii) The essential rule for the whole sentence needs further work. For

instance, in (20), clause (C-ii) made reference to the illocutionary mode of the sentence and (C-iii) explicated that mode (in this case assertion), but surely it should not be left at that: These are connected in a way that should be **reflected** in the resulting essential rule for the whole sentence but is not. Searle never discusses such complex examples, so it is not clear what he intends the essential rule(s) to look like.

4. SOME ALTERNATIVE SOLUTIONS

In the face of these problems, one may well be tempted to seek alternatives. Since our difficulties arise mainly from trying to combine Searle's rule-based theory of speech acts with an *Aspects* grammar, there are at least two dimensions of freedom here. We could keep just one of these, the speech act theory or the syntax, and modify the other; or we could modify them both. Consider the first alternative.

One way around at least some of these problems—especially (ii) and (iii) —would be to modify our syntax so as to provide the right constituents in the right order for pragmatic interpretation and projection. This would have the advantage of retaining the following virtues of Searle's theory:

(i) By postulating the internalization of constitutive rules, Searle has a potential explanation for the fluent speaker's ability to perform and recognize speech acts.

(ii) The theory contains the most plausible and principled taxonomy of illocutionary acts in print so far.

(iii) The theory has a fairly natural extension from direct to indirect speech acts.

Recalling Searle's semantic–pragmatic schemata listed at the very beginning of this paper, and the order required by (**AR**), we might generate structures of the form:

(21)

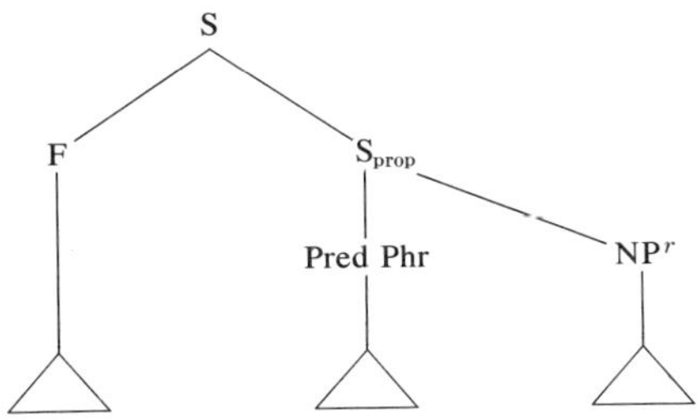

We could then assign lexical items and speech act rules to each terminal node, and finally amalgamate cyclically using the following right-to-left version of (**AR**):

(RtLAR) If R_1 is to the right of R_2, and R_1 is being composed with R_2, then the result of amalgamating R_1 with R_2 is the sequence: U-rules of R_1 + U-rules of R_2 + C-rules of R_1 + C-rules of R_2. Diagrammatically:

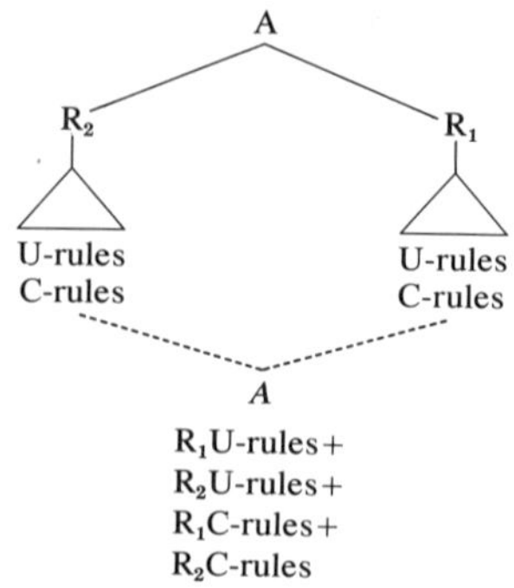

In the case of structures like (21) we would get sequences like the following:

(22)

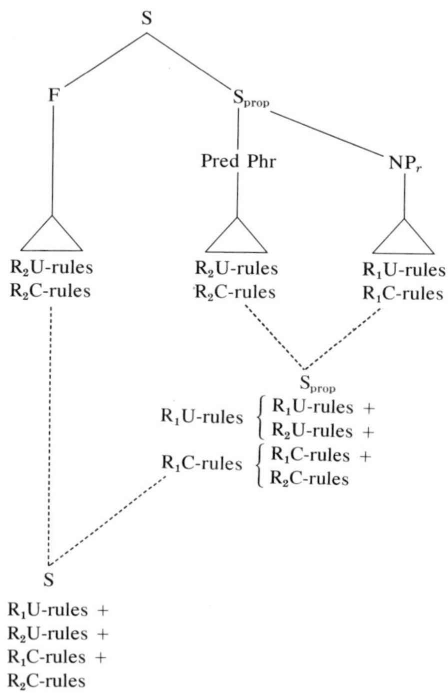

This gives us exactly what we want for a wide range of simple cases. Unfortunately, there are a host of problems with this proposal, not the least of which is its total lack of syntactic motivation.[24]

The syntax closest in "geometry" to the above seems to be the original $\bar{X}$ system of Chomsky 1970 (p. 210). The relevant structures can be schematized as in (23):

(23)

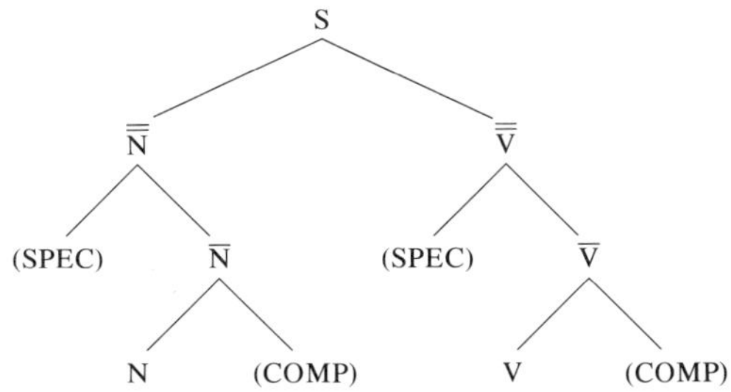

But such a structure would require considerable modification if it were to serve pragmatic purposes. $\bar{\bar{N}}$ and $\bar{\bar{V}}$ would have to be reversed, with force indicators added. The most obvious move in this context is to regard the force indicator as a kind of sentence specifier (of illocutionary force). This suggests the following:

(24)

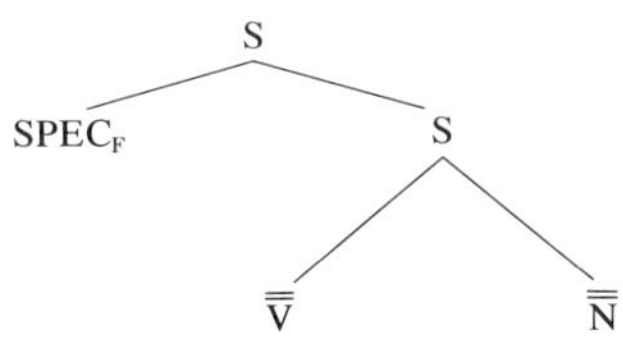

But besides being totally ad hoc, (24) fails to relate the performative prefix *I promise* (*that*) to the verb *promise*. Thus, no solution seems in sight by simply altering the syntax. We may well have to develop an alternative pragmatics.[25]

[24] The question of the weighing of independent syntactic evidence becomes tricky when a language is conceived of as a system of sounds (or shapes) with fixed meaning, and a grammar is conceived of as the principles of pairing. On this conception it would seem natural to strive for overall simplicity of the *system of pairing* as well as for independent evidence for each component. But what if these goals come into conflict? How strong must syntactic (semantic, phonological) evidence be to outweigh overall simplicity?

[25] In Bach and Harnish (1979) an approach to speech acts is developed in considerable detail which has a bearing on our problem. The theory has two major components: an account of the structure of linguistic communication, and a principled taxonomy of illocutionary acts with associated definitions. What is important for present purposes is that the theory does not posit a level of representation containing "constitutive rules" for performing speech acts. Instead, it utilizes a system of inferences based on (conventional) linguistic meaning to achieve the same ends. With a level of semantic representation (proposition) available, the theory does not have to solve the projection problem for speech acts with just the syntax, and so the sorts of problems we have encountered with Searle's theory do not arise.

REFERENCES

Alston, W. (1964). *Philosophy of Language*. Englewood Cliffs, New Jersey: Prentice-Hall.

Alston, W. (1974). Semantic rules. In M. Munitz and P. Unger (Eds.), *Semantics and Philosophy,* New York: New York University Press.

Bach, K. and Harnish, R. (1979). *Linguistic Communication and Speech Acts*. Cambridge, Massachusetts: MIT Press.

Chomsky, N. (1965). *Aspects of the Theory of Syntax*. Cambridge, Mass.: Cambridge.

Chomsky, N. (1970). Remarks on nominalization. In R. Jacobs and P. Rosenbaum (Eds.), *Readings in English Transformational Grammar*. Waltham, Mass: Ginn.

Davidson, D. (1964). Theories of meaning and learnable languages. In *Proceedings of the 1964 International Congress for Logic, Methodology and Philosophy of Science*. Amsterdam: North-Holland.

Davidson, D. (1974). Semantics for natural languages. In G. Harman (Ed.), *On Noam Chomsky*. New York: Doubleday.

Frege, G. (1923). Compound thoughts. *Mind,* January, 1963, 1–17.

Grice, H. P. (1968). Utterer's meaning, sentence-meaning, and word-meaning. In J. R. Searle (Ed.), *The Philosophy of Language,* New York and London: Oxford University Press.

Harman, G. (Ed.) (1974). *On Noam Chomsky,* New York: Doubleday.

Harnish, R. (1976a). Logical form and implicature. In T. Bever, J. Katz, and T. Langendoen (Eds.), *An Integrated Theory of Linguistic Ability*. New York: Crowell.

Harnish, R. (1976b). Two consequences of transparent subject position. *Philosophical Studies,* **30,** 11–18.

Katz, J., and Fodor, J. (1963). The structure of a semantic theory. In J. Fodor and J. Katz (Eds.), *The structure of language,* Englewood Cliffs, New Jersey: Prentice-Hall.

Lehrer, K. (1972). Reference, predication, and semantic interpretation. *Proceedings of the 11th International Congress of Linguistics*. The Hague: Mouton.

Schiffer, S. (1972). *Meaning*. New York and London: Oxford University Press.

Searle, J. (1965). What is a speech act? In M. Black (Ed.), *Philosophy in America,* Ithaca, New York: Cornell University Press.

Searle, J. (1968). Austin on locutionary and illocutionary acts. *Philosophical Review* **LXVII,** 4, 405–424.

Searle, J. (1969). *Speech Acts*. Cambridge, Mass: Cambridge University Press.

Searle, J. (1972). Chomsky's revolution in linquistics. In G. Harman (Ed.), *On Noam Chomsky*. New York: Doubleday, 1974.

Stalnaker, R. (1970). Pragmatics. *Synthese,* **22,** 272–289.

Wallace, J. (1970). On the Frame of Reference. *Synthese,* **22,** 117–150.

ON SOME GENERAL PRINCIPLES OF SEMANTICS OF A NATURAL LANGUAGE

H. HIŻ

1. INTRODUCTION

The problem of the relation which language bears to reality is old, venerable, and afflicted with the hereditary disease of obscurity. At least from the time of Frege, it has been a central problem for philosophers, although linguists have tried to keep away from it as a rule. In recent years, however, several influential theories of semantics of a natural language have been proposed, and some linguists as well as philosophers have taken part in their advancement. One of the theories which is gaining some momentum, mainly through the work of Richard Montague, assigns objects of set theory to various phrases of a natural language like English or Italian. If somebody doubts the nature or the existence of these objects, he may choose to follow those who assign logical formulas to sentences of a natural language, and claim that a sentence says the same as the corresponding formula. This is practiced by linguists who call themselves "generative semanticists." A somewhat similar idea is found in the genotype language of S. K. Shaumian (1971). Furthermore, if one doubts that whatever is expressible in a natural language is expressible in logic as well, he may side with the interpretivists, according to whom there are interpretive rules for a language and these give

SYNTAX AND SEMANTICS
VOLUME 10

ISBN: 0-12-613510-X

rise to the semantics of the sentences. If I understand what these interpretive rules are, what I have to say later does not depart very far from the interpretive theory, though I will state it in a new way.

Linguists, as a rule, practice a far more restrained semantics than is found in the theories like Montague's, and I will side with that tendency. In this respect it is instructive to reflect on the attitude of Harris toward semantics. In his work, he uses semantics both explicitly and in a syntactic guise. Contrary to general belief, in his early work (1950), Harris does not eliminate explicit semantics. Rather, semantics is reduced to a single, simple, and testable question: Are two utterances repetitions of each other, or do they contrast? *Book* and *hook* contrast and this is a fundamental semantic datum. From it, by formal manipulations, the linguist reconstructs two different phones /b/ and /h/. To contrast means to not be a repetition; to say or to mean something else. This rudimentary, explicit semantic element was never eliminated from Harris's work and—I may add—is assumed in all linguistic efforts, even if that fact is not always recognized. It is present in phonetics, in syntax, in discourse analysis, in field methods, and in comparative studies. In addition, Harris uses semantics implicitly. The entire effect of Harris's syntax, including the latest work (1976), is oriented toward rendering semantic differences by syntactic means. His syntax is always semantically motivated, in spite of the changes in form through the years. It is not that the result of the syntax—the derived sentences—will later receive semantic interpretation, but that each syntactic step reflects or records a semantic property, a paraphrase being a particular case of such a property.

2. TRUTH SEMANTICS

A restrained semantics might use only truth as a link between language and the world. A still more restrained semantics would not talk about truth but about relations of equivalence (simultaneous truth) or of consequence (preservation of truth), and would not say which sentences are true but only that if a sentence is true, then another sentence is true also. It is clear that linguists traditionally do not openly engage in a strong semantics. But it is less clear whether a very weak semantics suffices for the purposes of a linguist. It may be that to construct an insightful syntax and discourse analysis one has to know that some sentences are true or, at least, that some sentences are considered true by the speech community whose language is under examination. In what follows, I will use the concept of truth as primitive. I will not define it but I will use it in axioms and those axioms will to some extent demonstrate how I use that concept, and how it relates to other concepts used in the axioms. But this information is only very partial. The axioms are, of course, far from complete. They are only some starting principles for understanding the semantics of a language.

There will be two kinds of axioms for semantics: those which are independent of particular phrases, of a particular vocabulary of the language; and those which do relate the general semantic concepts to some particular vocabulary of a particular language. The first kind of axioms are therefore general principles of semantics of any language. I will discuss here six such principles. Some of the axioms are due to Tarski; others are slight extensions and modifications of his ideas. All of them use the concept of truth, the concept of consequence, or both. They of course also use the concept of a sentence. Each of the three concepts is to be taken with proper caution. To avoid misunderstanding, I must say that I take truth to be a property of some sentences (like Tarski) and of some finite strings of sentences.

Not all sentences of a natural language are true or false if taken out of context. Some of them are like sentential functions of a formal language, like $x = 5$ which is neither true nor false: It has the form of a sentence but is not a sentence, since it contains a variable, a blank. In a natural language, variables do not occur but other devices are used which make a sentence not a suitable argument for truth or falsity. For instance *It was John* or *They revolted* contain referential phrases (*it* and *they*) which refer the listener to occurrences of other phrases in the same utterance. Only after knowing to what occurrence of a phrase *it* refers, is it reasonable to ask whether *It was John* is true in that context. For instance, if the text is *Jane did not give me the book. It was John,* we conclude that *It was John* contains a cross-referential to *gave me the book* and that what we are asked to judge the truth of is *John gave me the book.* One may still ask what book is spoken about in this sentence. *The book* is a referential to some previous phrase. One must distinguish, therefore, between the truth of a single sentence and the truth of that sentence as a segment in a string of sentences. The former may be improper whereas the latter may apply correctly. We may then easily say that a sentence with a cross-referential is true in a text, if the text which also contains the referend is true.

Similar complications are encountered when one analyzes the concept of consequence. *John gave me the book* does not follow from *It was John,* but it does follow from *Jane did not give me the book. It was John.* One has to distinguish between a sentence α being a consequence of the string of sentences $\beta_1 ⁀ \beta_2 ⁀ \beta_3$ (where the arc is concatenation), from α being a consequence of the set of which β_1, β_2, and β_3 are members. Thus, if $\alpha \in \mathrm{Cn}(\{\gamma_1 ⁀ \gamma_2\})$, then it is not necessarily the case that $\alpha \in \mathrm{Cn}(\{\gamma_1\} \cup \{\gamma_2\})$. Here, as usual, $\cup$ is a sign of logical sum, "$\in$ Cn" denotes "is a consequence" and the braces indicate a unit set.

3. AXIOM OF SENTENCEHOOD

$$\textbf{(A1)} \qquad \alpha \in \mathrm{Cn}(X), \qquad \text{then} \quad (\alpha \in S \text{ and } X \subset S)$$

If α is a consequence of X, then α is a sentence and X is a set of sentences.

As the relation of consequence rarely holds between a sentence and a single sentence, we rather say that $\alpha \in \mathrm{Cn}(\{\beta\} \cup X)$ in cases where α is a consequence of β with the help of sentences of X (presuppositions, obvious truth, general sentences known in the community, logical rules).

4. AXIOMS OF CONSISTENCY AND COMPLETENESS

The following two axioms are clear and are the expression of the facts that the set T of all true sentences is consistent and complete:

(A2) $\quad$ $not(T = S)$

(A3) $\quad$ $(\alpha \in S\ and\ not(\alpha \in T))$, then $\mathrm{Cn}(T \cup \{\alpha\}) = S$

5. AXIOM OF CONJUNCTION

One may be tempted to claim that if $\alpha \in \mathrm{Cn}(\{\gamma_1\} \cup \{\gamma_2\})$, then $\alpha \in \mathrm{Cn}(\{\gamma_1 \frown \gamma_2\})$. But this does not necessarily hold. For, in a natural language, it is not the case that any two sentences can be conjoined by a pause (represented by a period in the written form). The most we can claim is that any two sentences can be conjoined by at least one of the many available conjunctive phrases and that the resulting text implies each of the conjuncts (but is not necessarily equivalent to the set of them).

Both a sentence, even if it contains an unresolved cross-referential, and a string of sentences are here called sentences. A string of sentences is therefore understood with a pause (period) between consecutive sentences. The term *sentence* is ambiguous. Sometimes it is used for an intonational unit (between pauses), at other times for a syntactic unit (a terminal string of a generative grammar), or for a logical or semantic unit. These (and other) understandings of the term are usually confused with each other. Here, only the third is intended. The first argument of the relation Cn (that is, a consequence) is a sentence. Also every member of the second argument of that relation (that is, a premise) is a sentence. Also, a part of a premise that gives a consequence of the premise after proper replacement of cross-referentials by their referends is a sentence. In the previous examples, *It was John* is a sentence in the premise, *Jane did not give me the book. It was John.* So *John gave me the book* is a consequence. (It takes some mechanism to describe the change from *It was John* to *John gave me the book.* One way may be to consider that in *It was John,* there is a zero occurrence of a referential at the end, so that replacement of the zero occurrence by the referend gives *It was John who gave me the book.* Here, *who* enters as a necessary adjustment of grammatical categories. From *It was John who gave me the book,* one obtains *John gave me the book* by a usual consequence transformation.

The following is the Axiom of Conjunction:

(A4) If $\alpha_1, \alpha_2, \beta_1, \beta_2, \alpha_1\frown\alpha_2, \beta_1\frown\beta_2$ are sentences and X is a set of assumptions, then there is a sentence γ such that $\alpha_1\frown\alpha_2 \in \mathrm{Cn}(\{\alpha_1\frown\gamma\} \cup X)$ and $\beta_1\frown\beta_2 \in \mathrm{Cn}(\{\beta_1\frown\gamma\} \cup X)$ and $\mathrm{Cn}(\{\alpha_1\frown\gamma\} \cup X) \neq S$ and $\mathrm{Cn}(\{\beta_1\frown\gamma\} \cup X) \neq S$.

The axiom of conjunction is a very useful tool for linguistic study. It allows one to consider one sentence γ instead of two sentences α_2 and β_2; then γ says at least as much as α_2 and β_2 together. The other factors α_1 and β_1, are to provide the contexts in which α_2, β_2 and γ appear, so that the principle applies to sentences with anaphoric referentials whose referends occur in the contexts. If α_2 is *John goes to school* and β_2 is *two plus two equals five,* then γ can be, for example, *John goes to school in order to learn that two plus two equals five*. Both *John goes to school* and *two plus two equals five* follow from γ. But γ does not follow from these two sentences, and their conjunction is not an acceptable English sentence. Here, *in order to learn that* is a conjunctive functor which forms γ from α_2 and β_2. In other cases it may take some ingenuity to find a proper conjunctive functor. The last two conditions in the axiom of conjunction insure that γ be chosen in a way consistent with what is said and known. Otherwise the axiom will be trivially true. The axiom of conjunction permits one to reduce grammatical problems which deal with a (finite) set of sentences to problems with a single sentence. Thus, when Harris and others derive a sentence by a binary transformation from two kernel sentences, it could also have been derived from a single sentence. It is important in practice to find as weak a γ as possible. The best is *and* (or a pause), for then we may have, barring some complications, in addition $\alpha_1\frown\gamma \in \mathrm{Cn}(\{\alpha_1\frown\alpha_2\} \cup X)$ and $\beta_1\frown\gamma \in \mathrm{Cn}(\{\beta_1\frown\beta_2\} \cup X)$. A γ which is equivalent to the set composed of α_2 and of β_2 does not always exist. Often, a conjunction says something more than the two conjuncts (like in the example with *in order that*).

6. AXIOM OF COMPACTNESS

For logic, and for other mathematical systems, one often asserts the principle of compactness which says that if a sentence is a consequence of a set of sentences, then it is a consequence of a finite part of the set:

(A5) If $\alpha \in \mathrm{Cn}(X)$ then there is a Y such that $Y \subset X$, Y is finite, and $\alpha \in \mathrm{Cn}(Y)$.

For some systems of logic this principle fails. Whether or not the principle of compactness holds for a natural language has not been sufficiently examined. The situation in this respect is not exactly the same as for logic. One reason for this is that a natural language contains its own metalanguage and allows reasonings going from premisses to metalinguistic conclusions which in turn may give consequences in the object language. In practice the compactness axiom is often used. However, any time it is used, it should be noticed that a certain risk is taken, or a restriction is imposed.

7. AXIOM OF INTERPRETATIONS

The next general principle connects the concept of consequence to that of interpretation. Interpretation is a three-place relation between a phrase, a sentence, and a phrase in that sentence. There is a similarity, and a fundamental difference, between interpretation and satisfaction. A sequence of objects satisfies a sentential function (a matrix), if the replacement of the variables by names of the successive objects of the sequence (the conforming free variables being replaced by names of the same object) results in a sentence which is true. We can generally say that the sequence $\langle \alpha_1, \alpha_2, \ldots \rangle$ satisfies $(f(x_1, x_2, \ldots, x_n))$ if and only if $f(\alpha_1, \alpha_2, \ldots, \alpha_n)$, provided that α_1 is the kind of thing which is named by phrases which the variable (x_1) represents, α_2 the kind of thing which is named by phrases which are represented by (x_2), etc. In such Tarskian constructions we assume that we know which kinds of things are named by which kinds of phrases represented by this or that variable. It is an assumption which for a natural language is hard to make. Take one example. *Blind* is usually labeled as an adjective, and, in set-theoretical terms, is viewed as a name of a property, a mapping from individuals to individuals, from a man to a blind man (or from a class of men to a class of blind men). But in the sentence *This is a home for the blind, blind* is used as a substantive and therefore as a name of a class rather than a mapping. French requires a plural substantive: *C'est la maison pour les aveugles*. In Polish, both the singular and the plural form can be used: *To dom niewidomego* and *To dom niewidomych* (both in the genitive, but the plural can also be with *dla*, 'for': *To dom dla niewidomych*). More important, the set-theoretic model for a language requires that there be a set of individuals; all relations are between the individuals or between relations between individuals, and so forth. But in natural language there is often no obvious way to base interpretation on a single set of individuals. For one thing, mass nouns (*water, wine, flesh*) constitute a separate basic set, and not a subset of count nouns (*a man, a book, a house*). (About the problems presented by the axiom of foundation in set theory see my forthcoming paper, "The Pretense of Existence").

Instead of satisfaction, one may take a different relation, namely, interpretation, which holds between some linguistic phrases and does not involve any other objects. It relates to the world only by truth. *A phrase* α *interprets the sentence* β *at* γ means, roughly, that if α replaces γ in β, the result is true. The result of the replacement of $\gamma_1, \gamma_2, \ldots$ by $\alpha_1, \alpha_2, \ldots$ in β will be represented by $repl(\beta, \gamma_i/\alpha_i)$. Thus, the word *sonnet* interprets the sentence *Petrach wrote a novel* at the occurrence of *novel* because it is true that Petrach wrote a sonnet. The formulation is only a rough one, for the phrase which interprets a sentence at a place must play the same grammatical role in the result of the replacement as the replaced phrase did. Otherwise, the matter will get out of hand and the theory will be useless. For instance, if we replace *novel* in the sentence *Petrarch wrote a novel* by *sonnet and was considered a great poet*, we obtain a true sentence, but *sonnet and was considered a great poet* is not of the same grammatical category as the substantive *novel* was—if indeed of *any* category. We must make our theory of interpretations relative to a theory of grammatical categories. Also, of course, one has to generalize the concept of interpreting to cover the case of a sequence of phrases to interpret a sentence or a set of sentences at proper places. Generally, f is an interpretation of the set of sentences X if f is a sequence of pairs $\langle \alpha_i, \gamma_i \rangle$ and every sentence in X results in a true sentence after replacing each γ_i by α_i. That is: $f \in \mathrm{Int}(X)$ if and only if $X \subset S$ and there is a sequence $\langle \alpha_i, \gamma_i \rangle$ such that $f = \langle \alpha_i, \gamma_i \rangle$ and for all β, if $\beta \in X$ then $repl(\beta, \gamma_i/\alpha_i) \in T$.[1]

With the help of the concept of interpretation, one can formulate the Axiom of Interpretations:

(A6) $\alpha \in \mathrm{Cn}(X)$ if and only if for all f, if $f \in \mathrm{Int}(X)$, then $f \in \mathrm{Int}(\{\alpha\})$

A sentence α is a consequence of the set X of sentences exactly when every interpretation of X is also an interpretation of α. This axiom looks like a definition of consequence in terms of interpretation.[2] But that is an illusion. For the concept of interpretation involves the concept of grammatical category, and grammatical category, I would claim, requires, in turn, the concept of consequence. For, a grammatical category is just a role that a fragment of a sentence plays in consequences. The structure of a sentence is imposed by the fact that consequences are drawn from the sentence. An adjective, approximately, is a phrase α in the following rule of inference, where "Det" stands for "determiner":

$$\psi(\text{Det } \alpha\ \beta) \rightarrow \text{Det } \beta \text{ is } \alpha$$

(For example, *I wear a blue sweater* → *A sweater is blue*.) Of course one can build a formal system in which the formula used here as the axiom of interpretations will be a definition, and interpretation will be accepted as a primitive concept and grammatical categories will be given in a lexicon.

[1] For more about interpretation see Hiż, 1969.

[2] Note that from the definition of interpretation it follows that if $X \subset Y$ and $f \in \mathrm{Int}(Y)$, then $f \in \mathrm{Int}(X)$. From this and the Axiom of Interpretations, if $X \subset Y$, then $\mathrm{Cn}(X) \subset \mathrm{Cn}(Y)$. From the Axiom of Interpretations it follows also that if $X \subset S$, then $X \subset \mathrm{Cn}(X)$ and that $\mathrm{Cn}(T) \subset T$. Therefore, if α and β are in T, so is $(\{\alpha\} \cup \{\beta\})$. But none of **(A1)**–**(A5)** follows from **(A6)**. **(A6)** departs from the spirit of Tarski's work, where, instead of interpretation of linguistic phrases by linguistic phrases, satisfaction of formulas by extra-linguistic entities is used.

Such a system may be elegant and instructive. But in empirical work consequences are given, and categories (even the category of a sentence) are constructs on the basis of a mass of material on consequences.

8. SOME LANGUAGE-SPECIFIC AXIOMS

Besides the general axioms (**A1–A6**), which are independent of any particular language, there should be axioms which connect the concept of consequence, and therefore other concepts appearing in the general axioms, with the specific phraseology of a language under study. For that purpose one may choose some phrases of the language as (grammatical) constants. The choice of constants is, in principle, arbitrary but some choices are better than others. For the language of logic, ordinarily, *if, not,* and *for every* are chosen. Sometimes *is a* is added; or, alternatively, *set,* or *class* and *set*. One may add *may be* or *but* or *although,* each time obtaining a variety of modal or relevance logic. The chosen constants are not subject to the reinterpretation referred to in the Axiom of Interpretations. If α is a (grammatical) constant, then the only interpretation allowed for it in (**A6**) is identity; if α is a constant, then for the purpose of (**A6**) $\mathrm{Int}(\{\alpha\}) = \alpha$. If a language contains a negation, N, then it would be reasonable to add

(**A7**) If $X \subset S$ and $\alpha \in X$ and $N(\alpha) \in X$, then $\mathrm{Cn}(X) = S$

For languages with the conditional (*if, then,* symbolically: $\supset$) one adds

(**A8**) If $X \subset S$ and $\alpha \in X$ and $(\alpha \supset \beta) \in X$, then $\beta \in X$

or

(**A8′**) $\beta \in \mathrm{Cn}(\{\alpha\} \cup \{\alpha \supset \beta\})$

Also, the rule of deduction is added (or derived) for languages with the conditional:

(**A9**) If $X \subset S$ and $\alpha \in \mathrm{Cn}(X \cup \{\beta\})$, then $(\beta \supset \alpha) \in \mathrm{Cn}(X)$

For English, one may choose as constants those words which play specific roles in those transformations that have the import of consequence rules. These are words which appear on one side of a transformation only, that is, phrases added or deleted by a transformation. Typically, there should be a number of negations (*no, not, few, neither, hardly, by no means, un-*, etc.), determiners, prepositions, affixes (*-s, -ed, -en, re-*), auxiliary and modal verbs. But the list of grammatical constants may vary greatly depending on which rules are to be described and with what degree of generality.

Some writers[3] consider paraphrase to be the basis for grammatical analy-

[3] The present writer included, in Hiż, 1964 and Bellert and Hiż, 1965.

sis, but consequence is handier. Paraphrase is a rare phenomenon, if taken very strictly, and it admits degrees of accuracy which are hard to notice. It is similar to measurement in the physical sciences. We say that two things are of the same length, if measurements fall within the allowed accuracy. But for paraphrase the accuracy is difficult to grasp. We know that there are closer and looser paraphrases, but we do not have a measure of the closeness. In principle, paraphrase is consequence both ways. But in practice we are often much more certain of the consequence one way than the consequence the other way around. From *It was at five o'clock that Jane arrived from Boston* follows the sentence *Jane arrived from Boston at five o'clock.* But the first sentence gives some other message. We expect that there is a problem as to when she arrived, or that she also arrived from a different direction at a different time.

One has to remember that consequences, and paraphrases, are rarely drawn from a single sentence. More often they are obtained from a sentence together with a set of assumed sentences (for example, those which in a discourse appeared just before the sentence). Logicians are aware that two theorems may be equivalent, if there are some assumptions present, but not otherwise. For instance, Peirce's law (*CCCpqpp*) and (*CCCpqrCqr*) are inferentially equivalent if the two usual axioms of *C*-calculus are used: the syllogism (*CCpqCCqrCpr*) and the simplification (*CpCqp*). If one requires for a paraphrase that two different sentences are so related only if they are inferentially equivalent in all contexts, with any assumptions, then presumably there are no paraphrases. Two sentences nearly never satisfy so strong a requirement. In logic only sentences which lead to contradiction are so related. In a natural language even this is not certain. The conclusion, which we reach in transformational grammar, that paraphrase is an unusual phenomenon is parallel to the similar conclusion of distributional linguistics (for example, Hoeningswald, 1965) that synonymy is very rare. For, if you ask in what environment a given sentence can be used, then it will be clear that usually its paraphrases, even close ones, cannot occur in all the same environments. There are, therefore, three different criteria for paraphrase. According to a distributionalist, two sentences are paraphrases of each other if they can occur in exactly the same contexts. A logician may say that two sentences are related paraphrastically if they are inferentially equivalent no matter what other sentences are used for deduction. Finally, one may say—and this is the most useful concept linguistically—that two sentences are paraphrases of each other if, for whatever assumptions, they have the same consequences. It is probable that, in all three senses, paraphrase is an exceptional case, if it exists at all. Even so, people say "in their own words" what other people have said, and a linguist seems obligated to describe this phenomenon.

If one takes as the meaning of a sentence the set of its consequences, then the meaning varies with the assumptions. Therefore, the meaning of a sentence α with respect to a set X of sentences is the set of consequences of X augmented by α, minus all the consequences of X alone. In other words, the meaning of α is what α says that is new.

$$M(\{\alpha\}, X) = \mathrm{Cn}(\{\alpha\} \cup X) - \mathrm{Cn}(X)$$

When we repeat correctly, we approximate the meaning of what was said by a set of sentences whose set of consequences, with the common assumptions, is similar enough to the consequences of the original utterance.

REFERENCES

Bellert, I, and Hiż, H. (1965). Paraphrastic sets and grammatical analysis. *Transformations and Discourse Analysis Papers,* **59.**

Harris, A. S. (1950). *Methods in structural linguistics.* Chicago.

Harris, Z. S. (1976). On the theory of language. *The Journal of Philosophy,* **73,** 253–276.

Hiz, H. (1964). The role of paraphrase in grammar. *Monograph Series in Language and Linguistics,* **17** (ed. C. I. J. M. Stuart), 97–104.

Hiż, H. (1969). Aletheic semantic theory. *The Philosophical Forum,* **1,** 438–451.

Hoeningswald, H. (1965). Review of John Lyons' *Structural semantics. The Journal of Linguistics,* **1,** 191–196.

Shaumian, S. K. (1971). *Principles of structural linguistics* (J. Miller, Trans.). The Hague: Mouton.

PRINCIPLES OF SEMANTICS

HANS-HEINRICH LIEB

1. PRINCIPLES AND PROBLEMS

1.1. Introduction

In the present paper, I intend to give a rough outline of a general semantic theory that I have been developing over the past few years. I shall emphasize fundamental matters rather than technical detail; for this reason, my presentation will be rather informal.

My theory is a semantic theory for **natural** languages, not for constructed ones—I disagree with Richard Montague's position that both natural and constructed languages should be dealt with in essentially the same way. It is a **linguist's** theory, not a logician's, and this will explain why certain problems are emphasized rather than others. And it is a theory whose subject matter is provided by **natural languages,** not their descriptions: It is not a "theory of semantic description," which deals with languages only in an indirect way. Thus, I am placing myself outside the context of most semantic work in generative grammar as I understand it. My own proposals for semantic description will not be discussed in this paper. These proposals ac-

SYNTAX AND SEMANTICS
VOLUME 10

ISBN: 0-12-613510-X

cept neither the framework of generative grammar nor of Montague grammar; nor do they attempt to combine the two approaches.[1]

It would of course be foolish to try and start *ab ovo;* like most linguists, I gratefully accept whatever seems valuable in logical semantics or philosophical semantics in general. I do feel, however, that ultimately the linguist's problems can only be solved in linguistics, and that the linguist must insist on his own formulation for those problems—even if his philosophical colleague is horrified. (Some of my philosophical colleagues will be.)

In developing my semantic theory I adopt the following **principles:**

(1) A general semantic theory of natural languages should satisfy the following conditions:
  a. The theory takes into account the internal variability of natural languages: the fact that most natural languages have a great number of different varieties (dialects, styles, etc.).
  b. The theory is directly relevant to
    (i) reconstructing the relations (if any) between properties of a language and the way in which its speakers conceive or perceive the world,
    (ii) reconstructing the relations between thinking and speaking,
    (iii) reconstructing the relations between speaking and performing nonverbal actions.
  c. The theory can be used in semantic theories of individual languages for the derivation of theorems on those languages.
  d. The theory can be formally combined with nonlinguistic theories that are relevant in connection with (b), such as psychological or sociological ones.

[1] Both my semantic theory and my theory of semantic description are only beginning to be published. The following references are meant as a guideline (hopefully they will not be taken as a bad case of self-quotation). An outline of my **semantic theory** is given in Lieb, 1976a, (Sections 3, 4, and 5.1) see also Lieb, 1976e. Extensions into pragmatics are considered in Lieb, 1976a (Sections 6.1f). The present paper contains certain improvements on my original account which will be pointed out later (see Footnotes 8, and 18). My semantic theory is based on a morpho–syntactic one which develops a new conception of "surface grammar"; for a general characterization, see Lieb, 1977; for a summary, Lieb 1976a (Sections 2.1f); for a fuller treatment (excluding morphology), Lieb, 1976b; for applications, Lieb, 1975, and Eisenberg, 1976. A complete version of my morpho–syntactic and semantic theories will be given in Lieb, in preparation. My morpho–syntactic and semantic theories assume the theoretical framework for dealing with language variability that was developed in Lieb, 1970, with certain improvements (Lieb, 1976a, Section 6.3). My theory of **semantic description** is part of the framework for axiomatic grammars proposed in Lieb, 1974 and 1976a (see also Lieb, 1976c). Semantic description is treated in Lieb, 1976a (Section 5) and also in Lieb, 1976d. For my arguments against generative grammar, see, in particular, Lieb, 1974 (Section 1); 1975 (Section 1.4).

I propose that these principles be taken as **conditions of adequacy** for any general semantic theory of natural languages. Other conditions may, of course, be added.

I do not know any semantic theory that actually satisfies Condition (1a); it is typically neglected in theories that emphasize the concept of social conventions (beginning with de Saussure, and up to Kutschera's recent (1976) restatement of Lewis's theory).

The first two subconditions of (1b) introduce the problems of language and thought—to use the traditional label—as key problems of any semantic theory. Such a position has proved detrimental to progress in traditional semantics, and is hardly made more attractive by Chomsky's vague philosophizing on language and cognition. I still maintain that the problems of language and thought are avoided in semantics only at a very high price; one of the principal ties to reality is severed, and semantic theories are left to float on the other, namely, the relations between speaking and nonverbal actions. Subcondition (iii) requires that a general semantic theory *should* be relevant with respect to these relations—a requirement that is meeting with growing support.

Conditions (1c) and (1d) imply requirements on the form of general semantic theories. Condition (1c) guarantees applicability with respect to individual grammars. Condition (1d) is needed because the problems of "language, thought, and action" cannot be exhaustively treated within linguistic semantics, or linguistics in general, for that matter.

In recent work I have developed a concept of theory integration that also applies in the cases of (1c) and (1d). For the purposes of theory integration—obtaining a single theory from several given ones—a semantic theory has to be assumed in axiomatic form. For further details, I must refer the reader to my other work.[2]

Limited space also forbids me from giving a detailed account of how I intend to meet Condition (1a). In Lieb 1970 I develop an extensive framework for dealing with language variability, which I have recently modified in some respects (cf. Lieb, 1976a, Sect. 6.3; Lieb, 1976b, Sect. 2). My treatment of language variability justifies the following conception: **A general theory of language directly deals with properties of idiolect systems,** and the same is true for its subtheories including any general semantic theory. An idiolect is taken not as a speaker's complete share of a language, but as a homogeneous part of his share, and systems are assumed that determine the idiolect. From here on I shall refer to idiolect systems instead of complete languages, without further justification.

In the present paper I shall concentrate on those aspects of my theory which relate to Condition (1b).

Let me first indicate roughly how I conceive the subject matter of a general semantic theory of natural languages.

[2] Lieb, 1974 (Section 3.4); and 1976c. With application to semantics, see Lieb, 1976a (Section 5).

1.2. On the Subject Matter of a Semantic Theory

My conception is standard in some respects but must still be stated explicitly. I begin with some additional remarks on idiolects and their systems.

An idiolect is a set of abstract texts. Each text is a pair consisting of a phonetic form and a meaning. A system of the idiolect determines which form–meaning pairs belong to the idiolect. This is achieved in the following way: The idiolect system determines properties of form–meaning pairs which jointly identify those pairs which are texts of the idiolect.

Any system of an idiolect has various subsystems. Among these, there are the morphosyntactic and the semantic subsystems. The morphosyntactic subsystem determines certain formal properties of texts, which are called morphosyntactic properties. For this purpose, the morphosyntactic subsystem specifies, among other things, morphological and syntactic units and their structures. I have worked out a detailed theory of the morphosyntactic subsystem, which is at the basis of my semantic theory but cannot here be characterized any further.[3] Let me just add that my morphosyntactic theory assumes two parts of the morphosyntactic subsystem, a morphological and a syntactic one.

The semantic subsystem determines **semantic** properties of form–meaning pairs, that is, properties which concern either the meanings themselves or the correlation between forms and meanings. There are certain properties of the semantic subsystem by which it becomes possible to determine the semantic properties of texts. Thus, the subject matter of a semantic theory may be characterized as follows:

(2) **Assumption 1:** A semantic theory deals with those properties of the semantic subsystem of any idiolect system which in any idiolect system have a share in determining the semantic properties of texts.

Given this conception, both traditional and current work in linguistic and philosophical semantics can be characterized as subscribing to the following assumption, either completely or in part:

(3) **Assumption 2:** Let *S* be a system of idiolect *C*. At least the following six properties of the semantic subsystem of *S* have a share in determining the semantic properties of texts in *C*:

a. The property that consists in determining, for each meaningful syntactic unit os *S* and morphosyntactic structure of the unit, the meanings of that unit in *S* given its structure;

b. The property that consists in determining the "semantic properties" of meaningful syntactic units of *S*, relative to a structure of a unit, that is, certain properties that

[3] Cf. the references given in Footnote 1.

can be attributed to a unit given its meanings in *S* for a certain structure;

c. The property that consists in determining the "semantic relations" between meaningful syntactic units of *S*, relative to structures of the units; that is, certain relations that may be said to hold between units, given their meanings in S for certain structures.

Formulations (d)–(f) are parallel to (3a)–(3c), but substitute "morphological" for "syntactic" and "morphosyntactic." In (3b), we include such properties as being ambiguous or unambiguous, analytic or nonanalytic, more vague or less vague; as well as having certain truth conditions, having certain reference conditions, etc. In (3c) we have relations such as the traditional meaning relations, for example, synonymy, but also relations of consequence, presupposition, etc. Formulations (3d)–(3f) are analogous with respect to morphological units.

From (2) and (3), we obtain the following:

(4) **Characterization of a semantic theory:** A general semantic theory of natural languages deals at least with properties (3a)–(3f).

Henceforth, I shall accept this characterization.

It is obvious from (3) that the semantic subsystem is in a sense based on the morphosyntactic subsystem: With respect to properties (3a)–(3c), it is based on the syntactic part of that subsystem, and with respect to (3d)–(3f) on its morphological part. Correspondingly, I assume two parts of the semantic subsystem, the **syntax-related part** and the **morphosemantic part.** At the level of theories we have a corresponding distinction between a **syntax-related semantics (syntactical semantics)** and a **morphosemantics,** both understood as theories, not disciplines. For obvious reasons, theories of the two types are characterized as follows:

(5) **Characterization of a syntax-related semantics and a morphosemantics:**

a. A general syntactical semantics of natural languages deals at least with properties (3a)–(3c).

b. A general morphosemantics of natural languages deals at least with properties (3d)–(3f).

I shall here restrict myself to the problems presented by any syntactical semantics.

1.3. Basic Problems for any Syntactical Semantics

Adopting (5a) and (3), we can immediately point out a number of basic problems that confront any syntactical semantics. First, there is a group of problems that arise in connection with (3a).

Let us informally distinguish between **syntactic meanings** and **morphological meanings** as follows:

(6) a. **Syntactic meaning:** meaning of a constituent of a syntactic unit.
b. **Morphological meaning:** meaning of a constituent of a morphological unit.

"Constituent" is to be understood in roughly the sense it would have in an immediate-constituent analysis based on surface structures (syntactic or morphological ones); any unit is assumed to be a constituent of itself.

The basic problems connected with (3a) may be grouped together as problems of the conception and determination of meanings. It may seem that only syntactic meanings need to be considered, but this is incorrect. Morphological units and syntactic units must be assumed to be related in such a way that the meanings of a syntactic unit depend on morphological meanings. Therefore, in constructing a syntactical semantics we must also clarify the concept of morphological meaning. We thus have the following problems in connection with (3a):

(7) **Group I of basic problems: Conception and determination of meanings:**
a. How are morphological meanings to be conceived?
b. How are syntactic meanings to be conceived?
c. How are morphological meanings to be related to syntactic ones?
d. How are meanings of simple syntactic constituents (that do not contain any other constituents) obtained?
e. How are meanings of complex (non-simple) syntactic constituents determined?

The following problems arise in connection with (3b):

(8) **Group II of basic problems: Semantic properties:**
a. Exactly what properties are to be allowed as "semantic properties" of syntactic units?
b. Assuming that (8a) has been answered, exactly how are the semantic properties to be conceived?

Analogously, for (3c):

(9) **Group III of basic problems: Semantic relations:**
a. Exactly what relations are to be allowed as "semantic relations" between syntactic units?
b. Assuming that (9a) has been answered, exactly how are the semantic relations to be conceived?

As a final problem, we have

(10) **The problem of interrelations:** How are properties (3a) to (3c) of the semantic subsystem to be interrelated?

As to (10), there is a traditional position in linguistics that amounts to taking (3a) as basic: It is tacitly assumed that (3b) and (3c) can be dealt with on the basis of (3a). However, there are well-known attempts, especially in philosophical semantics, to start from (3b) or (3c).[4] For my own theory, I adopt the traditional position of linguistic semantics.[5]

It then becomes imperative to provide a solution to the problems of Group I. In the rest of this paper I shall outline my solution to the Problems stated in (7a) and (7b): How are morphological meanings to be conceived? and how are syntactic meanings to be conceived? This will also involve some discussion of (7c) (the relation between morphological and syntactic meanings). My solution to (7d) (meanings of simple syntactic constituents) and (7e) (meanings of complex syntactic constituents) will only be hinted at.

1.4. Meanings and Determination of Meanings: General Ideas

In my conception of morphological and syntactic meanings, I deviate from the entire tradition in semantics in one important respect: I assume meanings of morphological units and meanings of syntactic units to be **entities of different type.** Traditionally, meanings of stems, word forms, and syntactic groups are all conceived as basically alike; it is only at the level of sentences that meanings in a new sense are introduced, for example, as propositions or states of affairs. On my conception of meanings, morphological meanings are radically different from syntactic ones, and all syntactic meanings are basically alike. Thus, I assign semantic significance to the fact that in any language the morphosyntactic subsystem of any idiolect system is bipartite.

As a second fundamental feature, my conception embodies the following idea: Morphological meanings are construed as concepts in a psychological sense, and syntactic meanings are constructed so as to represent necessary conditions for the normal use of syntactic units, for example, sentences. The novelty here consists neither in proposing a psychological conception for morphological meanings, for this is traditional, perhaps shockingly so, nor in introducing some type of "meaning as use" conception. The novelty consists in using the two approaches simultaneously and relating them to the distinction between morphology and syntax.

As a third characteristic, I assume the following relation between morphological and syntactic meanings: Meanings of simple syntactic constituents

[4] Cf., in particular, Hiż, 1969, Davidson, 1967 and 1970; also Lyons, 1968 (Ch. 9f).

[5] A position which, it seems to me, is also shared by Montague (1974).

are "contextualizations" of concepts that are morphological meanings, in a specific sense of "contextualization."

Finally, meanings of complex syntactic constituents are obtained from meanings of simple constituents by a composition process that depends only on the structures of the given syntactic unit; these structures are **surface** structures in a reasonable sense. (An analogous statement can be made for complex morphological constituents and their meanings.)

In its four salient features, my conception of morphological and syntactic meanings is motivated by principle (1b), which I adopted as a condition of adequacy. My semantic theory is to be directly relevant to an analysis of "language, thought, and action," and my conception of meaning is to be in line with this aim.

In the remaining two sections of this paper I shall deal first with morphological meanings (Section 2), then with syntactic ones (Section 3).

2. MORPHOLOGICAL MEANINGS

2.1. Perceptions and Conceptions

Morphological meanings are to be understood as concepts in a psychological sense. Such a conception of meanings can be traced back to Aristotle. In this century, it has been rejected by many as completely unmanageable whenever a more precise formulation is attempted. In spite of discouraging precedent, I shall make another try. I proceed from what I call perceptions and conceptions, beginning with perceptions. My discussion will be summarized by a table at the end of this subsection.[6]

Suppose somebody *sees a house* in the sense that he has certain sensations which lead him to assume that he is confronted with a house. What is taking place in this person is a complex process of having sensations and processing them mentally; this process results in the conscious recognition of a certain object. We may say that the person has **perceived** a house. Any **perception** is to be understood as the (mental) result of a process that consists in having sensations and processing them mentally. This in a way assigns an ontological status to perceptions; it is not a definition of "perception."

Any perception has a certain **content.** Suppose somebody perceives a house. In doing so, he consciously or unconsciously attributes certain properties to what he sees; properties such as being at a certain distance, being

[6] The following outline (Sections 2.1f) basically belongs within cognitive psychology. It is meant to be compatible with what is known in psychology about perceptions, cognition, and concept formation. The form of the outline is, however, determined by the needs of my semantic theory, hence, by linguistic considerations.

viewed from a certain angle, causing a sensation of redness, etc. The set of these properties forms the **content** of the person's perception.

Suppose that the same person perceives the same house on a different occasion. We shall assume that the two perceptions are different for the very reason that they result from different processes of perceiving. Their contents might still be identical although this is unlikely. What **is** identical in the two cases are the "object sets" of the two perceptions. By the **object set** of a perception we understand the set of entities that have the properties included in the content of the perception. The content of the perception may be such that the object set has at most one element, as would be the case in our example.

We shall also allow for perceptions whose content consists of relations-in-intension between entities that can be perceived; the object set in this case is a corresponding relation-in-extension.

The content of a perception must be logically homogeneous; that is, all elements must be of the same logical type; otherwise, the object set might be empty for trivial reasons. The object set is empty only in the case of an illusion. Since it may be empty, expressions such as "a perception **of** a house" are misleading; they presuppose existence and may be used only when existence is guaranteed.

We now turn from perceptions to conceptions. Consider an architect who has just drawn up his plans for a house. We may say that the architect has **conceived** a house: As a result of a complex mental process, he has formed a certain **conception** (embodied in the final drafts). The **content** of his conception is a set of properties of physical objects, such as, existing in the future, being at a certain place, and having certain measurements. The **object set** of the conception is the set of entities with these properties. The content here is such that the object set has at most one element. It may also be empty—no corresponding house may ever be built. Since the object set of a conception may be empty, we must again be careful with expressions like "conception of a house." Unless we want to populate the universe with all sorts of conceivable entities, such expressions may be used only when existence (in the real world) is guaranteed.

It is not essential that the elements of the object set should exist only in the future—I may form a conception whose object set consists of a person I know.[7]

In the case of a perception, the object set must consist of perceivable entities. This is not true for conceptions. We may, for instance, form a concept whose object set consists of a property of perceivable entities, for example, the property of *being a house*. (This is, so to speak, a conception of what it

[7] In the case of actually existing objects we should perhaps distinguish between the object set of a conception and its "intended object set": I may form a conception that is intended as a conception of a certain person but may, as such, be "incorrect."

TABLE 1

Perceptions and Conceptions

	Ontological Basis	Content		Object Set	
		Characteristics	Existence	Characteristics	Existence
Perceptions	Perceivable entities only	A homogeneous set of properties or of relations-in-intension that are ultimately reducible to the ontological basis either of perceptions or of conceptions	Nonempty	The set of entities that have the properties (among which the relations-in-intension hold) which form the content of the perception or conception (thus, possibly, a relation-in-extension)	Nonempty except for illusions
Conceptions	Perceivable entities only?				Potentially empty

means to be a house.) The content of such a conception consists of properties of the property of *being a house*.

In this example, the content of the conception is ultimately reducible to perceivable entities. It is one of the deepest problems of both philosophy and psychology whether this should always be the case, and I respect it as such.

Our discussion of perceptions and conceptions is informally summarized in Table 1. It goes without saying that we are still far removed from a formal account; it is not only matters of detail that will have to be filled in. Still, we are in a position to take the next step: introduction of concepts.[8]

2.2. Concepts

We introduce concepts as **properties of perceptions or conceptions,** where "or" is to be understood in its nonexclusive sense.

Consider perceptions and conceptions whose contents contain the following properties of physical objects: being constructed for the purpose of being inhabited, having a roof, and certain others. Now consider the property of being a conception or perception with such a content. This property is identified with a concept, which in turn may be suggested as a meaning of the English stems /haus/ and /hauz/. A particular perception or conception with this property has the previously mentioned properties in its content; in addition, there may be many other properties such as being at a certain distance, appearing from a certain angle, etc. The concept remains unaffected by the fact that corresponding conceptions or perceptions may differ in content. This already avoids a whole range of problems that traditionally beset psychological conceptions of meaning.

For a more precise formulation, we must make up our minds about the formal status of the properties and relations that appear as perceptual or conceptual content. For simplicity's sake I shall here make the following assumptions: The only **individuals** needed for such properties and relations are perceivable entities, that is, entities that can initiate processes of perceiving; and the properties and relations may be of arbitrary finite type. (Actually, only a few levels should be needed, whose number must be established by empirical investigations.)

The following variables are introduced, together with some additional notation:

[8] Our account of conceptions and perceptions in the present subsection improves the conception in Lieb, 1976a (Sections 3.3ff), in an important respect: It clearly distinguishes between the content and the object set of a perception or conception and clarifies the status of both. In Lieb, 1976a, the distinction was collapsed by positing a single "content function" whose status was only partially clarified. It could not be decided what kind of entities should be assumed as values of the function. This was recognized as a major drawback: "This means that my outline of a semantic grammar is, at a crucial point, incomplete" (1976a, p. 81). Furthermore, the conception of concepts was also affected: It was not possible to reconstruct the important distinction between the "comprehension" and the "extension" of a concept (see next subsection).

(11) **Variables, constants, lambda operator**

a. $x, x_1, \ldots$: any human perception and ordinary human conception (that is, any conception whose formation does not involve a process of scientific investigation).

b. $b, b_1, \ldots$: any first-level property of entities x.

c. *Convention:* Any constant of the same type as "b" consists of an English word between raised dots (·house·).

d. $y, y_1, \ldots$: any perceivable entity.

e. $c, c_1, \ldots$: any property or relation-in-intension (of arbitrary finite type) that presupposes only individuals that are perceivable entities y.

f. $z, z_1, \ldots$: any argument of any value of "c".

g. An expression of the form $(\lambda v_1 \ldots v_n)(\ldots v_1 \ldots v_n \ldots)$, where $v_1 \ldots v_n$ are variables and the usual conditions for λ-expressions are satisfied, is translated by an expression of the form: the property of v_1 (n-place relation-in-intension between $v_1 \ldots v_n$) determined as follows: $\ldots v_1 \ldots v_n \ldots$.

Given this minimal formal apparatus, we may now introduce a more precise formulation—though still not a formal one—of our conception of concepts:

(12) **Assumption and definitions:**

For any concept b:

a. There are (properties or relations) $c_1, \ldots, c_n$ such that:

$$b = (\lambda x)(\{c_1, \ldots, c_n\} \subseteq \text{content}(x)).$$

b. **The Comprehension of** b (${}^c b$) = the set of properties or relations satisfying (a).[9]

c. b is n-**place** iff (i) or (ii):

(i) $n = 1$, and each $c \in {}^c b$ is a property.

(ii) $n > 1$, and each $c \in {}^c b$ is an n-place relation.

d. b is **relational** iff for some $n > 1$, b is n-place.

e. b **applies** to z iff, for all $c \in {}^c b$, cz.

f. **The extension of** b (${}^e b$) = the set of all z to which b applies.

(13) **Example:**

a. Define: house = $(\lambda x)(\{(\lambda y)(y$ has been constructed for the purpose of being inhabited$), \ldots\} \subseteq \text{content}(x))$.

b. Assume: ·house· is a concept.

[9] We may obviously assume that there is at most one such set.

c. ^{c}·house· = {(λy)(y has been constructed for the purpose of being inhabited), . . .}.
d. ·house· is one-place.
e. ^{e}·house· = the set of all y to which ·house· applies (all y that have all the properties in ^{c}·house·).

In formulating (12) we have to presuppose a logic that makes sense of properties and relations-in-intension. In the present context, I will simply assume such a logic, leaving its nature unspecified.

By means of (12a) we clarify the logical and ontological status of concepts understood in a psychological sense. By (12b)–(12f) we introduce a number of important auxiliary terms that are traditionally used in speaking about concepts.

Our conception would have to be further elaborated. To mention just two problems, we would have to show how we deal with "family resemblance" in Wittgenstein's sense, and how we account for vagueness. Both problems seem solvable but I cannot here go into the details.[10]

In a more elaborate version, it would be emphasized that any conception or perception is a conception or perception by somebody during a certain time, and notions such as "having a concept" would be explicated. Concepts may be shared by different persons; some may even be shared universally. This is an empirical question. Our approach is completely neutral with respect to such problems.

We are now in a position to relate concepts and morphological meanings. Let me stress from the outset what is not involved in establishing this relation: We do not assume that every concept a person has must be a morphological meaning for this person, and we do not assume that in the case of two different persons it is exactly the same concepts that are the meanings of a given morphological unit. But we do assume that morphological meanings are concepts.

2.3. Morphological Meanings as Concepts

In this subsection, I shall have to be very brief.[11]

I conceive morphological units as sequences of phoneme sequences; for example,

[10] "Family resemblance" may seem much less of a problem once we recognize the following fact: The properties or relations that form the comprehension of a concept may be of arbitrary complexity (the lambda operator may be prefixed to any appropriate sentential formula); in particular, we may obtain a property from a "disjunction of conditions." The problems of vagueness, formidable enough, are no more formidable than they always are, and may be attacked in one of the known ways. In particular, we may consider an application of "fuzzy set theory" in setting up comprehensions. A certain property or relation may belong to the comprehension of a concept "to a certain degree" (there may be empirical motivation for assigning the degrees).

[11] For the morphological notions involved, cf. Lieb 1976a, (Sect. 2.2).

(14) *sweet en ed* = $\{\langle 1, /\text{swi:t}/\rangle, \langle 2, /\text{en}/\rangle, \langle 3, /\text{d}/\rangle\}$

Each morphological unit has a morphological structure. The unit has **constituents** relative to the structure:

(15) $\{\langle 1, /\text{swi:t}/\rangle\}$; $\{\langle 2, /\text{en}/\rangle\}$; $\{\langle 3, /\text{d}/\rangle\}$; $\{\langle 1, /\text{swi:t}/\rangle, \langle 2, /\text{en}/\rangle\}$; and *sweet en ed*

The constituents are **occurrences** of morphological units in the given unit:

(16) $\{\langle 2, /\text{en}/\rangle\}$ is an occurrence of $\{\langle 1, /\text{en}/\rangle\}$ in *sweet* + *en* + *ed*.

I shall use the following variables:

(17) **Variables:**
a. S: any idiolect system.
b. $f, f_1, \ldots$: any relation between positive integers and phoneme sequences of any S.[12]
c. $s, s_1, \ldots$: any entity that has the same logical type as a morphological structure.

Morphological meanings are to be meanings of constituents of units of an idiolect system. Obviously, the concept of meaning of a constituent should be relativized to a morphological unit f, a structure s of the unit, and an idiolect system S. Consequently, we introduce a functor "m-mg" such that the arguments of the function m-mg are triples $\langle f, s, S\rangle$ and the values are relations-in-extension between meanings and constituents:

(18) "m-mg$(f, s, S)bf_1$" for "b is a morphological constituent meaning of f_1 relative to f, s, and S."

Here f_1 should be a constituent of f relative to s; and b is a property of perceptions or conceptions x.

We now formulate the key assumption:

(19) **Assumption:** If m-mg$(f, s, S)bf_1$, then b is a concept.

Because of (19), all concepts defined in (12) directly apply to morphological meanings. In particular, we may speak of the comprehension of a morphological meaning or of its extension.

To counter some obvious objections to (19), I will immediately add that meanings are assigned only to morphological word forms and their stems, for example, in (14), to $\{\langle 1, /\text{swi:t}/\rangle\}$, *sweet en* and *sweet en ed*. No semantic content is assumed for occurrences of **inflectional** affixes such as $\{\langle 3, /\text{d}/\rangle\}$,

[12] In particular, these variables can be used for morphological units (which are sequences of phoneme sequences, that is, certain relations between positive integers and phoneme sequences) and for constituents of units (which are subsets of the units).

whereas the semantic content of **derivational** ones such as {⟨2, /en/⟩} is not to be treated via meanings.[13]

Given an English idiolect system and an appropriate structure of *sweet en ed,* we might have the following meanings:

(20) a. ·sweet· as a meaning of {⟨1, /swi:t/⟩}
b. ·sweeten· as a meaning of *sweet* + *en* and of *sweet en ed*

The comprehension of ·sweet· (20a) should consist of properties of perceivable objects *y*, whereas ·sweeten· is a relational concept.

The source of morphological meanings in an idiolect is an assignment of concepts to morphological paradigms. This and many other things about morphological meanings stand in need of further clarification.[14]

I would like to conclude this section by a suggestion: It may be worthwhile to analyze the comprehension of concepts, or, generally, perceptual and conceptual content, by making prudent use of model-theoretical semantics. It may turn out that the primary usefulness of, say, Montague semantics is in precisely this area and not in syntactical semantics, where it has been sought. In any case, my own conception of syntactic meanings is quite different from the proposals made in model-theoretical semantics.

3. SYNTACTIC MEANINGS

3.1. Complete Syntactic Meanings: Preliminary Discussion

In Section 2 I outlined a solution to the problem in (7a): How are morphological meanings to be conceived? I shall now examine the problem in (7b): How are syntactic meanings to be conceived? In doing so, we must also briefly consider the problem in (7c), the problem of interrelating morphological and syntactic meanings.

According to (6a), a syntactic meaning is a meaning of a constituent of a syntactic unit. A few comments are needed to understand the syntactic notions involved.[15]

We previously conceived morphological units as sequences of phoneme sequences, (cf. (14)). Syntactic units—sentences and "word groups"—are

[13] Derivational affixes should be used to define particular morphological relations which may then be assigned a "semantic content." I consider as inadequate traditional attempts to assign this content to the affix instead of the relation. (Cf. also Lieb, 1976e.)

[14] Cf. Lieb 1976a, §3.

[15] For details, see Lieb 1976a (Sect. 2.1), and in particular, Lieb, 1976b.

understood in the same way; thus, we may have the following sequence as a syntactic unit:

(21) *the thief* = {⟨1, *the*⟩, ⟨2, *thief*⟩} = {⟨1, /ðe/⟩, ⟨2, /θi:f/⟩}

This unit may occur as a constituent in another unit, for example,

(22) a. (Unit) *send the thief to jail*
b. (Constituent) {⟨2,*the*⟩,⟨3,*thief*⟩}

A syntactic unit has constituents only relative to a syntactic structure of the unit. For such structures we assume:

(23) **Assumption:** Each syntactic structure of a syntactic unit is an ordered triple consisting of a constituent structure of the unit, the "marking structure" relative to the constituent structure, and an intonation structure relative to the constituent structure.

Syntactic structures and morphological structures are to be of the same logical type; thus, the variables "s", "s_1", . . . introduced in (17) may also be used for syntactic structures; and, obviously, f, f_1, . . . may be used for syntactic units and their constituents.

We now take up the general proposal for syntactic meanings that was formulated in Subsection 1.4. The following requirement is in agreement with that proposal:

(24) **Requirement:** Assume a constituent of a syntactic unit such that the constituent has "complete syntactic meanings." The complete syntactic meanings of the constituent jointly represent a necessary condition for normal utterances of the unit.

Constituents with complete syntactic meanings are, for instance, occurrences of noun phrases and of sentences. We shall concentrate on an explication of this concept, passing over the other types of syntactic meanings whose existence is implied by the qualifying adjective "complete."

Suppose we have a sentence f in which *the thief* occurs as a constituent. Assume a normal utterance of f. We may then assume that a condition of the following type is satisfied by utterance and speaker: The speaker is referring by the utterance to somebody of whom he presupposes that he is a thief; or he is referring to thieves in general; or . . . or . . . . We thus have a disjunction that formulates a necessary condition for normal utterances of f. The condition is that one of several **relations** must hold between utterance and speaker: either the speaker must be referring by the utterance to somebody of whom he presupposes that he is a thief, or . . . or . . . . **It is my basic idea to identify these relations with the meanings of** *the thief*. The complete syntactic meanings of *the thief* in f may then be said to jointly represent a necessary condition for normal utterances of f, in an intuitively obvious sense of "represent."

We have been considering utterances and speakers. A speaker is an object in space–time, and an utterance is an object or event in space–time. Let us identify such objects and events with the space–time regions they occupy. We may refer to arbitrary objects, events, and spatio–temporal parts of objects or events by means of the following variables:

(25) **Variables:** $V, V_1, \ldots$: any region in space–time.

These variables are helpful in the following reformulation of my basic idea for complete syntactic meanings:

(26) **Basic idea for complete syntactic meanings:** Assume a constituent of a syntactic unit such that the constituent has "complete syntactic meanings."

a. Each complete syntactic meaning is a relation-in-intension between objects or events V and persons V_1 that consists in V_1 producing V and making certain references by means of V, making certain presuppositions in producing V, and the like.

b. For any V and V_1 such that V is a normal utterance by V_1 of the syntactic unit, V and V_1 are related by one of the complete syntactic meanings of the constituent.

This idea can, of course, be related to the tradition originating from Strawson's classic paper (1950). I shall make it more precise by further discussing constituents like *the thief* in (22b), that is, occurrences of units that consist of the definite article followed by the singular form of a noun.[16] Such constituents will be called **definite singular noun occurrences,** or **definite noun occurrences,** for short.

I shall now introduce my main conceptual tool for the semantic analysis of definite noun occurrences.

3.2. The Concept of Reference Basis

Using our previous example (22a), we assume:

(27) a. f = *send the thief to jail*

b. $f_1 = \{\langle 2, the\rangle, \langle 3, thief\rangle\}$

c. $f_2 = \{\langle 3, thief\rangle\}$

In producing an utterance of f, the speaker may be referring to a certain per-

[16] For the following discussion we may need further restrictions on the type of noun allowed; these will be left unspecified.

son. The person is, as it were, selected from the set of all thieves. It seems natural to take this set as the set of potential reference objects (potential referents). There is a certain concept,

(28) ·thief·

that is a morphological meaning of the stems {⟨1,*thief*⟩} and {⟨1,*thiev*⟩}. If this concept is somehow associated with the constituent f_2, we may identify the set of potential referents with the extension of ·thief·.

This account, which fully agrees with classic conceptions, contains fundamental mistakes. In producing a normal utterance of *f*, the speaker selects for singular reference an object that satisfies two conditions:

(29) a. The speaker believes that the concept ·thief· applies to the object.
b. In producing {⟨3,*thief*⟩}, the speaker is willing to predicate the concept ·thief· (or rather its comprehension) of the object.

I shall take up the second condition first.

Let V_2 be that part of the utterance *V* which corresponds to {⟨3,*thief*⟩}. In producing V_2, the speaker V_1 considers certain objects as being under discussion, say, two men, one woman, and a book. Suppose V_1 is convinced that the woman but not the men stole the book. Then the speaker is willing, at this point, to predicate the comprehension of ·thief· of the woman—by asserting that she is a thief—**and also of the men**—by denying that they are thieves. He is not willing to predicate the comprehension of ·thief· of the book: It would be strange for him either to assert or deny that the book is a thief. Nor is he willing to predicate the comprehension of ·thief· of all the thieves in the world, for a different reason: They are not under discussion. The two men and the woman together form what may be called the **reference basis** relative to {⟨3,*thief*⟩}, the utterance *V*, the speaker V_1, and the concept ·thief·. The following definition may be given:

(30) **Definition:** For any *n*-place concept *b*, the (*n*-place) **reference basis** for *f* relative to *V*, V_1, and *b* [$\text{reb}^n(f, V, V_1, b)$] = the set of $\langle z_1, \ldots, z_n\rangle$ such that:
a. V_1 produces *V*.
b. There is exactly one V_2 such that:
(i) V_2 is a part of *V*
(ii) V_2 corresponds to *f*
(iii) V_1 is willing to predicate ${}^c b$ of $\langle z_1, \ldots, z_n\rangle$ in producing V_2[17]

[17] For $n = 1$, $\langle z_1\rangle = z_1$.

The concepts of producing, correspondence, willingness, and predication would of course have to be further clarified.[18]

The reference basis functions as a **momentary universe of discourse,** a universe that is relativized to a certain point of the utterance and a single concept. At the next point another universe may have to be assumed. Thus, the basis for the interpretation of quantifiers may change from one point of the utterance to the next. This extreme versatility of natural languages is not reconstructed in model-theoretical semantics, and there are strong reasons to believe that it cannot be reconstructed by any known version of the model-theoretical approach. This would mean a fundamental inadequacy of those versions, and a strong additional argument in favor of my own conception.[19]

In our example, the speaker selects an object not simply from the reference basis [Condition (29b)]: The object must be such that the speaker **believes** it belongs to the extension of the concept [Condition (29a)]. His belief may well be false and he may actually choose an object to which the concept does **not** apply. Thus, there are two independent reasons why it would be totally inadequate to identify the set of potential referents with the extension of the concept.

It has to be admitted, though, that there are other uses of definite noun occurrences in which a potential referent must indeed be in the extension of the concept. For a realistic construal of the meanings of definite noun occurrences, the semantic analysis of our example has to be extended.

3.3. The Uses of Definite Noun Occurrences

Consider the following four situations for normal utterances of f = *send the thief to jail:*

[18] Definition (30) differs from definition (47c) in Lieb, 1976a (p. 48) in essentially the following respect: The reference basis is not constructed from perceptions and conceptions but from entities to which concepts apply (cf. (12e)), that is, from entities that appear in the object sets of conceptions or perceptions (cf. Sect. 2.1). Correspondingly, it is no longer the concept that is predicated but its comprehension (that is, each element of the comprehension). These changes are due to our improved version of conceptions and perceptions (cf. Footnote 8).

[19] The reasons just mentioned are as follows. Any known version of model-theoretical semantics that tries to account for something like a reference basis uses "points of reference" for this purpose. But Östen Dahl has argued—convincingly to my mind—that there is no principled way of choosing "points of reference" so as to identify something like a reference basis (Dahl, 1975, my summary). Model-theoretical semantics, working with "points of reference" or "indices" (n-tuples of "coordinates"), attempts to isolate factors that determine potential reference objects. My own approach essentially operates with a single factor, the willingness of the speaker to predicate the comprehension of a concept. It is left open what influences the speaker's willingness to predicate. Model-theoretical semantics seems to assume that the influences on the speaker can be represented, once and for all, by a fixed n-tuple of coordinates (relevant factors). But why should this be true? Indeed, it seems to be false.

(31) a. Jones is charged with stealing Smith's wallet. At one point during the court trial, Smith says, with a meaningful look at Jones, "Send the thief to jail."

b. Smith informs the police: "I noticed a man stealing my wallet but too late. Hope you'll get him. Send the thief to jail."

c. Smith informs the police: "My wallet got stolen, I don't know by whom. Hope you'll find out. Send the thief to jail."

d. Smith on law and order: "Hang the murderer. But thieves? Send the thief to jail."

The linguist would say that in (31a) to (31c) the article occurs in its "specific sense," and in (31d) in its "generic sense." The philosopher would further distinguish between the "referential" use of *the thief* in both (31a) and (31b), and its "attributive" use in (31c).[20] Our analysis in the previous subsection assumes the referential use.

The four cases can be distinguished in terms of two criteria: reference and presupposition.

(31a) **Reference:** There is exactly one object such that Smith is referring to it by means of {⟨2,*the*⟩,⟨3,*thief*⟩} in his utterance. Furthermore, this object is in the reference basis for {⟨3,*thief*⟩} relative to Smith, the utterance and the concept ·thief·. Finally, a pointing gesture (performed, in this case, simply by looking) is directed towards the object.

Presupposition: Smith presupposes that, for any object to which he is referring, the object is in the extension of ·thief·.[21]

(31b) **Reference:** This is the same as for (31a) except for the pointing gesture. This is replaced by a doxastic condition: Smith believes that the hearer **knows of** the object Smith is referring to, in the sense that the hearer has reason to believe the object exists.

Presupposition: This is the same as in (31a).

(31c) **Reference:** Smith is referring to exactly those objects in the reference basis that are also in the extension of ·thief·.

Presupposition: There is exactly one such object.

(31d) **Reference:** This is the same as in (31c). (It may be argued, though, that we are here dealing with a different concept: a concept which applies to professional thieves, not to persons who occasionally steal something.)

[20] To use Donnellan's terms. See Donnellan, 1966.

[21] The precise interpretation of "presupposes" may be left open. In spite of recent criticism, I would still not reject a concept that involves beliefs of the hearer in, say, the following form: "The speaker believes that *p* and believes that the hearer has reason to believe that *p*." (Here and elsewhere I disregard the problem of pretended beliefs.)

TABLE 2

Uses of Definite Singular Noun Occurrences

Uses	Specific			Generic
	Referential		Attributive	
	Gestural	Doxastic		
Reference condition	Exactly one referent: it is in reference basis; a pointing gesture is directed toward it.	Exactly one referent: it is in reference basis; speaker believes that hearer knows of the referent.	Referents: exactly those elements of the reference basis to which the concept applies.	
Presup-position	The concept applies to any referent		Exactly one element of the reference basis to which the concept applies	Several elements of the reference basis to which the concept applies

Presupposition: Smith presupposes that there are several objects in the reference basis that are also in the extension of ·thief·.

Generalizing, we may distinguish four uses of definite noun occurrences: **specific-referential-gestural, specific-referential-doxastic, specific—attributive, and generic.** They are characterized in a summary fashion in Table 2.

As our next step we give a more formal account of the four uses.

3.4. The Meanings of Definite Noun Occurrences

We first provide an explicit solution to a problem that in previous subsections was tacitly taken as solved: how do we know which morphological meanings should be introduced in connection with syntactic constituents? Where did we get the concept ·thief· from?

Compare:

(32) a. $\{\langle 1, /\theta\text{i:v}/\rangle, \langle 2, /\text{z}/\rangle\}$
b. $/\theta\text{i:vz}/$

The word form in (32a) is morphological. The phoneme sequence in (32b) is **the phonological form** of (32a), in an easily defined sense.[22]

To (32a) we assign morphological meanings, given a certain morphological structure. These are transferred to (32b) as **morphophonological meanings** in the following sense:

(33) **Variables:** p, p_1, . . .: any phoneme sequences.

(34) b is a **morphophonological meaning** of p relative to s and S (mp-mg$(s, S)bp$) iff there is an f such that:
a. p = the phonological form of f.
b. m-mg$(f, s, S)bf$.

If we now have a one-element syntactic constituent like $\{\langle 3, \textit{thieves}\rangle\}$, we can make use of the morphophonological meanings of the phoneme sequence *thieves,* with one proviso: a certain morphological structure has to be presupposed. Such structures may not be unique for a given phoneme sequence. Therefore, for any syntactic unit f we need a function that relates morphological structures to one-element constituents of f.

Using the following variables:

(35) **Variables:** v, v_1, . . .: any relation between entities f and s.

we characterize the concept of **morphosyntactic structure** by means of the following assumption:

(36) **Assumption:** Let $\langle v, s\rangle$ be a morphosyntactic structure of f in S. Then s is a syntactic structure of f in S; and v is

[22] Cf. Lieb, 1976a (p. 21, fn. 47).

a function that assigns, to each one-element constituent of f relative to s, a morphological structure of a morphological unit whose phonological form is a member of the constituent. (If there are no one-element constituents, v is empty.)

In addition to the concepts of morphophonological meaning and morpho–syntactic structure, we need concepts of utterance, referring, pointing, believing, knowing of, and presupposing. The first three are introduced as follows:

(37) a. V is a normal utterance by V_1 of f, v, and s in S: $\mathrm{NU}VV_1fvsS$.
b. V_1 is referring by f in V_1 to z: $\mathrm{Ref}V_1fVZ$.
c. V_1 is indicating z at f in V_1: $\mathrm{Ind}V_1zfV$.

Normal utterances are relativized not only to syntactic units f but to their morphosyntactic structures $\langle v, s\rangle$. In the case of indicating, any pointing gesture is allowed, and z must be something that can be pointed at. The epistemic concepts (believing, etc.) will not be made formally explicit, which means that the following formulations remain partly informal. We first introduce an abbreviation (38), which we then use to formulate our assumption on definite singular noun occurrences (39).

(38) **Abbreviation:** $S, f, v, s, k, \mu, I, f_1, f_2, n, p$, and p_1 **are as required** iff:
a. S is a system of an English idiolect.
b. $\langle v, s\rangle$ is a morphosyntactic structure of f in S.
c. $s = \langle k, \mu, I\rangle$.
d. I "represents a normal declarative intonation of f in S."
e. f_1 is a "definite singular noun occurrence relative to f, k, and μ."
f. $f_1 = \{\langle n,p\rangle, \langle n + 1, p_1\rangle\}$.
g. $f_2 = \{\langle n + 1, p_1\rangle\}$.
h. There are morphophonological meanings of p_1 relative to $v(f_2)$ and S, and any such meaning is one-place.[23]

(39) **Assumption regarding normal utterances of units in which definite singular nouns occur:** Suppose that $S, f, v, s, k, \mu, I, f_1, f_2, n, p$, and p_1 are as required. Then, for all V and V_1:

[23] (38h) may be dropped if the following plausible assumptions are correct: the first part of (38h) is implied by (38e), and in case of a relational meaning, reference always is to the first component of an n-tuple to which the meaning applies. (The second assumption would require obvious reformulations in (39).)

If V is a normal utterance by V_1 of f, v, and s in S, there is a b that is a morphophonological meaning of p_1 relative to $v(f_2)$ and S such that at least one of the following conditions (a)–(d) is satisfied:

a. **Specific-referential-gestural:**
   (i) **Reference:** There is exactly one z such that $\text{Ref}V_1 f_1 Vz$; and for all z, if $\text{Ref}V_1 f_1 Vz$, then $\text{reb}^1(f_2, V, V_1, b)z$ and $\text{Ind}V_1 z f_1 V$.
   (ii) **Presupposition:** V_1 presupposes that, for all z, if $\text{Ref}V_1 f_1 Vz$, then ${}^e bz$.[24]

b. **Specific referential-doxastic:**
   (i) **Reference:** As in (a-i), with "$\text{Ind}V_1 z f_1 V$" replaced by "V_1 believes that any hearer of V knows of z."
   (ii) **Presupposition** = a-ii.

c. **Specific-attributive:**
   (i) **Reference:** For all z, $\text{Ref}V_1 f_1 Vz$ iff $\text{reb}^1(f_2, V, V_1 b)z$ and ${}^e bz$.
   (ii) **Presupposition:** V_1 presupposes that there is exactly one z such that $\text{reb}^1(f_2, V, V_1, b)z$ and ${}^e bz$.

d. **Generic:**
   (i) **Reference:** c-i.
   (ii) **Presupposition:** V_1 presupposes that there are several z such that $\text{reb}^1(f_2, V, V_1, b)z$ and ${}^e bz$.[25]

Assumption (39) represents an analysis whose importance goes far beyond the case of definite singular noun occurrences. Suppose that we generalize the auxiliary concept "are as required" in (38) by allowing in (38e) not only combinations of definite article and noun but also of any "definite" pronoun (possessive, demonstrative, etc.) plus noun. Then (39) remains essentially unchanged. The same holds if we include personal pronouns in (38e). In the case of indefinite articles and pronouns, (39) is replaced by a similar assumption of essentially the same type. Moreover, plural forms are easily analyzed along the same lines, taking into account the generic case for the singular forms and its possible elaboration (see Footnote 25).

[24] In Lieb, 1976a (Section 4), the specific uses (39a) and (39b) were not yet analysed.

[25] (39) may be incomplete with respect to the generic case: Compare (a) *send the thief to jail* with (b) *the thief is everywhere in our society* and (c) *the thief is a recent development in our society*. In (b) and (c), the referents are not individual persons. Calling (a) an example of **distributive use,** we might consider (b) as an example of **collective–extensional use:** there is exactly one referent, namely, the intersection (a set!) of the reference basis and the extension of the concept, and it is presupposed that there are several elements in the intersection (that is, the presupposition is the same as in the distributive case). (c) provides a (doubtful) example of **collective–intensional use:** The referent seems to be the comprehension of the concept, and the presupposition the same as before. It seems feasible, though, to account for these differences by differences of sentential context.

On the basis of (39) we now specify our conception of complete syntactic meanings as previously outlined in (26).

From (39a), we obtain the following lambda-expression:

(40) $(\lambda V V_1)$(There is exactly one z such that $\mathrm{Ref} V_1 f_1 V z$; and for all z, if $\mathrm{Ref} V_1 f_1 V z$, then $\mathrm{reb}^1(f_2, V, V_1, b)z$ and $\mathrm{Ind} V_1 z f_1 V$; and V_1 presupposes that, for all z, if $\mathrm{Ref} V_1 f_1 V z$, then ${}^e bz$).

For fixed f_1, f_2, and b, for example, $\{\langle 2, \mathit{the}\rangle, \langle 3, \mathit{thief}\rangle\}$, $\{\langle 3, \mathit{thief}\rangle\}$, and ·thief·, (40) denotes a relation-in-intension between arbitrary V and V_1. **This relation is identified with a complete syntactic meaning of the definite noun occurrence** $\{\langle 2, \mathit{the}\rangle, \langle 3, \mathit{thief}\rangle\}$.

For complete syntactic meanings, we thus arrive at a concept of the following form:

(41) **Variables:** u, u_1, . . .: any relation-in-intension between entities V and V_1.

(42) u is a complete syntactic meaning of f_1 relative to f, v, s, and S: $\text{cs-mg}(f, v, s, S)uf_1$.

Suppose there are m morphophonological meanings b for a fixed definite singular noun occurrence f_1. Assuming (39) we then have $4m$ complete syntactic meanings, and it is a necessary condition for normal utterances of the entire unit f that utterance and speaker be related by at least one of these relations. Thus, our proposal for complete syntactic meanings of definite singular noun occurrences is in agreement with (26).

My basic idea for complete syntactic meanings has been exemplified for an important case. It is, of course, anything but obvious how syntactic meanings can be subjected to a composition process. My solution to this problem—given in (7e)—is outlined in Lieb 1976a (Section 4.3f); it has to be somewhat adjusted to account for the improvements introduced in the present paper.

REFERENCES

Dahl, Ö. (1975). On points of reference. *Semantikos,* **1**(1), 45–61.

Davidson, D. (1967). Truth and meaning. *Synthese,* **17,** 304–333.

Davidson, D. (1970). Semantics for natural languages. In B. Visentini (Ed.), *Linguaggi nella società e nella tecnica.* Milano: Edizioni di Comunità. Pp. 177–188.

Donnellan, K. (1966). Reference and definite descriptions. *Philosophical Review,* **75,** 281–304.

Eisenberg, P. (1976). *Oberflächenstruktur und logische Struktur. Untersuchungen zur Syntax und Semantik des deutschen Prädikatadjektivs.* Tübingen: Niemeyer (=*Linguistische Arbeiten* **36**).

Hiż, H. (1969). Aletheic semantic theory, *The Philosophical Forum,* **I** (new series), 438–451.

Kutschera, F. von (1976). Conventions of language and intensional semantics. *Theoretical Linguistics,* **2,** 255–283.

Lieb, H. (1970). *Sprachstadium und Sprachsystem: Umrisse einer Sprachtheorie.* Stuttgart, etc.: Kohlhammer.

Lieb, H. (1974). Grammars as theories: The case for axiomatic grammar (Part I). *Theoretical Linguistics,* **1,** 39–115.

Lieb, H. (1975). Universals of language: Quandaries and prospects. *Foundations of Language,* **12,** 471–511.

Lieb, H., (Ed.) (1975). *Oberflächensyntax, und: Syntaktische Konstituentenstrukturen des Deutschen.* LAB (Linguistische Arbeiten und Berichte) Berlin (West) 4, Berlin: Freie Universität, Fachbereich 16.

Lieb, H. (1976a). Grammars as theories: The case for axiomatic grammar (Part II). *Theoretical Linguistics,* **3,** 1–99.

Lieb, H. (1976b). *Vorlesungen zur Sprachtheorie: Syntax, Semantik, Morphologie.* Wintersemester 1975–1976: Einleitung: Allgemeine Orientierung; Teil I: Syntax, Berlin: Freie Universität, Fachbereich 16. (Mimeographed.)

Lieb, H. (1976c). Zum Verhältnis von Sprachtheorien, Grammatiktheorien und Grammatiken. In D. Wunderlich (Ed.), *Wissenschaftstheorie der Linguistik.* Kronberg: Athenäum (Athenäum Taschenbücher Sprachwissenschaft). Pp. 183–199.

Lieb, H. (1976d). Deskriptive oder rekonstruktive semantische Analyse? Kommentar zu Schnelle. In D. Wunderlich (Ed.), *Wissenschaftstheorie der Linguislik.* Kronberg: Athenäum. Pp. 244–247.

Lieb, H. (1976e). Grammatische Bedeutung in näturlichen Sprachen, *Sprache im Technischen Zeitalter,* **57,** 16–24.

Lieb, H. (1977). Surface syntax as a basis for semantics. In R. J. DiPietro, E. L. Blansitt, Jr. (Eds.), *The Third LACUS Forum 1976.* Columbia, S.C.: Hornbeam Press, Pp. 405–412.

Lieb, H. (in prep.) *Integrational Linguistics.* Amsterdam: Benjamins.

Lyons, J. (1968). *Introduction to Theoretical Linguistics.* Cambridge: Cambridge University Press.

Montague, R. (1974). *Formal Philosophy: Selected Papers of Richard Montague,* edited and with an introduction by R. Thomason. New Haven: Yale University Press.

Strawson, P. F. (1950). On referring. *Mind,* **69,** 320–344.

CONTENTS OF PREVIOUS VOLUMES

Volume 4

Volume 5

Volume 6

Contents of Previous Volumes

Contents of Previous Volumes

Further Notes on Logic and Conversation, *H. Paul Grice*
Remarks on Neg-Raising, *Laurence R. Horn*
DTHAT, *David Kaplan*
Conversational Implicature and the Lexicon, *James D. McCawley*
Two Types of Convention in Indirect Speech Acts, *J. L. Morgan*
On Testing for Conversational Implicature, *Jerrold M. Sadock*
Synonymy Judgments as Syntactic Evidence, *Susan F. Schmerling*
Assertion, *Robert C. Stalnaker*